Membrane Processes in Biotechnology and Pharmaceutics

Membrane Processes in Biotechnology and Pharmaceutics

Catherine Charcosset

ELSEVIER

AMSTERDAM • BOSTON • HEIDELBERG • LONDON
NEW YORK • OXFORD • PARIS • SAN DIEGO
SAN FRANCISCO • SINGAPORE • SYDNEY • TOKYO

Elsevier
The Boulevard, Langford Lane, Kidlington, Oxford OX5 1GB, UK
Radarweg 29, PO Box 211, 1000 AE Amsterdam, The Netherlands

British Library Cataloguing in Publication Data
A catalogue record for this book is available from the British Library

Library of Congress Cataloging-in-Publication Data
A catalog record for this book is available from the Library of Congress

ISBN: 978-0-444-56334-7

For information on all **Elsevier** publications
visit our web site at elsevierdirect.com

Printed and bound by CPI Group (UK) Ltd, Croydon, CR0 4YY

Transferred to Digital Print 2012

To my parents,
To Jean-Philippe, Laurane and Sébastian

Contents

Preface

One of the growing applications of membrane processes is the downstream processing and purification of pharmaceuticals, enzymes, antibiotics and therapeutic proteins. The purpose of this book is to provide a general overview of membrane processes used in the biotechnology and pharmaceutical industry. Ultrafiltration, microfiltration, virus filtration, membrane chromatography and membrane emulsification are examined in detail. Other processes include liquid membranes, membrane reactors, solvent-resistant nanofiltration and membrane crystallization. For each process, membrane material and devices, process parameters, control of fouling and general rules on modelling are discussed. A broad range of applications is presented for each membrane process.

Chapter 1 deals with general aspects of membrane processes which will be discussed in the following chapters. Chapter 1 examines some general rules on membrane materials, including organic and inorganic membranes. The device is also a key point of a membrane process, developed for both cross-flow and dead-end configurations. Membrane characterization techniques include chemical structure characterization, permeability measurement, measurement of solute rejection, liquid displacement techniques, as well as microscopic methods. Membrane fouling is an essential and complex process because of the enormous range of inorganic, organic and biological components that can interact with the membrane and the different mechanisms by which these interactions can occur. As a consequence, membrane cleaning is an essential component of almost all membrane processes including physical and chemical cleaning.

Chapter 2 provides a general overview of ultrafiltration (UF), a pressure-driven separation process which first came into use in the 1960s. UF is extensively used for the concentration, diafiltration and separation of biological solutions, both for the final product formulation and as conditioning of feed streams prior to other separation processes. UF membrane materials, especially developments to improve membrane properties such as selectivity and permeability and reduce fouling, are examined. Various methods have been proposed to limit the negative effects of fouling and concentration polarization by improving the hydrodynamics of the cross-flow over the membrane surface such as dynamic filtration which uses a rotating disk, or by rotating or vibrating the membrane, backflushing and backpulsing, Dean vortices, gas sparging, corrugated surfaces and electro-UF. UF can be operated in various modes including diafiltration, high-performance tangential flow filtration (HPTFF) and affinity UF. Some applications of UF are discussed including processing of antibiotic broths, DNA purification, monoclonal antibodies purification, whey fractionation and lysozyme purification from chicken egg white. In the last section, theoretical backgrounds of UF are considered.

Chapter 3 focuses on various aspects of microfiltration (MF) for biotechnologies and pharmaceutical applications. MF is an older process than UF which traces its roots back to Germany in the beginning of the twentieth century. MF is a pressure-driven separation process, which is widely used in cell recycle and harvesting, separation of recombinant proteins from cell debris and purification of process streams. In this chapter, MF membrane materials are briefly examined, such as developments to improve permeability and reduce fouling. Like for UF, hydrodynamic methods have been proposed to reduce the negative effects of fouling and concentration polarization. The techniques include dynamic filtration, vortex systems, backflushing and backpulsing, and electric field. Other techniques are proposed such as constant filtrate flux operation, flocculation, gas sparging, secondary membranes and turbulence promoters. Common applications of MF are discussed, including bacterial removal (sterile filtration), concentration and washing of cultures of single-cell organisms, recovery of intracellular molecules produced from fermentation broth including proteins, antibiotics, lactic acid and polysaccharides, and purification of nanoparticle suspensions. Finally, some results on MF modelling are examined.

Chapter 4 deals with virus filtration, whose principles are very similar to those of UF and MF. Virus capture is needed in the biotechnology industry for two main goals: viral clearance and purification of viral vectors and vaccines for gene therapy applications. Virus clearance is an essential component in the production of biopharmaceutical products derived from human or animal origin, as manufacturers are required to validate virus clearance before regulatory authorities approve the product. Commercial membranes are specifically designed for virus clearance. Purification methods of viral vectors and vaccines are aimed at eliminating contaminants originating from host cells or culture media and producing large volumes of concentrated, biologically active viruses. The main features of virus filtration are presented including membranes and configurations, fouling, and applications related to virus clearance of plasma products and monoclonal antibodies, and purification of viral vectors and vaccines.

Chapter 5 provides an overview of membrane chromatography, which dates back to the 1980s with a major paper published by Brandt et al. in 1988 who reported the use of Protein A hollow fibre membranes for the purification of fibronectin from blood plasma and purification of IgG. Membrane chromatography is now a competitive technique which has the potential to maintain high efficiencies both at high flow-rates and for use of large biomolecules with small diffusivities, reducing biomolecules degradation and denaturation. Interaction mode include affinity interaction, ion exchange, hydrophobic interaction, reversed-phase and multistage chromatography. Chapter 5 highlights the principles, membrane materials and devices, membrane activation and interaction modes, operational parameters and some general modelling aspects of the purification process. The large range of applications is examined, including purification of monoclonal antibodies and DNA, and viral clearance. A specific section is also devoted to monolith chromatography.

Membrane emulsification, which is discussed in Chapter 6, was introduced more than 20 years ago by Nakashima et al. at the annual Meeting of the Society of Chemical Engineers in Japan in 1988. Its use has been increasingly reported these last 10 years. In membrane emulsification, the dispersed phase is pressed through the pores of a microporous membrane, while the continuous phase flows along the membrane surface. Droplets grow at pore openings until they detach. The resulting droplet size is controlled primarily by the choice of the membrane and not by the generation of turbulent droplet break-up. Thus, emulsions with narrow droplet size distributions are produced and shear-sensitive ingredients can be used. Chapter 6 provides a general overview of membrane emulsification focusing on membrane materials and devices, process parameters, optimization and some general aspects of modelling. Many applications are reported in the literature which may be of use in the pharmaceutical and biotechnological industries, such as simple emulsions, multiple emulsions, polymeric and lipid nano- and microparticles, and liposomes.

In Chapter 7, some other membrane processes related to biotechnologies and pharmaceutical industry are considered. Liquids that are immiscible with the source (feed) and receiving (product) phases can serve as a liquid membrane. Liquid membranes are reported for concentration and separation of biochemical compounds like amino acids, organic acids and antibiotics. Another field of application of liquid membranes is chiral resolution. In the first section of this chapter, the principles, configurations, applications and some theoretical backgrounds of liquid membranes are examined. In the second section, membrane reactors are presented. Membrane bioreactors are alternative approaches to classical methods of immobilizing enzymes, microorganisms and antibodies. The biocatalysts are suspended in solution and compartmentalized by a membrane in a reaction vessel or immobilized within the membrane matrix itself. In the last section, membrane bioreactors are considered including membranes and configurations, applications such as chemical synthesis and protein hydrolysis, and general rules on modelling.

Chapter 8 provides an overview on two recent membrane processes. The first section of the chapter focuses on solvent-resistant nanofiltration (SRN). Nanofiltration has recently emerged from its traditional application area with the development of SRN. The extraction of drugs and natural compounds become possible using these new membranes. Membranes and devices, governing phenomena, configurations and applications are summarized. The second section describes a recent membrane process which uses membranes to improve crystallization efficiency of various pharmaceutical compounds. Several configurations are discussed such as osmosis and reverse osmosis, evaporative membrane crystallization and membrane contactor. For each configuration, membranes, devices, governing phenomena and applications are presented.

1 Principles on membrane and membrane processes

Chapter Outline

Membrane Processes in Biotechnology and Pharmaceutics. DOI: 10.1016/B978-0-444-56334-7.00001-0

1.1 Membrane materials

1.1.1 General characteristics

Depending on their internal structure, membranes can be classified as symmetrical or asymmetrical. Symmetrical membranes show uniform pore sizes in cross section. The pores of asymmetric membranes are usually smaller on the membrane surface. Composite membranes combine two different structures into the same membrane. The different layers can be either symmetrical or asymmetrical, with a distinct pore-size distribution, aspect ratio (ratio of pore sizes on the two faces of the membrane) and thickness. Multi-layer membranes have different membranes layered together, each of which is cast separately with the desired pore size and surface characteristics. The first layer is typically used as a pre-filter while the pore size of the second layer depends on the application.

From their morphological point of view, membranes can be divided into two large categories: dense and porous. Membranes are considered to be dense when the transport of components involves a stage of dissolution and diffusion across the material constituting the membrane. A membrane is denominated as porous when permeate transport occurs preferentially in the continuous fluid phase which fills the membrane pores.

Membranes are usually classified accordingly to their average pore sizes (Figure 1.1). Microfiltration (MF) membranes typically have pore sizes on the order of 0.1–10 μm. Ultrafiltration (UF) membranes have pore sizes in the range of 0.001–0.1 μm and are capable of retaining species in the molecular weight range of 300–10,00,000 Da. Membranes designed specifically for virus filtration fall between these limits. Reverse osmosis (RO) membranes retain solutes, such as salts and amino acids, with molar mass below 1000 Da. Nanofiltration (NF) membranes retain solutes, such as small polypeptides, in the range of molar mass between 1000 and 3000 Da.

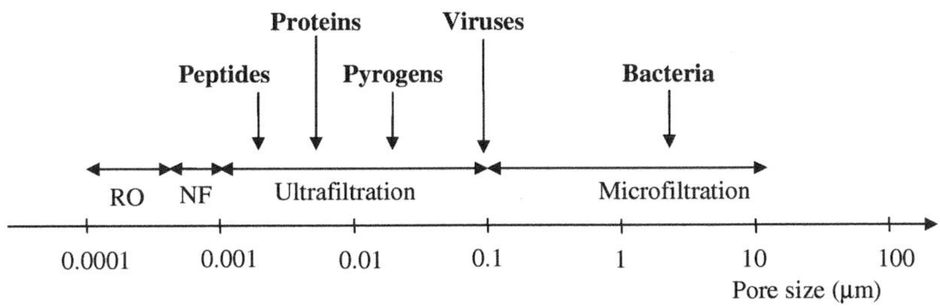

Figure 1.1 Approximate pore size ranges of different types of membranes, compared to dimensions of some components separated by membrane processes [1].
This article was published in Comprehensive Biotechnology, Second Edition, Vol. 2, C. Charcosset, Downstream Processing and Product Recovery/Membrane Systems and Technology, pp. 603–618, Copyright Elsevier (2011).

1.1.2 Organic membranes

Common polymers used in commercial applications such as UF, MF, membrane chromatography and virus filtration are listed in Table 1.1. In addition, many polymers are grafted, custom-tailored, blended or used in the form of copolymers, to improve membrane properties, such as lower protein adsorption, higher flux, higher flux recovery ratio and lower membrane fouling. Common polymeric membranes are obtained by casting technologies, including air, immersion and melt casting. Other common techniques include track-etched membranes, membranes made by controlled stretching of films, composite membranes and nanofibrous membranes.

Table 1.1 Common polymers used in commercial membrane manufacture

Polymer	UF	MF	MC	VF
Cellulose diacetate and triacetate (CA, CTA)	×	×		
Cellulose nitrate (CN)	×			
CA/CN blends	×			
Cellulose and regenerated cellulose	×	×	×	×
Polyacrylonitrile (PAN)			×	
Polyamide (aromatic and aliphatic)	×	×		
Polysulfone (PS)	×	×		
Polyethersulfone (PES)	×	×	×	×
Polycarbonate (track-etched)	×	×		
Polyethylene terephthalate (PET) (track-etched)	×	×		
Polyimide		×		
Polyethylene (PE)	×			
Polypropylene (PP)	×			
Polytetrafluoroethylene (PTFE)	×			
Polyvinylidene fluoride (PVDF)	×	×		×
Polyvinylchloride (PVC)	×	×		

UF: ultrafiltration, MF: microfiltration, MC: membrane chromatography, VF: virus filtration

They are briefly described below. Details on these various techniques are given in various references [2,3].

Casting technologies

Polymeric membranes are usually manufactured by a phase-inversion process. This technique involves preparing a casting solution consisting of one or more polymers in an appropriate solvent or solvent blend and possibly one or more non-solvents, surfactants and other additives such as inorganic salts. The polymeric membrane is prepared by causing this casting solution to undergo phase inversion, which involves the transformation of a homogeneous solution in which the polymer molecules are dispersed in the mixture of solvent(s) and non-solvent(s) into a porous membrane in which the polymer forms an interconnected matrix. The phase-inversion process is obtained (1) by removing solvent from a casting solution that contains non-solvent (wet casting); (2) by simultaneously removing solvent while adding non-solvent (dry casting) or (3) by cooling a casting solution that contains a latent solvent displaying only a limited ability to dissolve the polymer (thermally induced phase-separation [TIPS]) [4].

The wet casting process involves immersing the casting solution into a non-solvent bath that causes simultaneous loss of solvent and gain of non-solvent. The dry casting process entails evaporating sufficient solvent from a casting solution that initially contains some non-solvent. The TIPS process necessitates cooling a casting solution containing a solvent that can dissolve the polymer only at elevated temperatures. Combinations of these three fundamental phase inversion methods lead to hybrid processes. For example, wet casting may involve a precursor evaporation step that predisposes the interface of the casting solution to gel or vitrify. The thermally assisted evaporative phase separation process combines dry- and TIPS-casting, in which phase separation is caused by evaporative solvent loss coupled with temperature control. The vapour induced phase-separation process combines wet and dry casting whereby the casting solution is initially contacted with humid air (i.e. water is a non-solvent) followed by immersion into a non-solvent bath. Membranes from polymers with excellent chemical and thermal resistance can thus be produced, such as polyolefins, polyfluorocarbons and poly(ether ether ketone).

In recent years, several polymer blends have been used for the development of novel membranes with improved properties [5–7]. In addition, organic or inorganic additives as the third component to the blend polymers have been used to control the morphology and performance of membranes.

Studies were conducted by adding additives such as polyethylene glycol (PEG), hyperbranched polyglycerol (HPG) and polyvinylpyrrolidone (PVP) in the casting [8,9]. Su et al. [8] prepared a series of amphiphilic poly(ethylene glycol)-*graft*-polyacrylonitrile (PEG-*g*-PAN) UF membranes with various molecular weights of PEGs by wet phase precipitation copolymerization using ceric(IV) ammonium nitrate as an initiator. All prepared PEG-*g*-PAN UF membranes have lower BSA adsorption, higher flux for protein solution, higher flux recovery ratio and lower membrane fouling during protein UF in comparison with the control PAN membrane. The authors concluded that these improved properties endow PEG-*g*-PAN membranes with potential applications in protein separation and purification. Sivakumar et al. [9] investigated the effects of PVP on CA/PS blend UF membranes and showed that an increase in the concentration of PVP in casting solution resulted in improved performance. Arthanareeswaran et al. [10] introduced sulfonated poly(ether ether ketone) (SPEEK) to modify a cellulose acetate (CA) membrane in order to obtain a CA/SPEEK blend UF membrane with improved performance. The same authors prepared PS and SPEEK blend membranes and characterized their UF performance [11]. Susanto and Ulbricht [12] prepared polyethersulfone (PES) UF membranes by a non-solvent-induced phase separation method using different macromolecular additives: PVP, poly(ethylene glycol) (PEG) and poly(ethylene oxide)-*b*-poly(propylene oxide)-*b*-poly(ethylene oxide) (Pluronic®). Pluronic showed the best results in water flux and rejection of BSA.

Track-etched membranes

Track-etched membranes are made from thin polymeric films of polycarbonate, poly(ethylene terephthalate [PET]) or polyimide. Tracking is produced by bombardment of the film (10–20 μm thick) with a beam of high energy nuclear particles generated in a nuclear reactor. The interaction of the high energy beams leads to an array of linear, perpendicular tracks. The subsequent etching operation in a bath containing typically a hot NaOH solution, or H_2O_2, enlarges the tracks into cylindrical pores. The range of commercially available pore diameters is approximately 0.05–20 μm. UF and MF membranes of this type are manufactured by several companies, such as Millipore, Osmonics and Whatman.

Stretching

The controlled stretching technique is based on a controlled uniaxial or biaxial stretching of the film made of a homogeneous semi-crystalline polymer such as poly(tetrafluoroethylene) (PTFE) or polyolefins. This technology produces openings in the film in the form of slits or fissures in the size range of 0.02–20 μm. The resultant filter structures often consist of nodes of solid polymer connected by filaments. Membranes made by controlled stretching of films are commercialized by companies such as Millipore, Osmonics and Sartorius.

Composite membranes

The method of producing composite membranes is based on a very thin selective layer deposited on top of a highly permeable substrate that provides the necessary mechanical support. Such methods are reported to include (1) casting of the selective layer separately, followed by lamination to the substrate; (2) dip coating of the substrate with a precursor solution, followed by curing with heat or radiation; (3) gas-phase deposition of the selective layer from a glow-discharge plasma and (4) interfacial polymerization of reactive monomers on the surface of the substrate. The defect-free characteristics of the selective layer allowed, for example, the development of virus retentive membranes (Viresolve membranes, Millipore).

Nanofibrous membranes

Electrospinning is a simple and versatile method for producing fibres with diameters ranging from several micrometres to tens of nanometres from a variety of materials [13,14]. In electrospinning, an electrostatic field is applied between a nozzle and a collector. The polymer solution is ejected from the nozzle towards the collector due to electric force. The solvent evaporates from the polymer jet, but the jet also undergoes instabilities, which are mainly responsible for the nano-sized fibre diameters. The diameter of electrospun nanofibre depends on a number of parameters: (1) properties of the polymer solution (molecular weight, concentration, surface tension, viscosity, conductivity, etc.) and the environment (temperature, humidity, pressure, etc.); (2) operational parameters including applied voltage, flow rate and distance between the nozzle and the collector. The resultant fibre diameter determines the properties of the electrospun fibre membrane, such as mechanical, electrical and optical properties.

The electrospun nanofibre membranes have found applications in filtration, drug delivery carrier, tissue engineering, wound dressing, nano-sensors and enzyme immobilization due to their features such as high bulk porosity (up to 80%) with fully interconnected pore structures, and high surface area-to-volume ratio. When used as supports for enzyme immobilization, nanofibrous membranes offer many attractive features, such as the large surface area for the attachment of enzymes, and macroporous structure for improvement of the mass transfer rate of substrate [15,16].

Morphology

The morphology of polymeric membranes is complex, diverse and irregular. Typical scanning electron micrographs of polymeric membrane surfaces are shown in Figure 1.2. Membranes prepared by casting technologies show usually one of these three morphologies: opened-cellular (a percolating foam), lacy (a fibrous network) or nodular (a layer of packed nodules) [2]. The opened-cell morphology is typical for many MF membranes, prepared by any of the three main casting technologies (air, immersion and melt casting). The pore space is formed by coalesced or communicating quasi-spherical hollow domains (Figure 1.2a). Only one side of a membrane may exhibit the opened-cellular morphology, while the opposite side has a lacy structure. Any of the three main casting technologies can generate lacy structured membranes (Figure 1.2b and c). The nodular morphology is typical for surface layers of UF membranes. With crystalline polymers, membrane structures can be obtained that consist entirely of large (5–30 μm) packed or impinged nodules. Any of the three main casting technologies can produce nodular membranes. In addition, membranes prepared by a track-etched process show uniform and cylindrical pores (Figure 1.2d).

1.1.3 Inorganic membranes

Inorganic membranes are of great interest in membrane technology because of their higher chemical, thermal and mechanical stability compared to most polymeric membranes. With inorganic membranes, processing at high temperature (up to

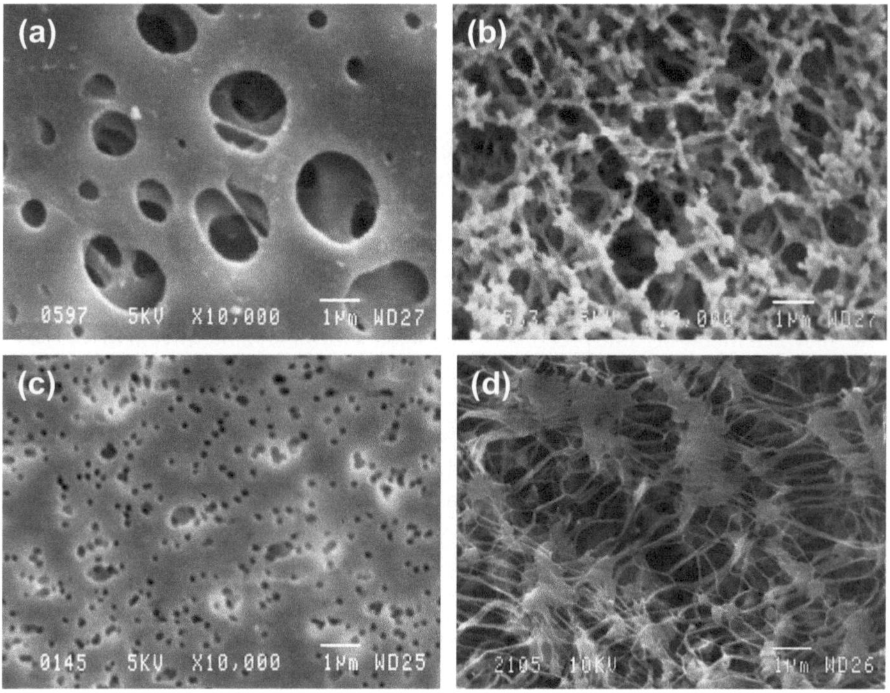

Figure 1.2 Scanning electron micrographs of the surface: (a) 0.16-μm polyethersulfone (PES) (Filtron, MA) membrane, (b) 0.22-μm mixed cellulose esters membrane (Millipore, MA), (c) 0.2-μm cellulose acetate (Sartorius, NY), and (d) 0.2-μm polycarbonate (PCTE) (Osmonics, CA) [29].
This article was published in J. Membr. Sci., Vol. 155, C.C. Ho, A.L. Zydney, Effect of membrane morphology on the initial rate of protein fouling during microfiltration, pp. 261–275, Copyright Elsevier (1999).

500 °C) and extreme pH-value (pH 1–14) is possible. In addition, inorganic membranes can be cleaned with aggressive chemicals, organic solvents or hot water stream.

Inorganic membranes can be classified into (1) porous and amorphous membranes, (2) porous and crystalline membranes and (3) dense membranes. Dense inorganic membranes are usually manufactured from metals such as palladium, nickel, silver, zirconium and their alloys. Porous inorganic membranes are produced from metallic oxides (alumina, titania, zirconia), carbon, silica (glass), metals and zeolite. Ceramic membranes with a wide range of pores sizes can be obtained by using different preparation techniques including slip casting, tape casting, pressing, extrusion, sol–gel process, dip coating, chemical vapour deposition and preparation of hollow fibre ceramic membranes. In addition, anodized aluminium membranes share common features with both the track-etched films (showing arrays of straight parallel pores of uniform size) and inorganic membranes (due to the nature of membrane material). The specific techniques of sol–gel, chemical vapour deposition and anodized aluminium are briefly described below. Details on these various methods can be found in Li [17].

Sol–gel technique

The sol–gel technique was first used for the preparation of ceramic UF membranes by Leenaars et al. [18] and developed further by several authors [19,20]. The advantage of the sol–gel technique is that the membrane pore size can be controlled, especially for small pores. The two main methods to prepare membranes through the sol–gel technique are (1) the colloidal route, in which a metal salt is mixed with water to form a sol, the sol is coated on a membrane support, where it forms a colloidal gel and (2) the polymer route, in which metal-organic precursors are mixed with an organic solvent to form a sol, which is then coated on a membrane support, where it forms a polymer gel [17].

Colloidal sols are colloidal solutions of dense oxide particles such as Al_2O_3, SiO_2, TiO_2 or ZrO_2. Leenaars et al. [18] developed γ-alumina membranes with pore diameters of 4–10 nm from a boehmite sol. As the membrane pore size was in the UF range, these membranes have been used in separation of colloidal particles and large-molecular-weight solutes or as a support where smaller pore size membrane can be further developed [17].

Chemical vapour deposition

The chemical vapour deposition technique modifies the properties of membrane surfaces by depositing a layer of the same or a different compound through chemical reactions in a gaseous medium surrounding the component at an elevated temperature. The gas mixture, consisting typically of hydrogen, nitrogen or argon, and reaction gases such as metal halides and hydrocarbons, is carried into a reaction chamber that is heated to the desired temperature. The deposition of coatings by chemical vapour deposition can be achieved in a number of ways such as thermal decomposition, oxidation and hydrolysis.

Anodized aluminium membranes

These membranes, with typically pore sizes in the 0.02–0.20 μm range, are produced by an electrolytic oxidation of aluminium. In the membrane formation process, the Al metal sheet is electro-oxidized in an acidic electrolyte. Initially, randomly distributed circular nuclei of Al_2O_3 are formed on the sheet surface. The pores grow subsequently from the nuclei where the current density is higher than in the rest of the sheet. The pore growth is limited by the presence of adjacent pores with their cross section being polygonal. Anodized aluminium membranes are commercialized by Whatman.

Morphology

The properties of ceramic membranes are strongly dependent on their microstructure [17]. The important features are the size and shape of the grains, the porosity, the pore size and the distribution of the pores. In order to achieve the desired properties during the fabrication of ceramic membranes, the microstructural features must be controlled.

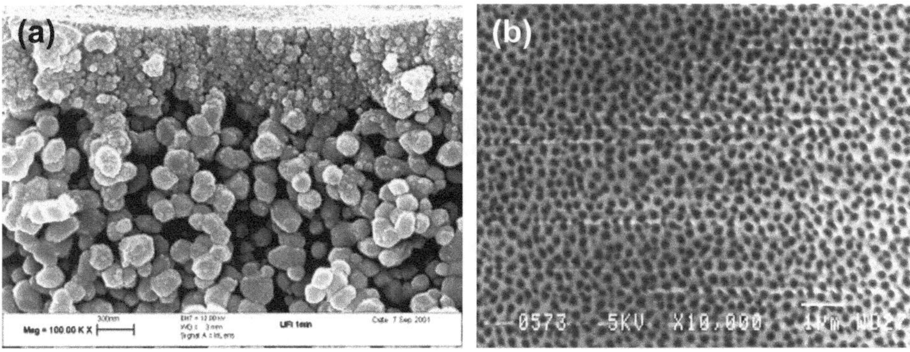

Figure 1.3 (a) FESEM image of a TiO_2 ultrafiltration membrane on alumina support [21]. (b) SEM of a 0.2-μm Anopore anodized aluminium membrane (Whatman, UK) [29].
Figure 3a: This article was published in Adv. Eng. Mater., Vol. 6, S. Benfer, P. Arki, G. Tomandl, Ceramic membranes for filtration applications – preparation and characterization, pp. 495–500, Copyright John Wiley and Sons (2004).
Figure 3b: This article was published in J. Membr. Sci., Vol. 155, C.C. Ho, A.L. Zydney, Effect of membrane morphology on the initial rate of protein fouling during microfiltration, pp. 261–275, Copyright Elsevier (1999).

For most applications, this means the achievement of a high density, a small grain size and a homogeneous microstructure. Figure 1.3a shows a field emission scanning electron microscopy (FESEM) image of a two-layer composite membrane, consisting of an α-alumina macroporous support and a TiO_2 separation layer. The membrane has pores in the UF range with a separation layer thickness around 400 nm [21]. Figure 1.3b shows typical regular and circular pores of an anodized aluminium membrane.

1.2 Membrane devices

The two standard modes of membrane operation are dead-end and cross-flow (Figure 1.4). In the dead-end mode, the fluid to be filtered is forced through the membrane pores usually by applying a pressure on the feed side. In the cross-flow mode, the feed flows parallel to the membrane surface and permeates through the membrane due to a pressure difference. The cross-flow reduces the formation of the filter cake to keep it at a low level. In both dead-end and cross-flow configurations, a large range of membrane devices is available.

1.2.1 Cross-flow configuration

In the cross-flow mode, the most commonly used membrane modules include hollow fibre, tubular, flat plate and spiral wound devices [2,22]. Hollow fibre and spiral wound modules have the highest membrane packing densities since they have the thinnest channels. However, this makes them more susceptible to fouling and can make cleaning more difficult.

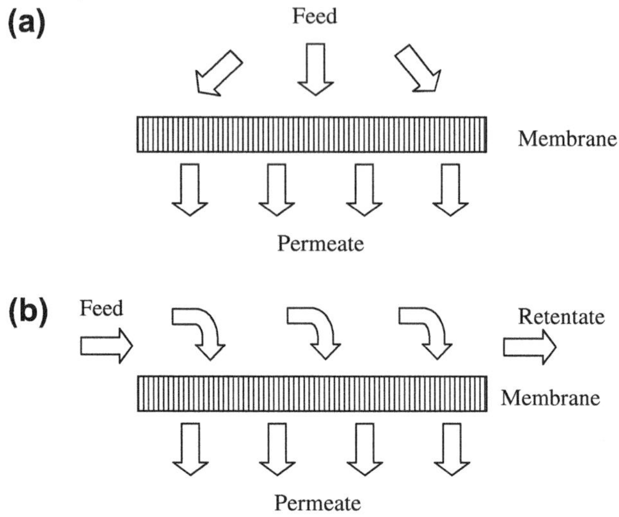

Figure 1.4 Dead-end (a) and cross-flow (b) configurations [1].
This article was published in Comprehensive Biotechnology, Second Edition, Vol. 2, C. Charcosset, Downstream Processing and Product Recovery/Membrane Systems and Technology, pp. 603–618, Copyright Elsevier (2011).

Hollow fibre modules

Hollow fibre modules consist of an array of narrow fibres (Figure 1.5). The membrane has a dense skin layer (usually at the lumen side of the fibre) that gives its selectivity while the porous substructure provides the necessary mechanical support. Hollow fibre devices have a very high packing density due to the very narrow diameters of the fibres and the ability to tightly pack the fibre bundles in the cylindrical cartridge. This

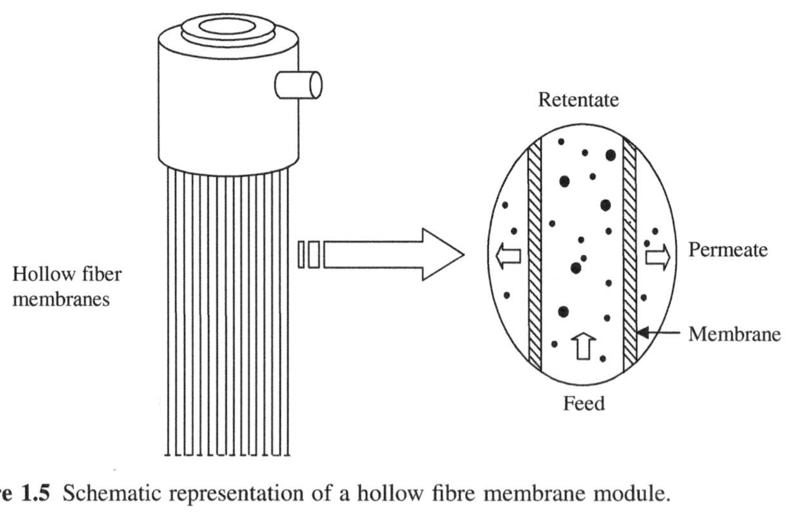

Figure 1.5 Schematic representation of a hollow fibre membrane module.

means that large volumes can be filtered, while utilizing minimal space, with low power consumption. One major disadvantage of the hollow fibre modules is that they are highly susceptible to particulate fouling.

Hollow fibre modules for bio-applications are commercialized by companies such as Koch Membrane Systems (USA), Asahi Kasei (Japan), GE Healthcare (United Kingdom) and Spectrum Laboratories Inc. (USA). Fibre diameters typically range from 200 to 2500 μm, with fibre thickness around 200 μm. Bundles of fibres contain 50–10,000 fibres, which are potted at the ends in an epoxy or polyurethane resin to form a tube sheet. Hollow fibre membranes are made from a variety of polymers, including regenerated cellulose, PES, polysulfone, polypropylene (PP), polyvinylidene fluoride and mixed cellulose esters. Ceramic hollow fibre devices are also commercialized by CEPAration (The Netherlands).

While hollow fibre modules are typically operated by a pump, small devices sold by Spectrum Laboratories Inc. offer the advantage of manual operation for processing small volumes. For concentration processing, two retentate syringes connected to the luer inlet/outlet ports are utilized to pass the sample back and forth through the membrane lumen while a third syringe collects the filtrate by way of one of the side-ports. In diafiltration, a fourth syringe can be used to either supply more sample for concentration or buffer for diafiltration via a luer T connector.

Tubular devices

Tubular devices are similar in design to the hollow fibre modules but employ much larger diameter tubes, typically between 0.3 and 2.5 cm (Figure 1.6). The tube section may be circular in shape or more elaborated. The tubular membranes are usually cast in place within a support tube made of fibreglass, ceramic, plastic or stainless steel. The individual tubes are placed inside a plastic or stainless steel sleeve to form a single tube cartridge, or they can be packed together in small bundles. Inorganic membranes are usually made of a tubular monolithic unit in which the membranes are arranged in a parallel array. The unit is housed in an appropriate plastic or stainless steel cartridge. One major advantage of the tubular modules versus hollow fibre devices is that they are more resistant to fouling. In addition, tubular membranes are easier cleaned using either chemical or mechanical methods. The main disadvantages of the tubular modules are the low packing densities associated with the use of the

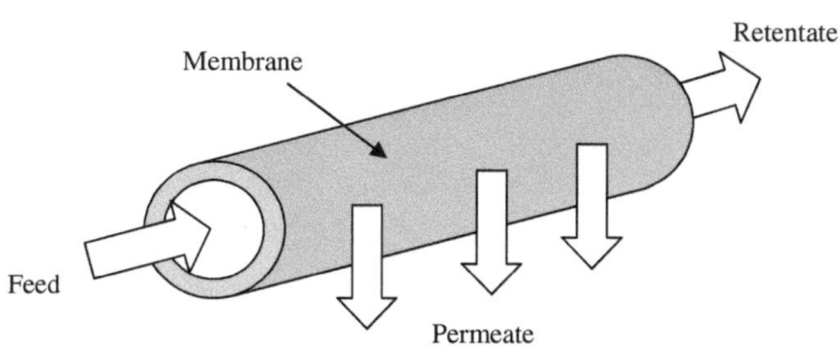

Figure 1.6 Schematic representation of a tubular membrane module.

large feed diameters, and the high hold-up volumes and high capital costs for these systems. Tubular devices for bio-applications are commercialized by companies such as Pall Corp. (USA), Koch Membrane Systems (USA), NovaSep Process (France), Tami Industries (France) and IBMEM (Germany).

Flat plate modules

Flat plate modules are among the first configurations to have been developed for large-scale commercial applications. They use multiple flat sheet membranes in a sandwich arrangement consisting of a support plate (which also defines the permeate flow path), the membrane and the channel spacer or separator (Figure 1.7). Several of these membrane elements are then stacked together and clamped to form a complete module. The feed channel is typically 0.03–0.1 cm high. The channel can employ an appropriate screen, such as a PP mesh, to promote local mixing and improve the overall mass transfer characteristics of the module. Flat sheet membranes are typically cast on a non-woven substrate and can have either a symmetrical or an asymmetrical structure. A variety of polymers is available, including polysulfone, PES, cellulose and hydrophilized polyvinylidene fluoride (PVDF). These materials are often surface modified to increase hydrophilicity and reduce fouling. The small channel height in the flat plate module leads to relatively high packing densities and low hold-up volumes, making them particularly attractive for recovering biological products. These devices are also susceptible to fouling, particularly in the tortuous flow path defined by the feed-side spacers. Flat plate modules are commercialized, for example, by Koch Membrane Systems (USA), Pall Corp. (USA), Sartorius (Germany), GE Healthcare (United Kingdom) and Microdyn-Nadir (Germany).

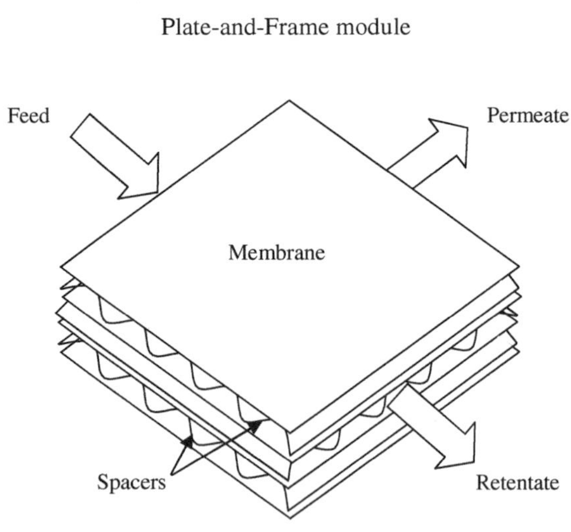

Figure 1.7 Schematic representation of a flat plate module.

Spiral wound modules

Spiral wound membranes are used intensively in wastewater and water treatment processes such as heavy metal removal, brackish water treatment and desalination. They are also found in the pharmaceutical and biotechnological industries. Spiral wound modules are constructed using flat sheet membranes in the form of a 'pocket' consisting of two membrane sheets separated by a permeable mesh that defines the region for permeate flow (Figure 1.8). The membranes are sealed along three of the edges to form a pocket. The open side of the pocket is attached to a central perforated tube that is used to collect the permeate flow. Several of these pockets are spirally wound around a single collecting tube using a feed-side mesh as a spacer between the pockets to establish the desired feed channel thickness. The spacer enhances turbulence in the process stream to reduce fouling. The edges of the membranes are bonded so that process solution cannot enter the channel where permeate is collected. Permeate moves spirally inside the membrane sandwich to the central collection tube, then exits through the header assembly.

Spiral wound modules have relatively high membrane packing densities and very effective mass transfer characteristics due to the feed-side spacers. The main disadvantage of the spiral wound modules is that they are much sensible to fouling due to the narrow and irregular flow path through the spacers. In addition, the presence of spacer elements in the narrow membrane channel tends to increase pressure drop

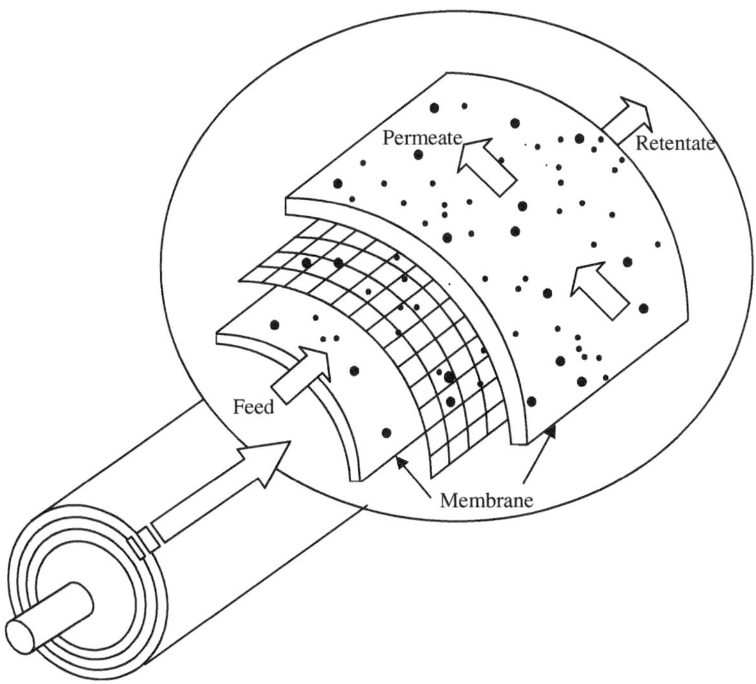

Figure 1.8 Schematic representation of a spiral wound module.

across the membrane channel. UF and MF spiral wound modules are commercialized, for example, by Koch Membrane Systems (USA), Microdyn Nadir (Germany), Millipore (USA) and Sepro Inc. (USA).

1.2.2 Dead-end configuration

In principle, any of the previous cross-flow devices can be used in the dead-end configuration, in which the retentate exit is closed and the feed solution is forced through the membrane and out the permeate ports. However, these configurations are susceptible to high fouling. Therefore, a number of devices have been developed specifically for dead-end applications [2]. Syringe-end filters, centrifugal membrane devices and some vacuum filtration devices are disposable systems. Their main advantage is to avoid a cleaning procedure. The disadvantages associated to these systems have minimal control over the applied pressure and flow rate, and inadequate control of polarization concentration.

Vacuum filtration

Vacuum filtration devices are mainly used for sterilizing tissue culture media, buffers, or other aqueous solutions. In vacuum filtration, a small flat sheet membrane disc is placed between an appropriate support plate and an open glass or plastic funnel (Figure 1.9). The feed solution is placed in the funnel, and drawn through the membrane into the collection flask by the application of vacuum. These systems typically combine a filter unit with a receiver flask for processing and storing volumes from 150 to 1000 mL. They are commercialized, for example, by Corning, Millipore and Nalgene.

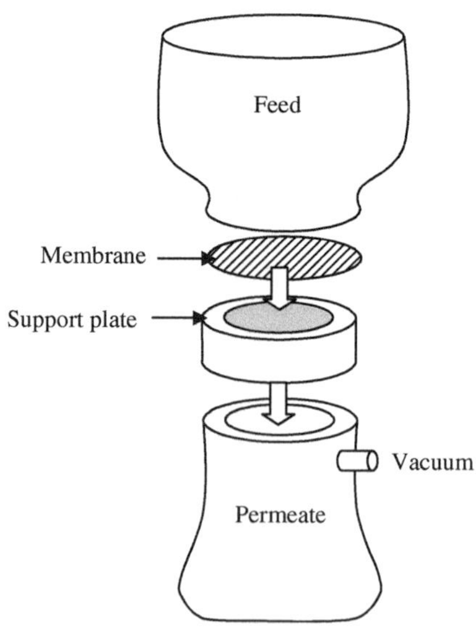

Figure 1.9 Schematic representation of a vacuum filter.

Disposable syringe filters

Disposable syringe filters are used for bio-applications such as removal of virus, biological sample preparation, capillary electrophoresis, filtration of colloidal material, HPLC sample preparation, sterile filtration and ion-exchange membrane chromatography. Disposable syringe filters use membranes bonded to a plastic support plate and sealed into a small disc-shaped unit equipped with luer-type connectors attached to the syringe (Figure 1.10). The solution to be filtered is manually pushed through the membrane by the syringe. Syringe filters are commercialized by companies such as Millipore, Nalgene, Whatman, Sartorius and Pall. A very large range of membrane materials is available including polytetrafluoroethylene (PTFE), polyvinylidene difluoride, polypropylene, PES, nylon, CA, cellulose nitrate (CN), regenerated cellulose, Anopore and glass microfibre/glass fibre. The diameter of the disc-shaped membrane is usually between 3 and 30 mm, corresponding to filtration area between 0.07 and 5.7 cm^2, and to sample volume to be filtered between 1 and 100 mL.

Centrifugal membrane devices

Centrifugal membrane devices are used for concentration of biological samples containing antigens, antibodies, enzymes, nucleic acids and microorganisms, for purification of macromolecular components, desalting, buffer exchange or diafiltration. In a centrifugal membrane device, a membrane cone or disk is placed inside a small centrifuge tube. The feed solution is driven across the membrane by the

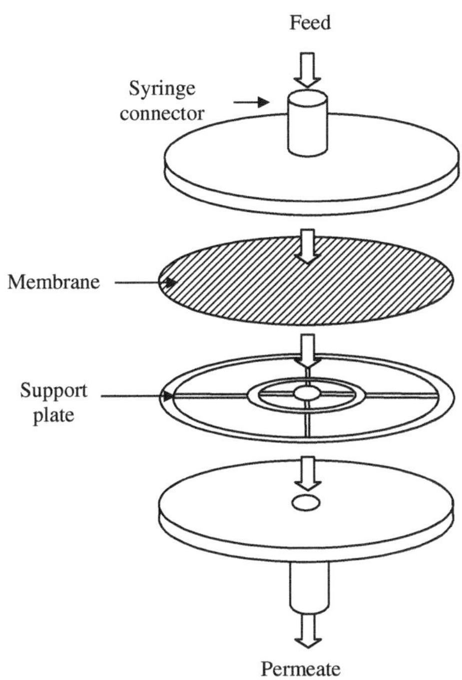

Figure 1.10 Schematic representation of a syringe filter.

centrifugal force. The concentrate is collected from the filter device sample reservoir using a pipettor, while the ultrafiltrate is collected in the provided centrifuge tube. Centrifugal membrane devices provide fast sample processing (typically between 15 and 60 min), high sample recovery (typically greater than 90% of dilute starting solution), for the processing of aqueous biological solutions in volumes ranging from 0.1 to 70 mL (Centricon Plus-70, Millipore). These devices are sold by companies such as Corning (USA), Millipore and Sartorius.

Stirred cell device

The stirred cell device has been used extensively in the 1980s to carry out theoretical studies on fouling and sieving of UF and MF membranes. It is still in use due to its facility and the ability to carry out experiments, for concentration, diafiltration and buffer exchange of macromolecule solutions, including proteins, enzymes, antibodies and viruses. In the stirred cell device, a flat sheet membrane is placed on an appropriate support plate and then sealed in place at the bottom of a plastic cylindrical housing. The fluid is agitated using a magnetic bar that is suspended from the top of the housing and remains just above the membrane surface. The stirred cell uses gentle magnetic stirring to minimize concentration polarization and shear denaturation of macromolecules. The device can be operated with air or nitrogen pressurization. It is commercialized by Millipore under the trade name Amicon and is available in different volume sizes to accommodate solutions from 3 to 400 mL.

1.3 Membrane characterization

The performance of membrane processes is directly associated with the overall membrane properties. Information on the chemical structure of the membrane, its permeation properties, its pore size, shape and distribution as well as its surface properties are of importance to both membrane manufacturers and users. It allows a useful prediction of separation performances of the membranes. Details on these methods can be found in many books [2,17].

1.3.1 Chemical structure of a membrane

Several analytical methods give information on the chemical structure of a membrane surface, its hydrophilicity and charge. The knowledge about the surface chemistry helps in a better understanding of membrane stability under different conditions and in the determination of fouling mechanisms and optimization of cleaning procedures. A very complete presentation of these complementary methods, their operation principles and their applications can be found in Ref. [23]. These analytical methods include Fourier transform infrared (FTIR) spectroscopy, Raman spectroscopy, energy-dispersive X-ray spectroscopy (EDS), electron spectroscopy for chemical analysis (XPS), Auger electron spectroscopy (AES) and secondary ion mass spectroscopy (SIMS). XPS, AES and SIMS are the most suitable methods for analysis of the skin layers of the membranes, while FTIR, Raman and EDS give information about both the skin layer and the bulk of the membranes. Raman, EDS, XPS, AES and SIMS give more detailed information on the chemical composition of the membrane surface than FTIR. In most of these methods, the samples have to be analysed in their dry state, because water produces strong signals that covers the information from the other compounds and/or because the samples have to be stable under vacuum conditions. A great advantage of Raman spectroscopy is the possibility of analysing samples in their wet state.

1.3.2 Transmembrane pressure and permeability

The transmembrane pressure (TMP) is the difference between the pressure on the feed side and on the permeate side. For a cross-flow device, the pressure on the feed side is evaluated as the mean of the pressures at the inlet and outlet of the device (Figure 1.11):

$$\mathrm{TMP} = \frac{(P_{\mathrm{in}} + P_{\mathrm{out}})}{2} - P_{\mathrm{f}} \tag{1.1}$$

where P_{in} and P_{out} are the pressure of the flowing bulk solution at the inlet and outlet of the device, respectively. The pressure on the filtrate side, P_{f}, is usually negligible.

The volumetric permeate flux through the membrane, J_{v}, is evaluated from experimental data by:

$$J_{\mathrm{v}} = \frac{M_{\mathrm{p}}}{\rho_{\mathrm{p}} A} \tag{1.2}$$

where M_{p} is the mass flow rate of the permeate, A the membrane surface area and ρ_{p}, the permeate density. The permeate flux is an essential parameter in the efficiency of the membrane process. Increasing the TMP increases permeate flux through the membrane.

An important indicator of the membrane functionality is its hydraulic permeability. The volumetric permeate flux, J_{v}, is related to the TMP:

$$J_{\mathrm{v}} = L_{\mathrm{p}} \mathrm{TMP} \tag{1.3}$$

where L_{p} is the hydraulic permeability of the membrane. The units for membrane permeability depend upon the units used for both permeate flux (usually ms^{-1}) and the TMP (usually bar, Pa or $\mathrm{N\,m}^{-2}$). Other definitions for the hydraulic permeability can be used to account for the effects of the solution viscosity and/or the membrane thickness.

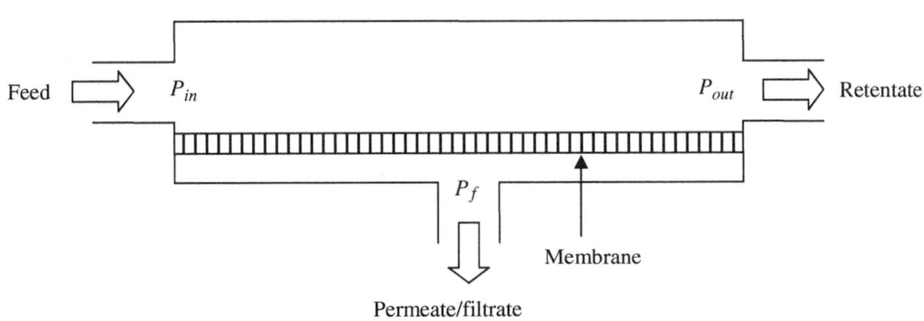

Figure 1.11 Transmembrane pressure in a cross-flow UF module.

The permeability defined as $L_p = \mu J_v/\text{TMP}$ has units of length (typically m or Å). The membrane resistance is equal to the reciprocal of the membrane hydraulic permeability.

For an idealized membrane composed of a parallel array of uniform cylindrical pores, L_p is typically expressed as [2]:

$$L_p = \frac{J_v}{\text{TMP}} = \frac{\varepsilon r^2}{8\mu\delta_m} \tag{1.4}$$

where ε is the membrane porosity, r the pore radius, μ the solvent viscosity and δ_m the membrane thickness. Equation (1.4) is only valid in the absence of osmotic pressure and/or solute rejection. It is often written with a tortuosity factor, τ, in the denominator to account for the tortuous and interconnected pore morphology of most membranes. The tortuosity factor provides a measure of the actual fluid path length through the membrane relative to the actual membrane thickness, δ_m. Typical values for τ are between 2 and 5 depending upon the porosity and morphology of the membrane.

The hydraulic permeability is determined by the membrane pore-size distribution, the membrane porosity and thickness as well as the solvent properties. Low ionic strength solutions will give slightly lower permeability due to the effects of counter-electro-osmosis associated with the non-zero streaming potential [24]. The permeability scales as the reciprocal of the solution viscosity. This effect can be significant with concentrated solutions.

1.3.3 Measurement of solute rejection

This method is a size exclusion technique. Depending on the pore sizes and surface properties of the membranes, different types of mixtures of solute/solvent solutions are employed. For UF membranes, macrosolute solutions such as dextrans and PEGs in water are often used. For characterizing RO or NF membranes, a mixture of mono- and multivalent salt solution is employed. The molecular weight cut-off (MWCO) is defined as the molecular weight above which molecules are at least 90% rejected by the membrane. For each solute, the intrinsic and the apparent solute rejections are, respectively, defined as:

$$R_i = 1 - \frac{C_p}{C_w} \tag{1.5}$$

and

$$R_a = 1 - \frac{C_p}{C_b} \tag{1.6}$$

where C_p is the species concentration in the permeate stream, C_w the species concentration at the membrane wall and C_b is the species concentration in the feed (or bulk) stream.

R_i is the intrinsic solute rejection, which takes into account the actual solute concentration at the membrane interface, which is not possible to measure. R_a is the observed solute rejection, which is measured by sampling the feed and permeate phases. The difference between R_i and R_a is due to the hydrodynamic resistance, which can be minimized by proper control of process parameters such as the concentration and the velocity of the feed solution. Details on the measurement of the MWCO can be found in Chapter 2.

1.3.4 Liquid displacement techniques

Methods for determining membrane pore sizes and pore size distribution by monitoring liquid permeation have been available for many years. The liquid displacement techniques are based on the displacement of a wetting liquid. In general, the membrane is first wetted by a liquid that is held in the pores by capillary forces. Air or a less wetting liquid used for displacement is then introduced on one side of the membrane. As the pressure of the air or the less wetting liquid is increased, the initial liquid held in the pores starts to displace from the largest pores until the liquid held in the smallest pores is eventually displaced. These techniques can be classified into two main methods: (1) bubble point method, that is use of air for the displacement of the initial liquid in the membrane pores; (2) liquid displacement method, that is use of a less wetting liquid for the displacement of the initial liquid [17].

Bubble point method

The bubble point method has been one of the most widely used methods for determining the membrane pore size and pore-size distribution. It is based on the principle that the pressure required to force an air bubble through a pore is inversely proportional to the pore size as described by the Laplace equation:

$$r_{\mathrm{p}} = \frac{2\sigma}{\Delta P}\cos\theta \tag{1.7}$$

where ΔP is the pressure difference across the membrane, σ the surface tension at the liquid/air interface, r_{p} the pore radius and θ the contact angle between the two liquids and the membrane pore wall.

After increasing the pressure over the capillary pressure, the liquid is expelled from the largest pores, allowing air to permeate. By successively increasing the pressure, smaller and smaller pores are opened for air permeation. The ideal flow versus pressure drop curve obtained is usually S shaped. This method is suitable for macropores characterization and can be applied to UF membranes with pore sizes up to 50 nm. The bubble point method is a very simple method to characterize membrane pore size and pore-size distribution. A disadvantage is that different results are obtained with different liquids. Moreover, the rate of pressure increase and the pore length may influence the result [17].

Liquid displacement method

This method is similar to the bubble point method. The difference is that instead of air a liquid is used to displace an initial liquid held in the pores of the membrane. The two liquids must be immiscible and the first one must easily wet the membrane, but not the second one. The principle is based on the Laplace equation (Equation (1.7)) and the displacement starts at the largest pores. By increasing the pressure, the liquid in smaller pores is displaced. In this way, the flow is obtained as a function of the pressure (or pore radius calculated using equation (1.7)). From this relation, the pore-size distribution is then obtained.

1.3.5 Other physical techniques

Mercury porosimetry

Mercury porosimetry can be used to evaluate the pore-volume distribution of the membranes. The method is based on the fact that mercury is a strongly non-wetting liquid on most materials. When mercury is forced into a dry membrane with the volume of mercury being determined at each pressure, a cumulative volume of mercury as a function of the applied pressure is established, from which the pore-size distribution is deduced. The relationship between the operating pressure and pore size of the membrane can be described by the Laplace equation (Equation (1.7)). This equation assumes cylindrical pores, which is generally not the case for most membranes; therefore, a morphological constant must be introduced to correct the results. During the experiment, the largest pores are filled with mercury at a certain minimum pressure. As the pressure increases further, smaller pores are filled until a maximum intrusion value is reached, that is all pores are filled. Thus, the pore-size distribution of the membrane can be determined because every pressure is related to one specific pore size.

Mercury porosimetry data can be difficult to interpret for membranes with significant pore throats since the volume behind the throat is measured at the operating pressure associated with the constricted diameter of the throat. In addition, the membranes must be characterized dry and mercury intrusion requires high pressures, both of which can alter the pore structure. Mercury porosimetry is most suitable for characterization of inorganic membranes with very stable structures.

Gas diffusion

The air diffusion method (also called the forward flow test) is performed at a pressure that is typically chosen at approximately 80% of the bubble point pressure (based on the maximum pore size or the size of the defects of greatest interest). The air flow rate through the wetted membrane is measured using an inverted graduated cylinder or an appropriate flow metre. Flow rates that exceed the manufacturer's specifications indicate the presence of defects that are large enough for the bubble point to be exceeded.

Liquid–liquid integrity tests

Liquid–liquid integrity tests are used to identify the presence of defects that could not be identified using bubble point measurements since the required pressures would exceed the pressure ratings of the membrane and/or device [25]. The method uses liquids with very low interfacial tension. The most common liquids are immiscible mixtures of alcohol and water or the two phases produced by a mixture of ammonium sulphate, PEG and water. The flow rate of the intrusion fluid is evaluated at a specified operating pressure, the membrane being thoroughly flushed with the wetting fluid. The module is then flushed with buffer, with the buffer flow rate evaluated at the same pressure. The CorrTest value (CTV) is defined as the logarithm (base 10) of the ratio of the buffer flow rate to the CorrTest fluid flow rate. The CTV below a critical level are indicative of membrane failure.

Streaming potential

The streaming potential provides a measure of the membrane surface charge. Data can be obtained for flow tangential to the membrane surface or flow through the membrane pores. The streaming potential is defined as the voltage that develops to balance the net convective flux of counter-ions through the pores (or along the surface) of an electrically charged membrane. The voltage can be measured by Ag/AgCl electrodes as a function of the TMP, with the slope of the resulting data used to evaluate the membrane zeta potential and surface charge density [26].

1.3.6 Microscopic techniques

Advances in the study of membrane structure have been made possible thanks to microscopic techniques such as scanning electron microscopy (SEM), transmission electron microscopy (TEM), atomic force microscopy (AFM) and confocal scanning laser microscopy (CSLM). Among these various techniques, the most widely used are SEM and AFM. Details on these various methods and others can be found in a very complete book focusing on monitoring and visualizing membrane-based processes by optical and electronic microscopic techniques, followed by electrical, laser and acoustic methods [27]. These microscopic methods applied to the characterization of membrane morphology were also presented recently in a book chapter by Kallioinen and Nyström [23].

Optical microscopy

Optical microscopy has been for a long time a routine tool for characterization of synthetic membranes [2]. The optical microscope is particularly useful for detection of large (>1 μm) membrane defects, macrovoids and foulant particles. During membrane preparation, it has been extensively used in quality control for detection of inhomogeneities, particles and gas bubbles in lacquers and melts. Polarized-light optical microscopy is also a powerful tool for observing changes in crystalline morphology of polymer systems such as spherulite formation. However, most membrane features are smaller than the wavelength, λ, of visible light ($\lambda = 0.4$–0.7 μm) and are not resolvable by optical microscopy.

Scanning electron microscopy

In SEM, a beam of electrons (with kinetic energy of 1–25 kV) is produced at the top of the microscope by heating a metallic filament [28]. The electron beam passes through electromagnetic lenses, which focus and direct it down towards the sample. Detectors collect the secondary or backscattered electrons, and convert them to a signal that is sent to a viewing screen. SEM has a resolution around 5 nm. SEM has been used for many years to characterize membrane structure and membrane fouling, for organic as well as inorganic membranes [29,30]. For example, Hwang and Lin [31] characterized by SEM three MF membranes with a mean pore size of 0.1 μm. They also observed the membrane fouling after filtration of a solution containing polymethyl methacrylate particles with a mean diameter of 0.4 μm.

For membranes with very small pores, SEM can only give qualitative information about the membrane morphology, because the resolution is not good enough at magnifications greater than 10^5 times. Field emission scanning electron microscopy (FESEM) has a better resolution of about 1.5 nm. For SEM and FESEM, membrane samples have to be dried and then coated with a charge conducting (usually gold) layer if they are non-conductive. The coated layer with a thickness of the order of a nanometre (which may be also the size of the smaller pores in the membrane) can cause problems in interpreting the images obtained. Typical images of organic and inorganic membrane surface obtained by SEM and FESEM are shown in Figures 1.2 and 1.3.

Transmission electron microscopy

In TEM, primary high-energy electrons are transmitted and diffracted through a thin specimen [28]. Therefore, before TEM, the dry membrane is embedded in a resin, and a very thin slice is cut using ultramicrotoming. TEM has a very high resolution (about 1 nm). However, due to the limitations in sample preparation, such as small cracks in samples caused by the cutting procedure, and the influence of the embedding resin on the membrane structure, resolution at 10 nm is often used to interpret the images obtained. Due to the limits of the technique, TEM has been much less used than SEM for membrane characterization. For example, Sheldon [32] examined the structure of polysulfone and cellulose UF membranes by performing C and C/Pt replicas and thin ultramicrotomed sections (70 nm). The method allowed to characterize the overall membrane structure including the finer structural elements such as surface pores in the range of 5–10 nm.

Atomic force microscopy

Atomic force microscopy (AFM) gives topographic images by scanning a sharp tip over a surface. In the contact mode, the tip is very close to the surface being imaged and is responding to very short range repulsive interactions with the sample. In the non-contact mode, the tip responds to attractive van der Waals interactions with the sample, the tip being generally at a distance of 5–10 nm away from the surface. This mode of operation is suitable for materials, such as polymeric membranes, which are soft or liable to mechanical damage as the forces used for imaging are lower than in the contact mode.

AFM was first used in the 1980s to study the structure of polymeric membranes [33]. The technique has been applied to organic and inorganic membranes, from RO to MF membranes [34–36]. AFM allows representing no conducting surfaces with a resolution of the order of the nanometre in dry and wet environments. AFM measurements give access to the surface roughness. The membrane surface roughness plays an important role in determining their fouling properties. Rough surfaces foul more easily because roughness increases surface area and affects other parameters, including water contact angle and zeta potential [37]. An increase in surface area increases the number of contact points between the foulants and the surface. Large protrusions can alter local flow fields and shield particles, such as bacterial cells from detachment. Using AFM, other data such as pore size, pore density and pore-size distribution can also be obtained.

Koyuncu et al. [38] examined two commercially available membranes with AFM: the HL NF membranes (GE Osmonics, USA) and the LFC-1 RO membrane (Hydranautics, USA). The dry membrane surfaces were imaged with AFM in tapping mode (Figure 1.12). Root mean square (R_q) roughness values were 21.2 and 105.3 nm for HL and LFC-1 membranes, respectively, at a scan-size of 100 μm^2. From AFM results, Wong et al. [37] established a fractal relationship relating roughness to length scale over several orders of magnitude for a series of MF, UF and NF membranes. The scale dependence of the root-mean-square roughness was first investigated. The fractal dimension was computed, and used to describe roughness at any length scale for which the fractal assumption was valid. The relationship between membrane roughness, MWCO and membrane zeta potential were also examined.

One drawback of the AFM technique is the limitation of the scanning depth due to the size of AFM scanning probe tips. Another limitation of AFM is the relatively small area that can be scanned, the maximum scan area being approximately 100 μm^2 [38]. This limitation is principally due to the operational set-up and the size of the cantilever used to produce the AFM images. Finally, AFM may distort membrane pore size due to rounded corners near pore entrance [39].

Confocal scanning laser microscopy

In CSLM, the sample is illuminated and observed so as to rid the resulting image of the light from out-of-focus planes. Imaging may be performed in the reflective or in the fluorescence mode. Therefore, CSLM allows accurate and non-destructive optical sectioning in a plane perpendicular or parallel to the optical axis of the microscope. A series of images can be recorded at different depths by changing the position of the focalization plane. These images can then be computer processed to present the information as a complete three-dimensional reconstruction.

Figure 1.13 shows an Immunodyne membrane (0.45-μm pore size, Pall, USA) observed by CLSM [40]. Prior to observation, the membrane was mounted in immersion oil. Optical sections were recorded at 1-μm intervals, from 0 to 60 μm. The pores located at 3 μm were much larger than the pores in the bulk of the membrane and had a hexagonal structure (Figure 1.13a–c: $z = 0$, 3 and 10 μm). At 60-μm depth (Figure 1.13d), the reinforced fibre structure of the membrane appeared clearly.

Three dimensional (3D) morphology characterization is a unique advantage of CSLM. CSLM requires only minimal specimen preparation. Unlike SEM and TEM, it does not require any metal coating or heavy-metal staining. CSLM provides the advantage that wet samples can be studied directly, like other methods of microscopy such as AFM, scanning probe microscopy and environmental SEM. CSLM has been

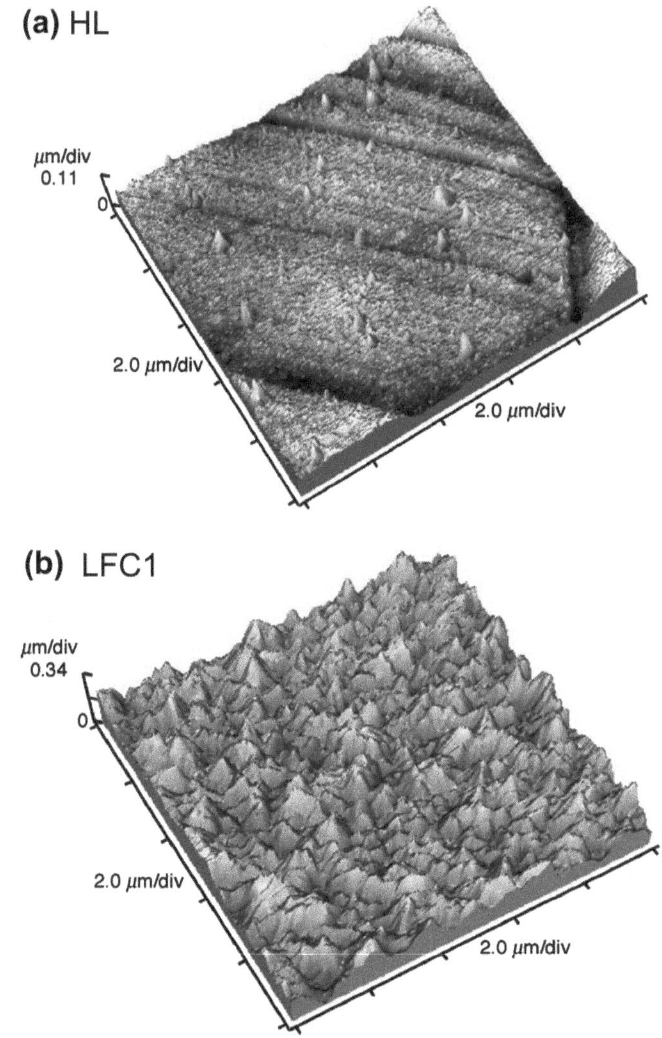

Figure 1.12 AFM images with a scan size of 100 μm^2: (a) HL and (b) LFC-1 membrane surfaces [38]. Original in colour.
This article was published in J. Membr. Sci., Vol. 278, I. Koyuncu, J. Brant, A. Luttge, M.R. Wiesner, A comparison of vertical scanning interferometry (VSI) and atomic force microscopy (AFM) for characterizing membrane surface topography, pp. 410–417, Copyright Elsevier (2006).

used to provide a 3D representation of MF membranes [41]. CSLM has also been applied for visualizing protein adsorption to ion-exchange membranes [42] and characterization of membrane fouling during filtration [43]. The main drawback of the technique is its low resolution, larger than 0.1 μm. CSLM is therefore only applicable to MF membranes.

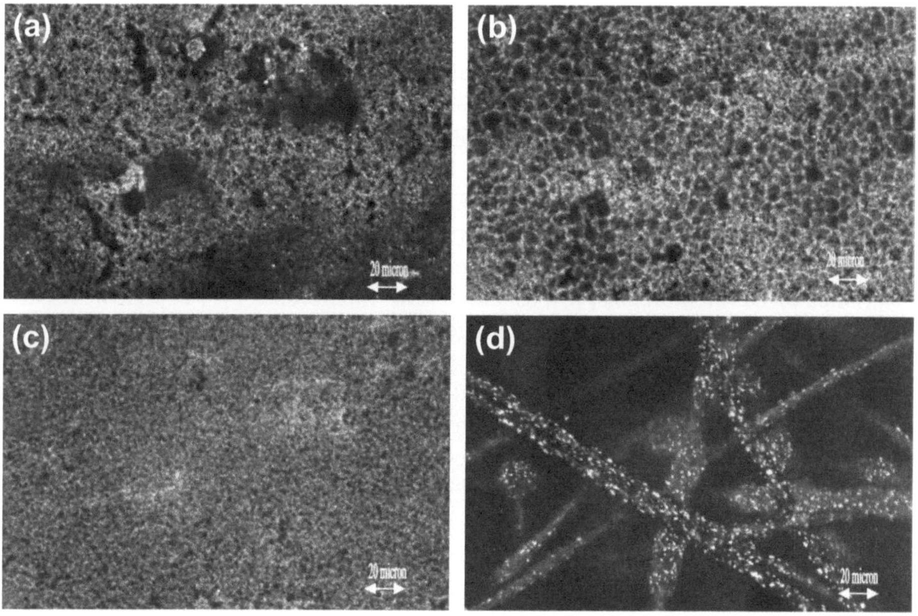

Figure 1.13 Confocal CSLM image of the Immunodyne membrane mounted in immersion oil. (a) $z = 0$ μm, (b) $z = 3$ μm, (c) $z = 10$ μm and (d) $z = 60$ μm (×63 oil immersion objective, numerical aperture 1.4) [40].
This article was published in Chem. Eng. Sci., Vol. 55, C. Charcosset, A. Cherfi, J.-C. Bernengo, Characterization of microporous membrane morphology using confocal scanning laser microscopy, pp. 5351–5358, Copyright Elsevier (2000).

Other techniques

SEM and AFM are the two most popular techniques for characterizing membrane structure and fouling. Other techniques such as optical interferometry, near infrared non-linear optical imaging and modern synchrotron radiation can be used. Koyuncu et al. [38] showed that the roughness values obtained by optical interferometry (scanned area = 64 μm^2 and 0.05 mm^2) were higher than those obtained by tapping mode AFM (scanned area = 100 μm^2). This was explained by the fact that membrane surface roughness increased with increasing scan size, until a critical scan size of 250,000 μm^2 was reached. The area that can be scanned by AFM is well below this critical value, so the results obtained using this technique can be misleading [38]. Another optical technique, two-photon femtosecond near infrared non-linear optical imaging, was developed by Hugues et al. [44] to give a 3D representation of flat membranes fouling by yeast suspensions. The use of synchrotron radiation computed microtomography also provides 3D visualization from 2D images. Remigy and Meireles [45] applied this method to describe the geometry of the pores and the 3D structure of polysulfone and PVDF-HFP hollow fibres. However, using 2D images to obtain this 3D representation requires quite advanced data processing software. Moreover, due to its low resolution, the technique is limited to the study of MF membranes.

Image analysis

Typical software for image analysis of membrane pore dimensions and shapes calculate parameters like porosity, pore-diameter distribution, pore-area distribution and pore-shape distribution on the membrane surface. The main difficulty in characterizing pore dimensions by SEM image analysis is the projection-type distortion. Projected pore images are smaller than the actual ones. This effect complicates measurements of the "throat-type" pores. The size of these pores evaluated from SEM images is further diminished due to their partial concealment by adjacent structures. The SEM image analysis method has been widely applied to studies of pore characteristics of UF membranes [46], as well as MF membranes [47–49]. However, the results obtained depend on the digital image processing program package used. Using two digital image processing program packages, Sun et al. [49] characterized different membranes, including polyacrylonitrile (PAN), polyaryletherketone with cardo (PEK-C), polyarylethersulfone with cardo (PES-C), polysulfone (PS), PVDF and poly(phthalazinone ether sulfone ketone) (PPESK) membranes. Comparing the analysis results of the two image analytical instruments, the porosity and the dispersion of distribution were found different.

1.4 Fouling

1.4.1 *Physical and chemical phenomena*

One of the major factors limiting the use of membranes in almost all applications is fouling, which is the irreversible alteration of the membrane caused by physical and/or chemical interactions between the membrane and components present in the feed solutions. Fouling typically manifests itself as a decrease in filtrate flux and/or membrane selectivity. This alteration requires cleaning or replacement of the membrane. The effects of fouling on filtrate flux and selectivity are very similar to those associated with concentration polarization. However, concentration polarization is a reversible phenomenon, which can be controlled by adjusting the device design and operating conditions. In contrast, membrane fouling occurs because of specific interactions between the various solutes and particles and the membrane. The changes in membrane properties can often be reversed by applying very high cross-flow velocities, backflushing or chemical cleaning.

Proteins

Membrane fouling, caused by the non-specific adsorption and deposition of proteins on the membrane surface and in the pores, results in a sharp decline in permeate flux (see also Chapter 2). Protein fouling is strongly influenced by the physical and chemical characteristics of the membrane. Monolayer protein adsorption is typically found throughout the internal pore surfaces of MF membranes. Ultrafiltration membranes with pores that are only slightly larger than the size of the protein are affected by 'steric' exclusion in the narrow pores. Protein adsorption to fully retentive membranes is primarily a surface phenomenon, since the membrane pores are virtually inaccessible to the proteins.

It is commonly accepted that aggregation, denaturation and deposition of proteins on the membrane are responsible for membrane fouling [50]. Protein fouling depends on feed properties (concentration, pH and ionic strength, prefiltration and the removal of aggregates, and component interactions), membrane materials (hydrophobicity, charge effects, surface rugosity and roughness, porosity and pore-size distribution) and process parameters (TMP, cross-flow velocity and turbulence promoters, back-flushing and temperature) [51–53]. Several techniques have been developed to provide visualization and monitoring of protein fouling, differentiation between adsorbed protein species, the detection of forces responsible for fouling, and the non-invasive in situ monitoring of concentration polarization [54]. Such techniques included SEM and TEM, electron paramagnetic resonance spectroscopy (EPRS), attenuated total reflection–Fourier transform infrared spectroscopy (ATR–FTIR), ellipsometry and small angle neutron scattering (SANS).

Bacteria

Bacterial fouling can also be a major problem in many membrane systems (see also Chapter 3). Although much of the bacterial cell surface is hydrophilic, many bacterial surfaces have hydrophobic properties and are able to attach to polymeric surfaces by strong hydrophobic interactions. In addition, many bacteria can adhere to different surfaces. Once bound to the membrane surface, these bacteria can grow and multiply using the nutrients present in the feed solution. Feed solutions with high levels of total organic carbon or high total biological oxygen demands are thus likely to cause severe biofouling. Hence, even 99.99% removal of microorganisms from a feed stream can still lead to the eventual formation of a fouling layer [55]. Differences in bacteria adhesion to surfaces over long times may result from biological factors such as the growth phase and nutritional condition of the microorganisms. In addition, the initial attachment is largely controlled by physicochemical factors such as solution chemistry, substrate surface properties and hydrodynamic conditions [55–58].

Inorganic species

Many inorganic compounds are naturally present in the feed solutions and process waters used in industry and can cause significant fouling of membranes. They include calcium sulphate, calcium carbonate, calcium phosphate, silica, metal oxides and hydroxides (particularly of iron and aluminium), colloidal sulphur and other inorganic particulates. Calcium salts are major inorganic foulants in brackish water, wastewater and agricultural drainage water processing [59,60]. Inorganic (salt) fouling can be also a problem in MF systems used for biotechnological applications. For example, Nagata et al. [61] have used a fermentation media containing both K_2HPO_4 and urea. These species reacted during steam sterilization (autoclaving) to form a $K_2NH_4PO_4$ precipitate, which was then deposited on the membrane surface during MF of the broth to harvest the cells. This problem could be eliminated by reformulating the fermentation medium, by separately sterilizing the different medium components to avoid this chemical reaction, or by using a different method of sterilization.

1.4.2 Modelling

The flux decline associated to fouling by macrosolutes and/or particles is generally explained by three separate mechanisms: pore blockage, pore constriction and cake filtration [2]. The effects of these fouling mechanisms can be understood using the Darcy's law modified to account for the osmotic pressure effect:

$$J_{\mathrm{v}} = \frac{\mathrm{TMP} - \sigma_{\mathrm{o}}\Delta\Pi}{\mu(R_{\mathrm{m}} + R_{\mathrm{c}})} \tag{1.8}$$

where σ_{o} is the membrane osmotic reflection coefficient, $\Delta\Pi$ is the osmotic pressure difference across the membrane and R_{m} and R_{c} are the hydraulic resistance of the membrane and particle cake, respectively. The membrane resistance can be evaluated from equation (1.9) assuming a uniform membrane pore size using the Hagen–Poiseuille equation for laminar flow:

$$R_{\mathrm{m}} = \frac{8\delta_{\mathrm{m}}}{N\pi r_{\mathrm{p}}^{4}} \tag{1.9}$$

where N is the number of pores per unit membrane area and r_{p} is the pore radius. The three fouling mechanisms associated with macrosolutes and/or particle deposition correspond respectively to a reduction in the number of accessible pores, N, a decrease in the effective pore radius, r_{p} and an increase in the resistance provided by a deposited particle cake on the upper surface of the membrane, R_{c}. Pore constriction is only possible for membranes with relatively large pores that are easily accessible to the macrosolutes and/or particles. Pore blockage and cake formation dominate when the pores are smaller than the particles.

Hermia [62] has reviewed the development of the pore blockage, pore constriction and cake formation models for dead-end filtration (Figure 1.14). The

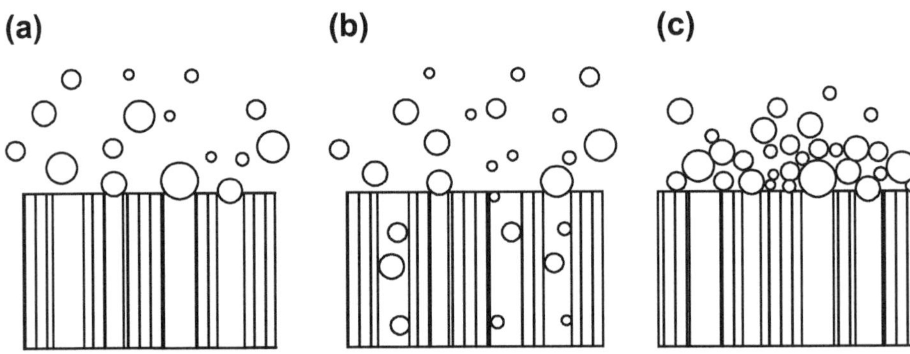

Figure 1.14 Pore blocking mechanisms: (a) complete pore blocking model $n = 2$, (b) intermediate blocking model $n = 1$ and (c) cake filtration model $n = 0$.

results are expressed in terms of filtration time t and cumulative filtrate volume V. Typically, the osmotic pressure term in equation (1.8) is assumed to be negligible due to the large size of the macrosolutes and/or particles. The rate of particles transport back into the bulk solution is assumed to be negligible due to the slow rate of Brownian particle diffusion and the absence of any shear-induced particle motion or inertial lift in the dead-end flow system. Long-range intermolecular (e.g. electrostatic and van der Walls) interactions are also assumed to be negligible, although this assumption is not valid for all systems.

In the pore blockage model, the rate of change in the number of open pores is assumed to be related to the rate of macrosolutes and/or particles convection to the membrane surface:

$$\frac{dN}{dt} = -\alpha_{\text{block}} AJ_v C_b \tag{1.10}$$

where A is the frontal area of the membrane and α_{block} provides a measure of the pore blockage efficiency. Cake formation is assumed to be negligible ($R_c = 0$ in equation (1.8)). Substitution of equation (1.10) into equations (1.8) and (1.9) yields, upon integration:

$$\frac{J_v}{J_0} = \exp\left(-\frac{\alpha_{\text{block}} AJ_0 C_b}{N_0} t \right) \tag{1.11}$$

where J_0 is the initial filtrate flux through the clean membrane and N_0 is the initial pore density.

In the pore constriction model, the rate of change in the pore volume (assumed to be cylindrical) is assumed to be proportional to the rate of particle convection to the membrane:

$$\frac{d}{dt}(\pi r_p^2 \delta_m) = -\alpha_{\text{pore}} AJ_v C_b \tag{1.12}$$

yielding:

$$\frac{J_v}{J_0} = \left(1 + \frac{\alpha_{\text{pore}} AJ_0 C_b}{\pi r_0^2 \delta_m} t \right)^{-2} \tag{1.13}$$

In the cake filtration model, the hydraulic resistance provided by the particle cake is assumed to be proportional to the cake mass, m_{cake}:

$$R_c = \frac{\alpha_{\text{cake}} m_{\text{cake}}}{A} \tag{1.14}$$

with the rate of particle deposition directly related to the rate of particle convection:

$$\frac{dm_{\text{cake}}}{dt} = AJ_v C_b \tag{1.15}$$

substitution of equations (1.14) and (1.15) in equation (1.8) yields:

$$\frac{J_v}{J_0} = \left(1 + \frac{2\alpha_{\text{cake}} J_0 C_b}{R_m} t \right)^{-1/2} \tag{1.16}$$

All of these classical models for flux decline can be described by a single mathematical expression:

$$\frac{d^2t}{dV^2} = K\left(\frac{dt}{dV}\right)^n \tag{1.17}$$

where t is the filtration time, V the cumulative filtrate volume, K a proportionality constant and n is an exponent which depends on the fouling model. For the cake filtration model, the exponent is $n = 0$. In the intermediate blocking model, $n = 1$. If fouling occurs by complete pore blocking, $n = 2$. An additional pore constriction model may be introduced, which assumes that foulants are deposited on the inner surface of the pore thereby constricting its diameter. In this case, the exponent is $n = 1.5$. The first derivative, dt/dV, is the inverse of the instantaneous filtrate flow rate, while the second derivative, d^2t/dV^2, is proportional to the rate of increase in the total resistance to filtration.

Recently, it has been demonstrated that the combining of these models lead to better agreement with fouling data [63–65]. For example, Ho and Zydney [63] developed a flux decline model accounting for initial fouling by pore blockage ($n = 2$) followed by cake formation ($n = 0$) to compare with experimental data of sterile filtration of bovine serum albumin (BSA) solutions through a track-etched polycarbonate membrane. The model was found to be in excellent agreement with experimental data obtained during the stirred cell filtration of BSA solutions through polycarbonate track-etched MF membranes over the entire course of the filtration (Figure 1.15). The model provides a smooth transition from the pore blockage to cake filtration regimes, eliminating the need to use different mathematical formulations to describe these two phenomena.

These models assumed that the membranes are composed of an array of straight-through cylindrical pores. Ho and Zydney [66] developed further a mathematical model explicitly accounting for fluid flow through composite membranes formed with two layers: an upper layer with non-interconnected pores and substructure with highly interconnected pores. The analysis of the effects of asymmetric membrane structure was also presented recently [67].

1.5 Cleaning

Membrane cleaning is an essential component of almost all membrane processes. The membrane fouls during operation, causing the membrane performance (i.e. flux and/or selectivity) to drop below a certain acceptable level. The foulants must then be removed from the membrane using an appropriate cleaning procedure. Cleaning requires membranes to be removed from the process set-up and thus has a significant impact on process operations and the commercial viability of the process.

Cleaning can be accomplished by physically removing the foulants, for example, by backflushing, and/or by using a specific cleaning solution containing

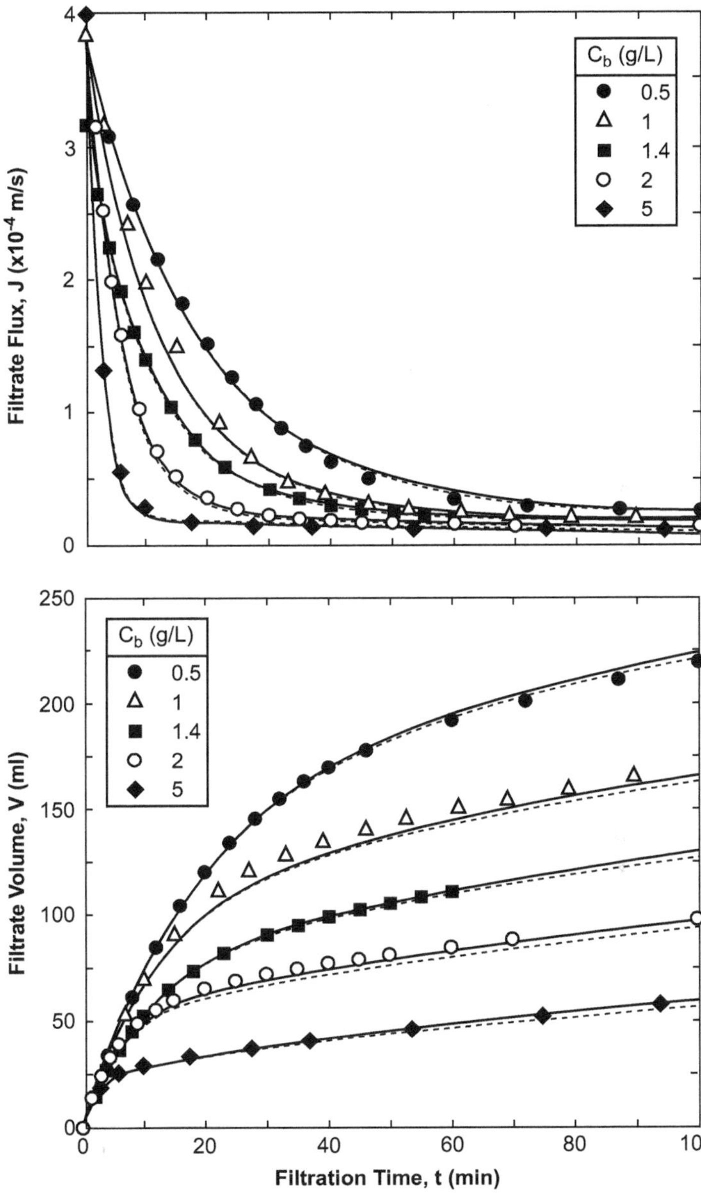

Figure 1.15 Filtrate flux (top) and total filtrate volume (bottom) for BSA filtration through the PCTE membranes at 14 kPa. Solid and dotted curves are model calculations using the complete model accounting for initial fouling by pore blockage ($n = 2$) followed by cake formation ($n = 0$) and the approximate analytical solution [63].
This article was published in J. Colloid Interface Sci., Vol. 232, C.C. Ho, A.L. Zydney, A combined pore blockage and cake filtration model for protein fouling during microfiltration, pp. 389–399, Copyright Elsevier (2000).

appropriate detergents and/or chemicals. The cleaning treatment must effectively remove and/or dissolve the foulants while not exceeding the mechanical or chemical limits of the membrane. The most common procedure for assessing the effectiveness of a cleaning treatment is measuring the recovery of water permeate flux. As different foulants may influence permeation characteristics, other methodologies, such as microscopic techniques or surface characterization, could also be considered.

1.5.1 Physical cleaning

Some foulants can be effectively removed from the membrane surface by physical cleaning [2]. Physical cleaning can be accomplished by removing the TMP and circulating the feed solution through the membrane module. The feed solution can be replaced with a rinse solution, which is then circulated through the membrane module at relatively high shear rates. This method is performed at zero TMP difference as it avoids the redeposition of suspended solids by the convective filtrate flux. However, this technique is usually not effective enough to restore 100% of the original flux.

Hollow fibre membranes and some flat sheet membranes with appropriate feed-channel (mesh) spacers can be cleaned by periodic backflushing in which a negative TMP is used. Backflushing is very effective at removing particle cakes from the membrane surface. It can also remove foulants from the membrane pores when a chemical cleaning solution is used. Rapid backpulsing, which involves reversing the TMP for a fraction of a second once every few seconds, has been investigated for various foulants, with up to 30-fold improvements in flux [68–70]. For example, Mores and Davis [70] investigated yeast cells deposited on CA membranes during cross-flow MF. The membranes were then cleaned using individual backpulses at varying backpulse durations, TMPs and shear rates. The recovered flux was observed to increase with increasing shear rate, backpulse pressure and backpulse duration, reaching up to 98% of the clean membrane flux. Ultrasonic cleaning is another potential physical cleaning method [71].

1.5.2 Chemical cleaning

The selection of a cleaning agent must take into account the fouling substances present, the process conditions, the nature of the membrane material and the formulation and concentration of cleaning agent [2]. Selected cleaning solutions must be chemically stable, safe, cheap and washable with water. Moreover, chemical cleaning should be limited to avoid changes in membrane performance and to minimize the amount of chemical reagents. Chemical cleaning methods depend on a number of different mechanisms, including (1) displacement of the foulants from the membrane (e.g. by competitive adsorption of appropriate surface-active agents); (2) solubilization of the foulants (e.g. by changing the foulant solubility or by providing an appropriate emulsifying, dispersing or peptizing agent) and/or (3) chemical modification of the foulants (e.g. saponification of fats and oils, oxidation or degradation of proteins, chelation of divalent cations, or reaction of metal oxides with appropriate acids). There are numerous cleaning solutions available, including acidic or alkaline solutions that may or may not contain enzymes and/or surfactants.

Cleaning solutions

Acid solutions are effective at removing calcium salts (carbonate and phosphate) and metal oxides. In both cases, these components are dissolved by reaction with the acid to form a soluble salt. Acid cleaners include hydrochloric acid (HCl), sulphuric acid (H_2SO_4) and nitric acid (HNO_3). However, acid solutions are usually very corrosive for membranes and modules. Phosphoric acid (H_3PO_4) is much less corrosive but also more expensive. Organic acids such as citric acid can also be used for cleaning metal oxides and calcium precipitates. They are much weaker, and thus cause less damage to membranes and modules.

Alkaline solutions are effective cleaning agents for silica, inorganic colloids, and many biological/organic foulants. Alkaline cleaners work by several different mechanisms, including neutralization of acidic materials (e.g. fatty acids and humic acids), saponification (hydrolysis) of fats and oils, and dispersion/emulsification of colloid materials. A large number of alkaline cleaners are available including hydroxides (caustic soda, NaOH), carbonates (soda ash, Na_2CO_3), silicates (sodium metasilicate, Na_2SiO_3) and phosphates (e.g. sodium tripolyphosphate, trisodium phosphate). NaOH is often used in membrane cleaning after bioprocesses [72]. For example, an expanded PTFE membrane flushed with caustic soda solution was found to be the most suitable and efficient membrane for MF of BSA with yeast suspension without significant reduction in flux and protein transmission, even after several filtration cycles [72,73]. Sodium hypochlorite (NaOCl) is also used as an alkali cleaner and is often recommended by membrane suppliers as cleaning agent. Depending on membrane materials and feed characteristics, NaOCl has been shown to provide a wide range of cleaning performances [74]. However, the oxidizing properties of NaOCl and its elevated working pH have the potential to damage membranes over their lifetimes.

Surfactants have part of the molecule soluble in water and the other part soluble in oil. Surfactants are effective cleaning agents by (1) displacing foulants from the membrane surface due to their strong surface adsorption; (2) emulsifying oils and (3) solubilizing hydrophobic foulants by incorporating them into surfactant micelles. Surfactants are available with a wide range of chemical structures. They can possess neutral (non-ionic), negatively charged (anionic), or positively charged (cationic) hydrophilic groups. For example, a non-ionic surfactant (Tween 20) was chosen for cleaning of a PES membranes after lysozyme processing [75]. The efficiency of the non-ionic surfactant to desorb lysozyme from PES was discussed according to contact angles measurements. A combination of SDS as an anionic surfactant and Triton X-100 or Triton X-114 as non-ionic surfactant has been found the most efficient cleaning agents for flux recovery of the PES UF membrane fouled by whey [76]. The cationic surfactants (CTAB, TTAB, DTAB) exhibited moderate effect.

Enzymatic cleaners can be used for membranes that cannot withstand elevated temperatures, strong chemicals, or pH extremes. Enzymatic cleaning has the benefits of operating in mild conditions, reduced chemical usage, lower energy costs due to lower cleaning temperatures, and biodegradable effluents. Enzymes are available that degrade proteins (proteases), starches (amylases), fats and oils (lipases) and cellulose (cellulases). Some of the most popular protein removers are neutral enzymatic detergents that contain complexing agents and a combination of organic and inorganic surfactants. Enzymatic cleaners are most effective when operated at a concentration that optimizes the cutting of the proteins [77]. Use of higher concentrations does not increase the enzymatic cleaning efficiency. The use of a water rinse during the cleaning procedure can be an effective method of removing loose foulant materials, if the rinsing is carried out at the same temperature as the chemical cleaning. For example, Petrus et al. [78] used single and binary protein solutions of BSA and beta-lactoglobulin as the feed solution for filtration process, and a proteinase as the enzymatic cleaning agent. Their experimental results indicated that optimum cleaning time and cleaning agent concentration was due to the competition between foulant removal and deposition of enzymes on the membrane during the cleaning process.

Parameters

The effectiveness of chemical cleaning is a function of chemical action and physical parameters. Chemical action is defined by the nature of the cleaning solution, its concentration, pH and ionic strength. Physical parameters include temperature, TMP and cross-flow velocity. The effect of cross-flow velocity may differ according to the cleaning solution used. For a PES UF membrane fouled by natural organic matter and cleaned with a 0.1 M NaOH solution, an increased in cross-flow velocity improved cleaning efficiency, whereas additional benefits of a higher cross-flow velocity were negligible for the high ionic strength cleaning solution 0.1 M NaCl [79]. Chemical cleaning is usually operated at a low TMP, lower to that used during the process. Under these conditions, the compressible fouling layer is relaxed and the additional irreversible fouling that arises from a high applied TMP can be decreased or eliminated [80,81].

Cleaning duration is also an essential parameter which is related to cleaning kinetics. A certain cleaning time is required to complete chemical reactions within fouling layers, while a longer time may lead to the re-fouling of the membrane by the cleaning reaction products previously released in the cleaning solution [82,83]. This effect was clearly demonstrated in a study where a 0.1-μm tubular ceramic MF membrane fouled by a 3.5 wt% whey protein concentrate suspension was cleaned using sodium hydroxide [81]. Hydraulic cleanliness was evaluated using three parameters: (1) percent flux recovery, (2) percent irreversible removed fouling and (3) a hydraulic cleanliness criterion representing the ratio of residual fouling resistance to initial membrane resistance. Figure 1.16 shows the influence of cleaning time on the hydraulic cleanliness criterion and the flux recovery at the following operating conditions: TMP = 0.5 bar, $v = 4\ \mathrm{ms}^{-1}$ (above which no improvement in flux recovery was observed). The hydraulic cleanliness criterion as well as flux recovery were not significantly affected by cleaning time, with average values equal to 0.26% and 79.7%,

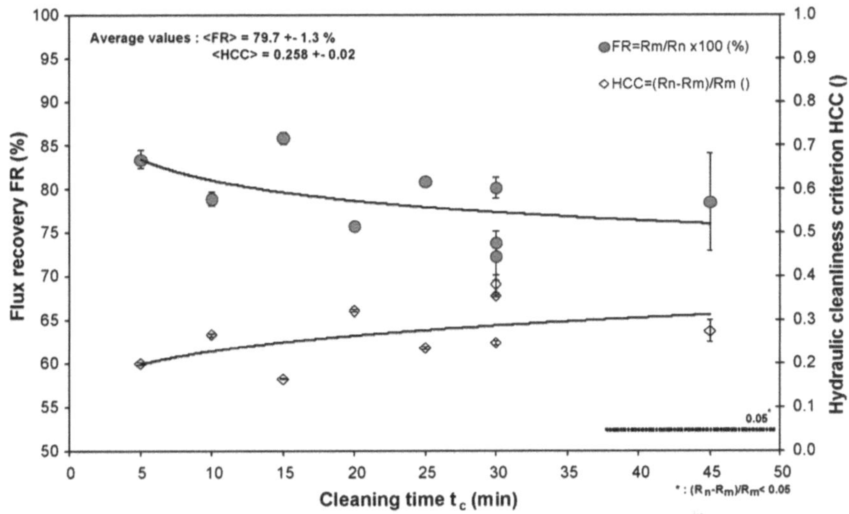

Figure 1.16 Effect of cleaning time on both percent flux recovery and hydraulic cleanliness criterion for NaOH cleaning. Operating conditions: 1 wt% NaOH, 55 °C, $TMP = 0.5$ bar, $v = 4\ ms^{-1}$. Microfiltration membrane fouled by a 3.5 wt% whey protein concentrate suspension [81]. Original in colour.
This article was published in J. Membr. Sci., Vol. 337, P. Blanpain-Avet, J.F. Migdal, T. Bénézech, Chemical cleaning of a tubular ceramic microfiltration membrane fouled with a whey protein concentrate suspension – Characterization of hydraulic and chemical cleanliness, pp. 153–174, Copyright Elsevier (2009).

respectively. Cleaning efficiency showed no significant improvement at cleaning times higher than 5 min which means that the major part of protein fouling was removed within the first minutes. At this time, cleaning was not effective anymore and a redeposition of foulant species previously released in the retentate occurred.

Microscopic techniques have been investigated to characterize cleaning results, such as SEM, streaming potential and FTIR measurements [84], multi-photon microscopy, a non-linear optical imaging technique capable of producing 3D images with sub-micron resolution [83], and confocal laser scanning microscopy [85]. For example, Väisänen et al. [86] studied the cleaning efficiency of selected cleaning agents on membranes using the combination of complementary surface characterization methods. Data from FTIR–ATR spectra allowed to determine the foulants present and the changes in foulant composition after the cleaning step. Using SEM pictures, the foulants were located primarily inside the membrane structure. Statistical factorial design was employed to identify the key factors as well as their interactions in both physical and chemical cleaning [87].

The use of CSLM to characterize mechanical cleaning of membranes was reported by Spettmann et al. [85]. CSLM in conjunction with image analysis was used for evaluating the efficiency of ultrasonic treatment by comparing fouled membranes before and after the removal of the deposit. Recently, Zator et al. [88] applied CSLM

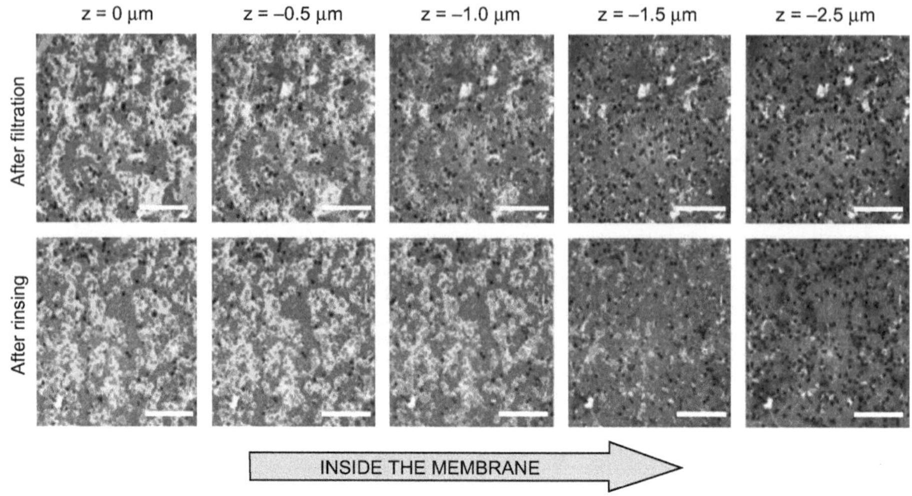

Figure 1.17 A 3D orthogonal reconstruction of CSLM images: membranes after the filtration of a 0.5 g/L BSA–FITC/dextran–TRITC 150 kDa solution; membrane after water rinsing. Images show the membrane (dark grey signal) and the protein (light grey signal) at different depths. Scale bar 10 μm [88]. Original in colour.
This article was published in J. Membr. Sci., Vol. 327, M. Zator, J. Warczok, M. Ferrando, F. López, C. Güell, Chemical cleaning of polycarbonate membranes fouled by BSA/dextran mixtures, pp. 59–68, Copyright Elsevier (2009).

to determine the efficiency of a water rinsing cycle and the use of a chemical cleaning agent for the removal of foulants deposited on polycarbonate membranes after the filtration of a model solution containing a protein (BSA) and a polysaccharide (dextran). A rinsing cycle, different cleaning times and different concentrations of the cleaning agent were used. Results on flux recovery, resistance reduction, amount of protein/polysaccharide detected inside the pores (calculated from CSLM images) and 3D reconstructions of CSLM images were used to compare the effect and efficiency of the cleaning protocols. In Figure 1.17, CSLM images illustrate the extent of membrane fouling caused by deposition/adsorption of BSA–FITC on the membrane surface and pore blockage at various membrane depths after the filtration of a BSA–FITC/dextran–TRITC 150 kDa solution and after rinsing the fouled membrane for 30 min with water. The protein (light grey signal) blocked most of the pores up to a depth of 1.5 μm inside the membrane after rinsing, while less protein was detected on the surface and inside the membrane after the filtration for the same depth.

References

[1] C. Charcosset, Downstream processing and product recovery/membrane systems and technology, in: M.Y. Murray (Ed.), Comprehensive Biotechnology, second ed., vol. 2, Elsevier, 2011, pp. 603–618.

[2] L.J. Zeman, A.L. Zydney, Microfiltration and Ultrafiltration, Principles and Applications, Marcel Dekker, New York, 1996.
[3] R.W. Baker, Membrane Technology and Applications, McGraw-Hill, New York, 2000.
[4] W.B. Krantz, A.R. Greenberg, D.J. Hellman, Dry-casting: computer simulation, sensitivity analysis, experimental and phenomenological model studies, J. Membr. Sci. 354 (2010) 178–188.
[5] M. Zeng, Z. Fang, C. Xu, Effect of compatibility on the structure of the microporous membrane prepared by selective dissolution of chitosan/synthetic polymer blend membrane, J. Membr. Sci. 230 (2004) 175–181.
[6] M. Wang, L.G. Wu, J.X. Mo, C.J. Gao, The preparation and characterization of novel charged polyacrylonitrile/PES-C blend membranes used for ultrafiltration, J. Membr. Sci. 274 (2006) 200–208.
[7] W.R. Bowen, S.Y. Cheng, T.A. Doneva, D.L. Oatl, Manufacture and characterisation of polyetherimide/sulfonated poly(ether ether ketone) blend membranes, J. Membr. Sci. 250 (2005) 1–10.
[8] Y.L. Su, W. Cheng, C. Li, Z. Jiang, Preparation of antifouling ultrafiltration membranes with poly(ethylene glycol)- *graft*-polyacrylonitrile copolymers, J. Membr. Sci. 329 (2009) 246–252.
[9] M. Sivakumar, D. Mohan, R. Rangarajan, Studies on cellulose acetatepolysulfone ultrafiltration membranes. II. Effect of additive concentration, J. Membr. Sci. 268 (2006) 208–219.
[10] G. Arthanareeswaran, K. Srinivasan, R. Mahendran, D. Mohan, M. Rajendran, V. Mohan, Studies on and sulfonated poly(ether ether ketone) blend ultrafiltration membranes, Eur. Polym. J. 40 (2004) 751–762.
[11] G. Arthanareeswaran, D. Mohan, M. Raajenthiren, Preparation, characterization and performance studies of ultrafiltration membranes with polymeric additive, J. Membr. Sci. 350 (2010) 130–138.
[12] H. Susanto, M. Ulbricht, Characteristics, performance and stability of polyethersulfone ultrafiltration membranes prepared by phase separation method using different macromolecular additives, J. Membr. Sci. 327 (2009) 125–135.
[13] M.M. Demir, I. Yilgor, E. Yilgor, B. Erman, Electrospinning of polyurethane fibers, Polymer 43 (2002) 3303–3309.
[14] Y.Z. Zhang, C.T. Lim, S. Ramakrishna, Z.M. Huang, Recent development of polymer nanofibers for biomedical and biotechnological applications, J. Mater. Sci. 16 (2005) 933–946.
[15] J.P. Chen, K.H. Ho, Y.P. Chiang, K.W. Wu, Fabrication of electrospun poly(methyl methacrylate) nanofibrous membranes by statistical approach for application in enzyme immobilization, J. Membr. Sci. 340 (2009) 9–15.
[16] Y. Wang, Y.-L. Hsieh, Immobilization of lipase enzyme in polyvinyl alcohol (PVA) nanofibrous membranes, J. Membr. Sci. 309 (2008) 73–81.
[17] K. Li, Ceramic Membranes for Separation and Reaction, John Wiley & Sons, Chichester, England, 2007.
[18] A.F.M. Leenaars, K. Keizer, A.J. Burggraaf, The preparation and characterization of alumina membranes with ultra-fine pores, J. Mater. Sci. 19 (1984) 1077–1088.
[19] A. Larbot, A. Julbe, C. Guizard, L. Cot, Silica membranes by the sol-gel process, J. Membr. Sci. 44 (1989) 289–303.
[20] M.A. Anderson, M.J. Gieselmann, Q. Xu, Titania and alumina ceramic membranes, J. Membr. Sci. 39 (1988) 243–258.

[21] S. Benfer, P. Arki, G. Tomandl, Ceramic membranes for filtration applications – preparation and characterization, Adv. Eng. Mater. 6 (2004) 495–500.

[22] R. van Reis, A. Zydney, Bioprocess membrane technology, J. Membr. Sci. 297 (2007) 16–50.

[23] M. Kallioinen, M. Nyström, Membrane surface characterization, in: N.N. Li, A.G. Fane, W.S.W. Ho, T. Matsuura (Eds.), Advanced Membrane Technology and Applications, John Wiley & Sons, Hoboken, NJ, 2008, pp. 841–878.

[24] A. Mehta, A.L. Zydney, Effect of membrane charge on flow and protein transport during ultrafiltration, Biotechnol. Prog. 22 (2006) 484–492.

[25] M.W. Phillips, A.J. DiLeo, A validatable porosimetric technique for verifying the integrity of virus-retentive membranes, Biologicals 24 (1996) 243–253.

[26] D.B. Burns, A.L. Zydney, Effect of solution pH on protein transport through ultrafiltration membranes, Biotechnol. Bioeng. 64 (1999) 27–37.

[27] C. Güell, M. Ferrando, F. López, Monitoring and Visualizing Membrane-Based Processes, Wiley-VCH Verlag GmbH, Weinheim, 2009.

[28] S.L. Fegler, J.W. Heckman, K.L. Klomparens, Scanning and Transmission Electron Microscopy: An Introduction, Oxford University Press, New York, 1995.

[29] C.C. Ho, A.L. Zydney, Effect of membrane morphology on the initial rate of protein fouling during microfiltration, J. Membr. Sci. 155 (1999) 261–275.

[30] S.M. Liu, K. Li, Preparation of TiO_2/Al_2O_3 composite hollow fibre membranes, J. Membr. Sci. 218 (2003) 269–277.

[31] K.-J. Hwang, T.-T. Lin, Effect of morphology of polymeric membrane on the performance of cross-flow microfiltration, J. Membr. Sci. 199 (2002) 41–52.

[32] J.M. Sheldon, The fine structure of UF membranes. I. Clean membranes, J. Membr. Sci. 62 (1991) 75–86.

[33] T.R. Albrecht, M.M. Dovek, C.A. Lang, P. Grütter, C.F. Quate, S.W.J. Kuan, C.W. Frank, R.F.W. Pease, Imaging and modification of polymers by scanning tunnelling and atomic force microscopy, J. Appl. Phys. 64 (1988) 1178–1184.

[34] W.R. Bowen, N. Hilal, R.W. Lovitt, P.M. Williams, Atomic force microscope studies of membranes: surface pore structures of cyclopore and anopore membranes, J. Membr. Sci. 110 (1996) 233–238.

[35] S. Singh, K.C. Khulbe, T. Matsuura, P. Ramamurthy, Membrane characterization by solute transport and atomic force microscopy, J. Membr. Sci. 142 (1998) 111–127.

[36] Y. Wyart, G. Georges, C. Deumié, C. Amra, P. Moulin, Membrane characterization by microscopic methods: multiscale structure, J. Membr. Sci. 315 (2008) 82–92.

[37] P.C.Y. Wong, Y.N. Kwon, C.S. Criddle, Use of atomic force microscopy and fractal geometry to characterize the roughness of nano-, micro-, and ultrafiltration membranes, J. Membr. Sci. 340 (2009) 117–132.

[38] I. Koyuncu, J. Brant, A. Luttge, M.R. Wiesner, A comparison of vertical scanning interferometry (VSI) and atomic force microscopy (AFM) for characterizing membrane surface topography, J. Membr. Sci. 278 (2006) 410–417.

[39] P. Dietz, P.K. Hansma, O. Inacker, H.-D. Lehmann, K.-H. Herrmann, Surface pore structures of micro- and ultrafiltration membranes imaged with the atomic force microscope, J. Membr. Sci. 65 (1992) 101–111.

[40] C. Charcosset, A. Cherfi, J.-C. Bernengo, Characterization of microporous membrane morphology using confocal scanning laser microscopy, Chem. Eng. Sci. 55 (2000) 5351–5358.

[41] C. Charcosset, J.-C. Bernengo, Comparison of microporous membrane morphologies using confocal scanning laser microscopy, J. Membr. Sci. 168 (2000) 53–62.
[42] U. Reichert, T. Linden, G. Belfort, M.-R. Kula, J. Thommes, Visualising protein adsorption to ion-exchange membranes by confocal microscopy, J. Membr. Sci. 199 (2002) 161–166.
[43] M. Zator, M. Ferrando, F. Lopez, C. Guell, Membrane fouling characterization by confocal microscopy during filtration of BSA/dextran mixtures, J. Membr. Sci. 301 (2007) 57–66.
[44] D. Hughes, U.K. Tirlapur, R. Field, Z. Cui, In situ 3D characterization of membrane fouling by yeast suspensions using two-photon femtosecond near infrared non-linear optical imaging, J. Membr. Sci. 280 (2006) 124–133.
[45] J.-C. Remigy, M. Meireles, Assessment of pore geometry and 3-D architecture of filtration membranes by synchrotron radiation computed microtomography, Desalination 199 (2006) 501–503.
[46] K.J. Kim, A.G. Fane, C.J.D. Fell, Quantitative microscopic study of surface characteristics of ultrafiltration membranes J, Membr. Sci. 54 (1990) 89–102.
[47] L. Zeman, L. Denault, Characterization of microfiltration membranes by image analysis of electron micrographs. Part I. Method development, J. Membr. Sci. 71 (1992) 221–231.
[48] R. Ziel, A. Haus, A. Tulke, Quantification of the pore size distribution (porosity profiles) in microfiltration membranes by SEM, TEM and computer image analysis, J. Membr. Sci. 323 (2008) 241–246.
[49] W. Sun, T. Chen, C. Chen, J. Li, A study on membrane morphology by digital image processing, J. Membr. Sci. 305 (2007) 93–102.
[50] S.T. Kelly, A.L. Zydney, Protein fouling during microfiltration: comparative behavior of different model proteins, Biotechnol. Bioeng. 55 (1997) 91–100.
[51] A.D. Marshall, P.A. Munro, G. Trägårdh, The effect of protein fouling in microfiltration and ultrafiltration on permeate flux, protein retention and selectivity: a literature review, Desalination 91 (1993) 65–108.
[52] Z.H. Syedain, D.M. Bohonak, A.L. Zydney, Protein fouling of virus filtration membranes: effects of membrane orientation and operating conditions, Biotechnol. Progress 22 (2006) 1163–1169.
[53] S. Loh, U. Beuscher, T.K. Poddar, A.G. Porter, J.M. Wingard, S.M. Husson, S.R. Wickramasinghe, Interplay among membrane properties, protein properties and operating conditions on protein fouling during normal-flow microfiltration, J. Membr. Sci. 332 (2009) 93–103.
[54] R. Chan, V. Chen, Characterization of protein fouling on membranes: opportunities and challenges, J. Membr. Sci. 242 (2004) 169–188.
[55] H.C. Flemming, G. Schaule, T. Griebe, J. Schmitt, A. Tamachkiarowa, Biofouling – the Achilles heel of membrane processes, Desalination 113 (1997) 215–225.
[56] R. Bos, H.C. Van Der Mei, H.J. Busscher, Physico-chemistry of initial microbial adhesive interactions – its mechanisms and methods for study, FEMS Microbiol. Rev. 23 (1999) 179–230.
[57] J.S. Baker, L.Y. Dudley, Biofouling in membrane systems – a review, Desalination 118 (1998) 81–89.
[58] S.T. Kang, A. Subramani, E.M.V. Hoek, M.A. Deshusses, M.R. Matsumoto, Direct observation of biofouling in cross-flow microfiltration: mechanisms of deposition and release, J. Membr. Sci. 244 (2004) 151–165.

[59] J. Gilron, D. Hasson, Calcium sulfate fouling of reverse osmosis membranes: flux decline mechanism, Chem. Eng. Sci. 42 (1987) 2351–2360.
[60] S. Bhattacharjee, G.M. Johnston, A model of membrane fouling by salt precipitation from multicomponent ionic mixtures in crossflow nanofiltration, Environ. Eng. Sci. 19 (2002) 399–412.
[61] N. Nagata, K.J. Herouvis, D.M. Dziewulski, G. Belfort, Cross-flow membrane microfiltration of a bacterial fermentation broth, Biotechnol. Bioeng. 34 (1989) 447–466.
[62] J. Hermia, Constant pressure blocking filtration laws: application to power-law non-Newtonian fluids, Trans. Inst. Chem. Eng. 60 (1982) 183–187.
[63] C.C. Ho, A.L. Zydney, A combined pore blockage and cake filtration model for protein fouling during microfiltration, J. Colloid Interface Sci. 232 (2000) 389–399.
[64] C. Duclos-Orsello, W.Y. Li, C.C. Ho, A three mechanism model to describe fouling of microfiltration membranes, J. Membr. Sci. 280 (2006) 856–866.
[65] G. Bolton, D. LaCasse, R. Kuriyel, Combined models of membrane fouling: development and application to microfiltration and ultrafiltration of biological fluids, J. Membr. Sci. 277 (2006) 75–84.
[66] C.C. Ho, A.L. Zydney, Protein fouling of asymmetric and composite microfiltration membranes, Ind. Eng. Chem. Res. 40 (2001) 1412–1421.
[67] W.Y. Li, C. Duclos-Orsello, C.C. Ho, Theoretical analysis of the effects of asymmetric membrane structure on fouling during microfiltration, AIChE J. 55 (2009) 1434–1446.
[68] S.G. Redkar, R.H. Davis, Cross-flow microfiltration with high-frequency reverse filtration, AIChE 41 (1995) 501–508.
[69] V.G.J. Rodgers, R.E. Sparks, Effect of solution properties on polarization redevelopment and flux in pressure pulsed ultrafiltration, J. Membr. Sci. 78 (1993) 163–180.
[70] W.D. Mores, R.H. Davis, Yeast foulant removal by backpulses in crossflow microfiltration, J. Membr. Sci. 208 (2002) 389–404.
[71] S. Muthukumaran, S. Kentish, S. Lalchandani, M. Ashokkumar, R. Mawson, G.W. Stevens, F. Grieser, The optimisation of ultrasonic cleaning procedures for dairy fouled ultrafiltration membranes, Ultrason. Sonochem. 12 (2005) 29–35.
[72] K.L. Tung, Y.L. Li, S. Wang, D. Nanda, C.C. Hu, C.L. Li, J.Y. Lai, J. Huang, Performance and effects of polymeric membranes on the dead-end microfiltration of protein solution during filtration cycles, J. Membr. Sci. 352 (2010) 143–152.
[73] H.L. Chen, Y.S. Chen, R.S. Juang, Flux decline and membrane cleaning in cross-flow ultrafiltration of treated fermentation broths for surfactin recovery, Sep. Purif. Technol. 62 (2008) 47–55.
[74] V. Puspitasari, A. Granville, P. Le-Clech, V. Chen, Cleaning and ageing effect of sodium hypochlorite on polyvinylidene fluoride (PVDF) membrane, Sep. Purif. Technol. 72 (2010) 301–308.
[75] M.C. Kaplan, A. Jegou, B. Chaufer, M. Rabiller-Baudry, M.C. Michalsky, Adsorption of lysozyme on membrane material and cleaning with non-ionic surfactant characterized through contact angle measurements, Desalination 146 (2002) 149–154.
[76] S.S. Madaeni, E. Rostami, A. Rahimpour, Surfactant cleaning of ultrafiltration membranes fouled by whey, Int. J. Dairy Technol. 63 (2010) 273–283.
[77] M.J. MunozAguado, D.E. Wiley, Enzymatic and detergent cleaning of a polysulfone ultrafiltration membrane fouled with BSA and whey, J. Membr. Sci. 117 (1996) 175–187.
[78] H.B. Petrus, H. Li, V. Chen, N. Norazman, Enzymatic cleaning of ultrafiltration membranes fouled by protein mixture solutions, J. Membr. Sci. 325 (2008) 783–792.

[79] H. Lee, G. Amy, J. Cho, Y. Yom, S.-H. Moon, I.S. Kim, Cleaning strategies for flux recovery of an ultrafiltration membrane fouled by natural organic matter, Water Res. 35 (2001) 3301–3308.

[80] G. Trägårdh, Membrane cleaning, Desalination 71 (1989) 325–335.

[81] P. Blanpain-Avet, J.F. Migdal, T. Bénézech, Chemical cleaning of a tubular ceramic microfiltration membrane fouled with a whey protein concentrate suspension – characterization of hydraulic and chemical cleanliness, J. Membr. Sci. 337 (2009) 153–174.

[82] M. Bartlett, M.R. Bird, J.A. Howell, An experimental study for the development of a qualitative membrane cleaning model, J. Membr. Sci. 105 (1995) 147–157.

[83] R. Field, D. Hughes, Z. Cui, U. Tirlapur, Some observations on the chemical cleaning of fouled membranes, Desalination 227 (2008) 132–138.

[84] H. Zhu, M. Nyström, Cleaning results characterized by flux, streaming potential and FTIR measurements, Colloids Surf. A 138 (1998) 309–321.

[85] D. Spettmann, S. Eppmann, H.C. Flemming, J. Wingender, Visualization of membrane cleaning using confocal laser scanning microscopy, Desalination 224 (2008) 195–200.

[86] P. Väisänen, M.R. Bird, M. Nyström, Treatment of UF membranes with simple and formulated cleaning agents, Trans. IChemE 80 (2002) 98–108.

[87] J.P. Chen, S.L. Kim, Y.P. Ting, Optimization of membrane physical and chemical cleaning by a statistically designed approach, J. Membr. Sci. 219 (2003) 2745.

[88] M. Zator, J. Warczok, M. Ferrando, F. López, C. Güell, Chemical cleaning of polycarbonate membranes fouled by BSA/dextran mixtures, J. Membr. Sci. 327 (2009) 59–68.

2 Ultrafiltration

Chapter Outline

Membrane Processes in Biotechnology and Pharmaceutics. DOI: 10.1016/B978-0-444-56334-7.00002-2

2.1 Principles

2.1.1 Introduction

UF is primarily a size exclusion-based pressure-driven membrane separation process. Operating pressures are usually in the range of 0.2–4 bar. UF membranes typically have pore sizes in the range of 10–1000 Å and are capable of retaining species in the molecular weight range of 300–10,00,000 Da. Typical rejected species include biomolecules, polymers and colloidal particles, as well as emulsions and micelles.

UF plays an important role in the biotechnology and pharmaceutical industries. UF has 'mild' operating conditions (relatively low temperatures, low pressures, no phase changes or chemical additives), which minimizes the denaturation, deactivation, and/or degradation of biological products. UF membranes are used extensively for the concentration, diafiltration and separation of biological solutions, both for the final product formulation and for conditioning of feed streams prior to other separation processes. UF is found throughout the downstream separation process for the purification of recombinant proteins, blood components, natural protein products and industrial enzymes. In addition, UF is used for the purification of antibiotics, amino acids and other small biological molecules.

2.1.2 Previous reviews

An important literature is available on UF presenting the main aspects of membrane materials, configurations, models, and the very broad range of applications, ranging from the processing of biological macromolecules to wastewater treatment. Books and book chapters have presented the major parameters and applications of this filtration process [1–3].

Other books and review articles have described more specific aspects. For example, van Reis and Zydney [4] have presented in detail protein UF and high-performance tangential flow filtration (HPTFF) including principles, membrane properties, modules and devices, process configurations and process design. Zydney and van Reis [5] have described the basic principles governing protein separations by HPTFF, with particular focus on the importance of the different phenomena in generating the high selectivity required for bioprocessing applications. Galaev and Mattiasson [6] have presented a detailed description of the operation mode of affinity UF (binding, washing, elution, regeneration) and the macroligands used (water-soluble macroligands, water-insoluble macroligands and other potential macro-ligands). Ghosh [7] has focused primarily on protein separation using UF including protein concentration, diafiltration, clarification and fractionation. He has provided an overview of recent developments in membrane biotechnology, focusing on the special characteristics of the membrane systems that are used for the commercial production and purification of recombinant protein products. Ultrafiltration was presented as well as sterile filtration, tangential flow microfiltration, depth filtration, virus filtration, membrane chromatography, HPTFF, membrane characterization and system design. Ultrafiltration was discussed according to these specific aspects: principles, membranes, modules, equipment, process configurations and diafiltration, and process optimization and control. Van Reis and Zydney [8] focused on protein UF, including theoretical principles, membrane materials, characterization and fouling, modules and devices, process configurations, process design, HPTFF and scale-up.

Koltuniewicz and Drioli [9] have given examples of UF in the pharmaceutical industries, including protein fractionation, purification of antibiotics, and affinity UF for enantiomer separation. Ghosh [10,11] focused on the application of UF in biopharmaceutical manufacturing. The author recalled that UF, MF and membrane chromatography were used for the purification of biopharmaceutical proteins such as monoclonal antibodies, plasma proteins, interleukins, interferon, growth factors, vaccines and hyperimmune antibodies. A detailed list of UF membranes and devices was given. Applications in protein bioseparation were detailed such as concentration (removal of solvent from protein solutions), desalting (removal of low-molecular weight compounds from protein solutions), clarification (removal of particles from protein solutions) and fractionation (protein–protein separation).

2.1.3 Basic definitions

Some basic definitions are given below including transmission and selectivity, the various modes to operate UF, and some general comments on lab scale and pilot scale. TMP and permeate (or filtrate) flux are also basic parameters for UF. Their definitions can be found in Chapter 1.

Transmission and selectivity

The transmission of a partially rejected species through an UF membrane is expressed by the intrinsic (or true) sieving coefficient, S_i:

$$S_i = \frac{C_p}{C_w} \tag{2.1}$$

where C_p is the species concentration in the permeate stream and C_w the species concentration at the membrane wall. The apparent sieving coefficient, S_a, is often more convenient to analyse experimental data:

$$S_a = \frac{C_p}{C_b} \tag{2.2}$$

where C_b is the species concentration in the feed (or bulk) stream.

S_i depends on the macromolecule-solvent system, the membrane, the physico-chemical conditions (i.e. pH, salt concentration and buffer type), the presence of other macromolecules and the permeate flux. S_a depends on the system hydrodynamics in addition to all parameters mentioned for S_i.

Rejection coefficients are defined by the intrinsic rejection coefficient:

$$R_i = 1 - \frac{C_p}{C_w} \tag{2.3}$$

and the apparent rejection coefficient:

$$R_a = 1 - \frac{C_p}{C_b} \tag{2.4}$$

The efficiency of binary species fractionation is commonly expressed in terms of the selectivity, ψ:

$$\psi = \frac{S_{a1}}{S_{a2}} \tag{2.5}$$

where 1 stands for the preferentially transmitted species and 2 for the preferentially retained species.

A throughput parameter can also be defined [12]:

$$J\Delta S = J(S_{a1} - S_{a2}) \tag{2.6}$$

where J is the filtrate flux. Effective separations require large values of both ψ and $J\Delta S$. The selectivity determines the separation capability of the process. The throughput parameter, which includes the filtrate flux, defines the practical limits on the actual purification.

Mode of operations

The main configurations to operate an UF process are dead-end and cross-flow (Chapter 1, Figure 1.4). Dead-end UF is used mainly for very small-scale and laboratory applications (i.e. for processing of less than 100 mL of feed solution). Most small-, medium- and large-scale UF processes are carried out in the cross-flow mode. The main advantage of cross-flow UF is the minimization of the accumulation of solute and particles near the membrane surface. The cross-flow arrangement also facilitates recirculation of the retentate stream to the feed tank, which may be associated to mixing with fresh feed. In cross-flow mode, the tangential flow is characterized by the flow rate value, the tangential mean velocity or by the shear rate at the membrane surface.

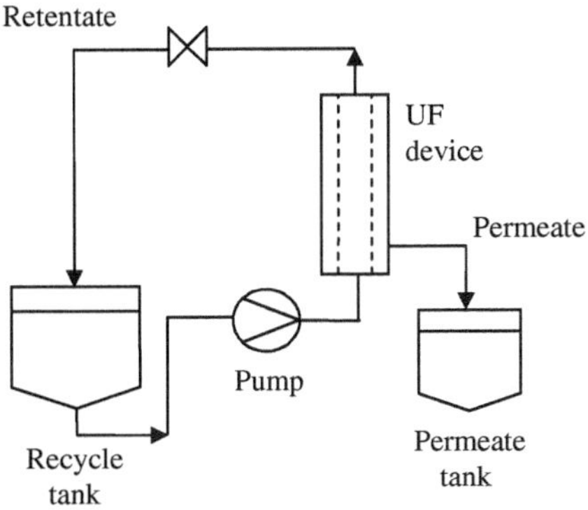

Figure 2.1 Batch-mode configuration.

UF can be operated in different modes, depending on the requirements of the process [4]. Commonly used modes of operation are batch, fed-batch, diafiltration and cascade systems. In a batch mode, the entire volume of feed is contained within the recycle tank (Figure 2.1). A fed-batch configuration uses an additional tank to feed into the recycle tank (Figure 2.2). Fed-batch configurations are commonly used in industry to obtain high concentration factors. Diafiltration is an operation in which a solvent or solution (typically a buffer) is added to the recycle tank simultaneously with filtration (Figure 2.3). The most common application is constant retentate volume diafiltration in which the buffer is added at the same flow rate as the filtrate flow rate. Cascade systems include both open-loop and closed-loop configurations. In both cases, the filtrate from the first stage is used as a feed to the second stage. In the open-loop configuration, the filtrate from the second stage is directed to drain while a replacement buffer is added to the first stage. In the closed-loop configuration (Figure 2.4), the filtrate from the second stage is used as the diafiltration buffer from the first stage, thereby reducing the amount of buffer required. Cascade systems can

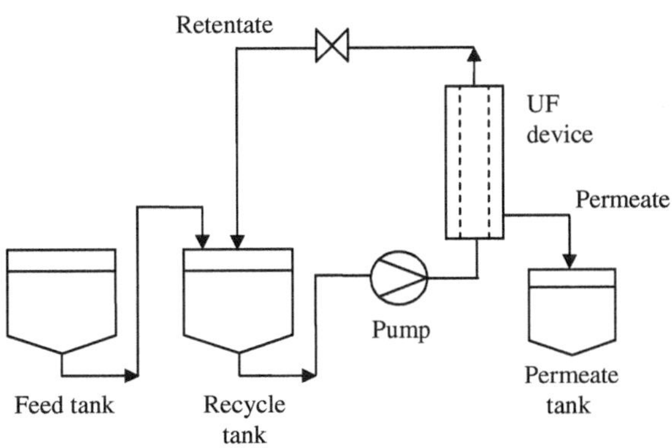

Figure 2.2 Fed-batch mode configuration.

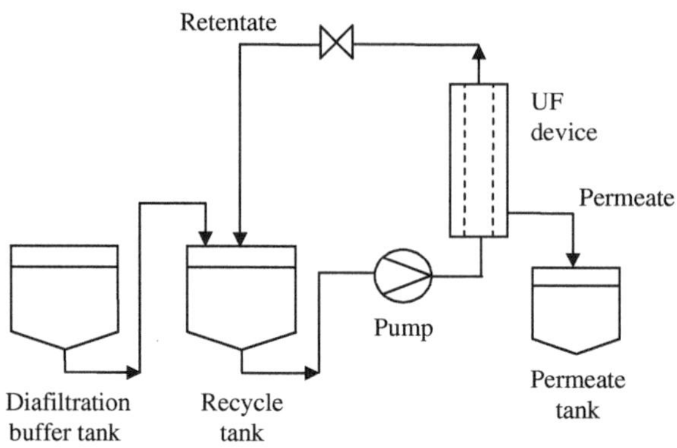

Figure 2.3 Diafiltration configuration.

be used to couple multiple steps into a single operation. For example, a MF step used to harvest protein from a culture can be cascaded with a second-stage UF step to achieve simultaneous harvest and protein concentration.

Lab scale and pilot scale

Once a process has been developed at lab scale, it must be translated to industrial scale. Scale-up considerations are given, for example, by Zeman and Zydney [3] and Allegrezza et al. [13]. The linear-scale technique is the simplest way to ensure translation of product yield and purity from lab scale to industrial scale. To linearly scale a cross-flow process [14], all fluid dynamic and membrane module parameters must be kept constant. Fluid dynamic parameters are set to be the user. They include the ratio of feed volume to membrane area, feed rate per membrane area, filtrate flux, and retentate and filtrate pressures. Membrane module parameters are inherent to the membrane device. They are membrane material and pore size, turbulence promoter,

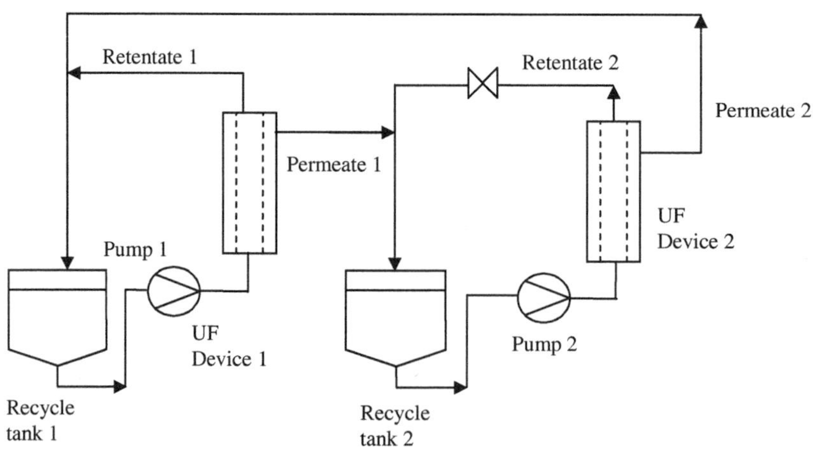

Figure 2.4 Two-stage closed-loop cascade configuration.

channel height, channel length, and feed and filtrate flow geometries. Linear scaling must be done with modules from the same supplier.

The typical steps of UF are set-up and pre-use cleaning, integrity and permeability testing, pre-use equilibration, processing, product recovery, post-use cleaning, testing and storage. The experimental set-up includes the membrane module(s). It also requires a recycle vessel, a feed pump, a retentate control valve, and pressure sensors for the feed and retentate lines. Many systems also include feed and filtrate flow-meters, a filtrate pressure sensor, and sensors for temperature, pH, conductivity or UV absorbance. Most cross-flow systems used for protein processing are operated in a sanitary manner, requiring sanitary fitting. In addition, the process parameters need to be adjusted during processing so that they remain constant. The process control is designed to maintain either a constant pressure drop or a constant cross-flow rate.

2.1.4 *Influence of parameters*

Transmembrane pressure and cross-flow velocity

UF is primarily a pressure-driven membrane separation process. The TMP range used is typically between 10 and 500 kPa. Due to the applied pressure, the solvent is forced through the pores. The solvent carries the solute molecules towards the membrane, and in certain cases through the membrane. Usually, a flux plateau is obtained in UF, where an increase in the applied pressure yields no permeate flux increase after a certain threshold (Figure 2.5). This phenomenon is usually explained by the gel

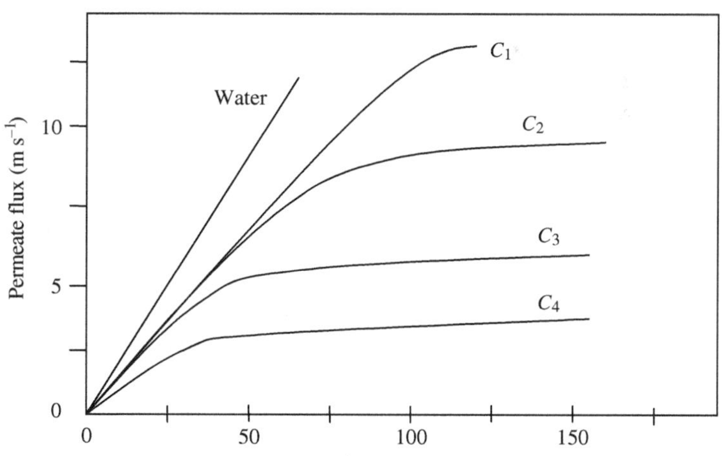

Figure 2.5 Typical permeate flux variation as a function of TMP for different solute concentrations ($C_1 < C_2 < C_3 < C_4$) [15].
This article was published in Comprehensive Biotechnology, Second Edition, Vol. 2, C. Charcosset, Downstream Processing and Product Recovery/Membrane Systems and Technology, pp. 603–618, Copyright Elsevier (2011).

model and the osmotic pressure model (see Section 2.6). It is usually admitted that optimum selectivity and throughput are obtained in the pressure-dependent flux regime, rather than in the pressure-independent flux regime [12]. To obtain reasonable flux and avoid formation of aggregates, the optimum *TMP* is usually chosen at the 'knee' of the flux curve [13].

A cross-flow configuration is usually preferred in UF and MF processes, especially for medium and large-scale systems. High tangential velocities (shear rates) and/or turbulence are needed in the immediate vicinity of the membrane in order to limit the build-up of a concentrated boundary layer of retained solutes. The shear rate at the membrane surface, γ_w, is calculated from the pressure drop due to flow through the device ΔP ($\Delta P = P_{in} - P_{out}$) using the following equation:

$$\gamma_w = \frac{d_H}{4\mu}\frac{\Delta P}{L} \tag{2.7}$$

where d_H is the hydraulic diameter of the flow channel, μ the solution viscosity and L the length of the membrane. The hydraulic diameter is equal to the tube diameter, d, for a tubular or hollow fibre device.

An important factor is the energy cost associated with pumping the feed (plus any recirculation flow) along the membrane surface. The energy associated with pumping is given by [3]:

$$E_{pump} = \Delta P Q \tag{2.8}$$

where ΔP is the pressure drop due to flow through the module and Q is the total volumetric flow rate. High mass transfer coefficients require high shear rates, but this must be done at low flow rate, Q, to minimize the energy cost associated with pumping. This is often achieved using very thin channels or by introducing turbulence promoters into the membrane device, for example, through the use of inserts, spacers, screens or obstructions in the flow channel.

Membranes

The effects of pore size on the performance characteristics of UF membranes are rather well understood. Membranes with large pores have usually high filtrate flux but low protein retention. On the contrary, membranes with small pores have lower filtrate flux but higher protein retention. In addition, pore geometry has an effect on the selectivity and permeability for UF membranes. Kanani et al. [16] presented experimental data for both commercial UF membranes and novel silicon membranes having slit-shaped nanopores of uniform size fabricated by photolithography. Their results showed that membranes with slit-shaped pores had higher performance, that is greater selectivity at a given value of the permeability, than membranes with cylindrical pores.

Membrane charge can have a large effect on the performance of many UF processes. The most common technique for evaluating the membrane surface charge is to determine the streaming potential as a function of the applied pressure. Data can be obtained with the flow either through the pores or along the membrane surface. The streaming potential develops because of the excess convective flux of oppositely charged counterions through the membrane pore [17]. The membrane zeta potential can be calculated using the following equation:

$$\zeta = \lambda \frac{\Delta U}{\Delta P} \tag{2.9}$$

where ΔP is the applied pressure (Pa), λ a system-dependent constant and ΔU the streaming potential.

The membrane charge effect on UF performance has been known from many years and is reported to alter both the extent of membrane fouling and the magnitude of the electrostatic interactions between proteins and membrane pores. Nakao et al. [18] used surface-modified polysulfone membranes, with a negatively charged version produced by sulfonation and a positively charged version produced by chloromethylation, followed by quaternization of the amino group. Protein rejection was found greatest under conditions where the membrane and protein had like-charge due to strong electrostatic repulsion [12,19,20]. Van Reis et al. [12] explored the effects of membrane charge, in combination with buffer pH, on protein separation. Protein separations were performed using Biomax™ 100 negative membranes and prototype Biomax™ 100 positive membranes (Millipore). Both membranes had nominal molecular weight cut-off values of 100,000 Da. Significant improvements in both selectivity and throughput were obtained by selecting the appropriate membrane charge at an optimum pH.

Properties of the feed solution

The properties of the feed solution influence the UF performance by a number of factors. In general, the permeate flux declines as concentration increases as seen in Figure 2.6 [21]. The viscosity of dilute solutions and suspensions are only weakly dependent of the concentration; however, the viscosity of more concentrated solutions can be significant. Non-Newtonian behaviour can also be observed at high

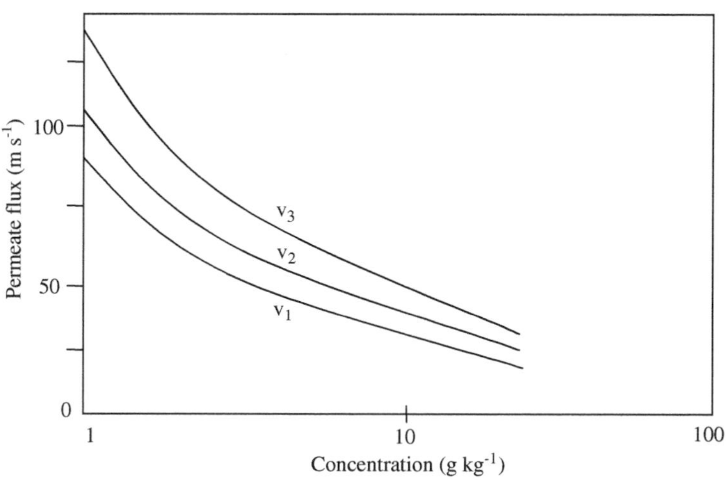

Figure 2.6 Typical permeate flux variation as a function of concentration for different cross-flow velocities ($v_1 < v_2 < v_3$).

concentrations which give unusual behaviour [22–24]. Using scleroglucan and concentrated skim-milk solutions, Aimar [22] showed that a pseudoplastic fluid could create better mass transfer than a Newtonian fluid which had the same viscosity at rest, because the pseudoplastic fluid had a higher shear rate and lower viscosity at the wall. Pritchard [23] proposed an expression for the laminar mass transfer coefficient, which incorporated the effect of pseudoplastic rheological behaviour. The model was a tool for understanding the cubic shape of the plot of flux versus log (xanthan concentration) rather than for a priori flux prediction. Charcosset and Choplin [24] reported UF experiments with newtonian and non-newtonian fluids (pectin, xanthan and polyacrylamide solution). For non-newtonian pseudoplastic fluids, the shape of the plot of flux versus log (bulk concentration) was different from the linear decline in flux which would occur for newtonian fluids.

Electrostatic interactions

Protein transport through UF membranes is strongly affected by electrostatic interactions. For example, Pujar and Zydney [25] showed that the transmission of BSA through a 100,000 molecular weight cut-off membrane decreased by nearly two orders of magnitude as the salt concentration was reduced from 150 to 1.5 mM. This large increase in protein rejection was attributed to the increased electrostatic exclusion of the charged BSA from the membrane pores at low ionic strength. The effect of solution pH on protein transmission was evaluated by several authors who indicated that the maximum transmission was attained near the protein isoelectric point; that is, at the pH where the protein has no net electrical charge [18,26].

The magnitude of the electrostatic interactions can be described using the theoretical analysis developed by Smith and Deen [27] for the partitioning of a charged sphere in an infinitely long charged cylindrical pore [28–30]:

$$S_a = (1-\lambda)^2 K_c \exp\left(\frac{-\psi_E}{k_B T}\right) \quad (2.10)$$

where S_a is the actual sieving coefficient. Equation (2.10) is valid at high filtration velocities where solute diffusion across the membrane is negligible, relative to the convective flux. The term $(1-\lambda)^2$ describes the steric (hard sphere) exclusion of the sphere from the region within one solute radius of the pore wall (with λ is equal to the ratio of the solute radius to the pore radius). K_c is the hindrance factor associated with convection and $(\psi_E/k_B T)$ is the dimensionless electrostatic energy of interaction [27]:

$$\frac{\psi_E}{k_B T} = A_s\sigma_s^2 + A_{sp}\sigma_s\sigma_p + A_p\sigma_p^2 \quad (2.11)$$

where A_s, A_{sp} and A_p are functions of the solution ionic strength, solute size and pore size [29], respectively. σ_s and σ_p are the dimensionless surface charge densities of the solute (protein) and pore. In Equation (2.11), the various terms represent the energy of interaction associated with the distortion of the electrical double layer around the solute, direct charge–charge interactions between the solute and the pore, and the distortion of the electrical double layer adjacent to the pore wall, respectively.

Table 2.1 Correspondence of membrane properties, protein properties and operating conditions on the impacts to protein–membrane interactions and mechanisms and rates of fouling [31]

Impact factor	Membrane	Protein	Operating conditions
Electrostatic interactions	Surface chemistry	PI, sequence	PH, ionic strength
Hydrophobic interactions	Surface chemistry	Hydrophobicity parameter	Temperature
Mechanism/rate of fouling	Porosity, morphology	MW, structure, conformational flexibility	Concentration, pressure, velocity, process volume

2.1.5 Protein fouling

Protein fouling occurs in membrane filtration processes such as UF and MF. In UF, the membrane MWCO is usually smaller than the molecular weight of the protein. In MF, the membrane pore size is usually greater than the molecule size; therefore, the protein deposit at the membrane surface may be considerably reduced compared to UF. Protein fouling presents a very complicated picture of electrostatic interactions, hydrophobic interactions and deposit. The influence of membrane properties, protein properties and operating conditions on protein–membrane interactions and mechanisms and rates of fouling is summarized in Table 2.1 [31].

Characterization

The characterization of membrane fouling by proteins has been the subject of extensive studies. A complete and recent review presents developments in membrane characterization applied to protein fouled membranes [32]. Surface deposition has been observed using techniques such as SEM [33,34], AFM [35,36], SEM coupled with energy dispersive X-ray (EDX) Spectroscopy [37] and confocal microscopy [38]. Pore and internal fouling by proteins has been examined by using standard infrared and X-ray photoelectron spectroscopy (XPS) [39], radiolabelling [40], streaming potential measurements [41], transmission electron microscopy (TEM) [42], electron paramagnetic resonance spectroscopy (EPRS) [43] and attenuated total reflection Fourier-transform infra-red spectroscopy (ATR-FTIR) [31,44,45].

Other recent methods have been employed to characterize protein fouling such as SANS [46], ELISA [47] and MALDI-MS [48]. Su et al. [46] developed an *in situ* method for monitoring the development of fouling under dynamic filtration conditions. The method was based on measuring the small angle neutron scattering (SANS) while monitoring the permeate flux under different conditions of flow. Since the ceramic materials silica and alumina investigated were fairly transparent to neutrons, SANS was able to probe the location of fouling on these materials under *in situ* filtration conditions. Yin et al. [47] applied the ELISA method to determine the amount of protein adsorbed on the membrane surface for various parameters. The target protein was firstly adsorbed onto the membrane surface; the primary antibody, secondary antibody–enzyme conjugate and substrate were then sequentially added and finally subjected to the photometric detection. Their model system was composed by human serum albumin (HSA) and a hydrophilic polysulfone MF membrane (HT Turffryn, Gelman Sciences, USA). An immunosensing system of HSA, in which rabbit IgG/anti HSA was used as primary antibody and alkaline phosphatase as enzyme label, and 4-nitrophenyl phosphate as substrate, was applied to the determination of HSA adsorbed on the membrane surface. Recently, matrix-assisted laser desorption ionization mass spectrometry (MALDI-MS) has been used to qualitatively analyse the composition of fouling deposits on regenerated cellulose and polysulfone membranes following the filtration of binary protein mixtures [48]. The method involved pre-adsorbing protein onto the membrane surface, then applying drops of a solution containing a laser-absorbing compound. The technique allowed the quantification of individually adsorbed proteins on membrane surfaces at levels as low as 50 fmol/mm^2. The high levels of compound specificity inherent to mass spectrometry make the method well suited to the quantification of individual components in mixed deposits.

Adsorption

A detailed presentation of protein adsorption in UF and MF can be found in Zeman and Zydney [3]. The very high affinity that proteins have for solid surfaces is due to the combination of the large overall size of the protein with the broad heterogeneity of the constituent amino acid residues. This allows the protein to participate in a wide range of electrostatic, hydrogen bonding, and van der Waals interactions with solid surfaces. Protein desorption is therefore highly unfavourable since it requires the simultaneous disruption of large numbers of interaction sites. Generally, if a protein molecule or aggregate is of opposite charge to a membrane surface, electrostatic attraction occurs and adsorption takes place [32]. If the protein and membrane have the same charge sign, then a high energy barrier must be overcome before adsorption may occur. Like proteins, membranes also have isoelectric points and therefore the net overall surface charge is also pH dependent. Protein adsorption in UF or MF membranes is generally characterized in terms of protein adsorption kinetics and isotherm. The static adsorption method consists in immersing a piece of membrane into the protein solution. It is generally employed to characterize the effect of protein adsorption on the filtration performance of UF and MF membranes. The dynamic adsorption method is performed during filtration, but can be affected by convective flow through the pores [49].

Protein adsorption occurs due to various physicochemical interactions between protein molecules and the membrane, such as electrostatic interaction, hydrogen bonding and/or hydrophobic interactions [49]. The electrostatic interaction can be estimated by the combination of the zeta potential of the pore surface and the protein. However, protein adsorption capacity is not completely explained in terms of electrostatic interactions. At high concentrations and alkaline pH, accumulation (non-adsorptive deposition) had also to be taken into account as confirmed by flux decay experiments and retention measurements [46,50]. Generally, the maximum adsorption capacity is observed at the isoelectric point (IEP) of the protein, where no electrostatic interaction is expected [3]. This maximum in adsorption capacity is due to the protein solubility which tends to be lowest at the IEP of protein, and to repulsive electrostatic interactions among the closely packed proteins in the adsorbed monolayer which are minimized.

Protein aggregation and precipitation

Protein aggregation and/or precipitation can limit the effectiveness of the process by increasing membrane fouling. Aggregation and precipitation of proteins are caused by a number of phenomena such as their exposition to air–water interfaces, adsorption to hydrophobic or hydrophilic surfaces and mechanical stress [51]. During UF, mechanical stresses include pumping, flow-induced shear stress and/or extensive contact to the membrane surface. In addition, extremely high concentrations can be reached in the gel or polarization boundary layer near the membrane which may also lead to aggregation [52]. The enhanced deposition of protein on the membrane surface is mainly due to protein–membrane and protein–protein interactions [53–55].

Protein aggregation during UF has been reported since many years ago [55–57]. For example, Meireles et al. [56] have carried out UF of BSA at various temperatures, cross-flow velocities, pressures, and concentrations with the aim of studying protein denaturation and its consequences on the process. Denaturation of proteins in solution was monitored by laser light scattering and size exclusion chromatography. It was shown that the rate of protein denaturation increased with temperature, cross-flow velocity and time. It was reported that under high concentration polarization, denaturation could occur in this boundary layer if the wall concentration exceeded 400 g/L.

Recently, new methods have been applied to the characterization of protein aggregation. Bee et al. [58] exposed two concentrated immunoglobulin-G1 (IgG1) monoclonal antibody ($>$100 mg/mL) formulations to shear rates between 20,000 and 250,000 s^{-1} for between 30 ms and 5 min using a parallel-plate and capillary rheometer, respectively. From their results, the authors suggested that in many cases, air-bubble entrainment, adsorption to solid surfaces (with possible shear synergy), contamination by particles or pump cavitation stresses could be much more important causes of aggregation than shear exposure. Van Reis and Zydney [52] have also pointed out that multiple passes through pumps and valves and concomitant micro-cavitation and air-bubble entrainment, rather than shear, cause protein aggregation during filtration. Ahrer et al. [59] applied dynamic light scattering to detect differences in the aggregate pattern during cross-flow filtration of human Ig preparations. The change of ionic strength of the process solution during diafiltration significantly influenced the aggregate pattern. Other causes for aggregation, such as flow-induced shear stress or high local protein concentrations due to the build-up of a gel layer, did not significantly influence the formation of aggregates. The use of weak buffered solutions suppressed aggregate formation and increased the permeate flux during diafiltration. Rosenberg et al. [60] monitored the formation of IgG aggregates during the UF of monoclonal antibody solutions by a number of techniques, including turbidity measurement, size exclusion high pressure liquid chromatography, light obscuration particle counting, dynamic light scattering and a microscopic method. These techniques allowed characterization of a wide range of sizes of soluble and insoluble aggregates that can occur during UF of monoclonal antibodies. FT-IR spectroscopy was performed to investigate the secondary structure of the aggregates, revealing different quantities of structurally perturbed protein depending on applied flow and pressure conditions.

2.2 Membranes

Numerous manufacturers provide UF membranes, such as Asahi Kasei (Japan), Koch Membrane Systems (USA), Novasep Process (France), Millipore (USA), Pall Corp. (USA) and Sartorius (Germany).

In the following section, main aspects of UF membranes are presented. The molecular weight cut-off is a major data used to define UF membrane performances. Methods used to obtain this parameter are detailed. In the next section, some classical membrane materials are listed. The surface modification of membranes contributes to enlarge the range of available supports. Finally, environment responsive membranes may be considered as an alternative to improve UF membrane performance. General rules on this technique are recalled in a specific section.

2.2.1 Nominal molecular weight cut-off

Commercially available UF membranes display generally a pore size distribution and contain tortuous interconnected pores. Consequently, UF membranes do not have a sharp molecular weight cut-off. In addition, the observed membrane rejection depends on more than just the membrane properties. The solute properties such as shape, degree of branching and deformability affect transport through the membrane pores [61], as well as the device configuration and operating conditions [62–64].

An UF membrane is typically characterized by its MWCO, defined as the equivalent molecular weight of the smallest species that exhibit 90% rejection. This value is usually referred to as R90. Although this choice is arbitrary, it is adopted by most of UF membrane users. However, the experimental conditions and systems used to evaluate R90 are not standardized. Since the MWCO depends on the solute species and operating conditions, it only applies to the test conditions specified by the manufacturer.

UF membranes have been first characterized by filtering solutions of individual proteins covering the molecular weight range of interest. However, the resulting data were difficult to analyse due to differences in adsorption, conformation and electrical charge between proteins. The use of neutral polymers like dextrans to characterize UF membranes dates back more than 25 years [65,66]. Under the experimental conditions used to determine rejection coefficients, dextrans may be modelled as rigid spheres. Because dextrans are neutral polymers of D-glucopyranose, electrostatic interactions

Table 2.2 Composition of the mixed dextran solution [63]

Dextran fraction	Concentration (g L^{-1})
T1	0.74
T4	1.22
T10	0.54
T40	0.74
T70	0.34
T500	0.27
T2000	3.65

among solute, solvent and membrane are avoided. In addition, dextrans do not adsorb onto most membranes or have limited adsorption [67].

Experiments typically use polydisperse dextrans over the whole molecular weight range of interest. For example, the composition of the dextran mixture used by Zydney and Xenopoulos [63] is given in Table 2.2. The concentration of the individual dextran fractions was selected to obtain a uniformly high concentration of dextrans over a wide MW range. Fractions T10 through T2000 were obtained from GE Healthcare, and fractions T1 and T4 were obtained from Serva (Germany).

The concentration of the different molecular weight fractions is evaluated using gel permeation chromatography [68,69]. A sieving (or rejection) curve can thus be generated from data obtained in the UF experiments (Figure 2.7). From these curves, R90 can be calculated, as well as the complete rejection profile. For typical

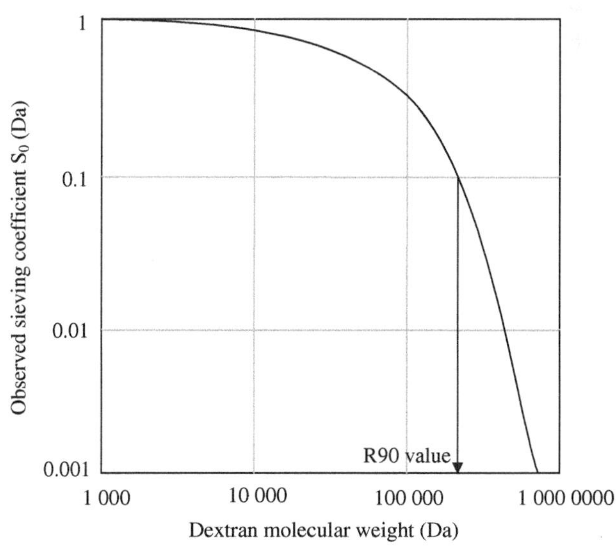

Figure 2.7 Typical dextran sieving curve (y-axis with a logarithmic scale).

biotechnology applications, where the product needs to be retained, R99 or R99.9 values can also be obtained (rejection of 99% and 99.9%, respectively). Recently, Wickramasinghe et al. [62] compared experimental rejection curves to calculated rejection curves based on FESEM imaging, assuming that the membrane pore size distribution might be described by the log-normal distribution function. Results showed relatively good agreement between calculated and experimental curves.

Other studies have extended the sensitivity of the dextran test by using fluorescently labelled dextrans, allowing detecting higher degrees of solute retention [70]. This 'Flex Test' was shown to provide sieving measurements down to $S = 0.001$ (corresponding to 99% yield) due to the greater sensitivity of the fluorescent detection. In addition, fluorescent-tagged charged dextrans can be used to take into account chemical membrane characteristics such as charge and hydrophobicity. For charged membranes, the sieving of a charged solute is usually measured with fluorescent-tagged dextrans (high sensitivity for low-sieving values), whereas the neutral dextrans may not require a fluorescent tag since they likely represent passing markers [8].

Some investigators have employed Ficoll [71] and polyethylene glycol [72], although dextrans remain the solute of choice for the characterization of UF membranes.

2.2.2 Membrane materials

UF membranes are cast from a variety of polymers, including polysulfone, polyethersulfone, cellulose acetate and regenerated cellulose. Most of UF membranes have an asymmetric structure, with the thin skin providing the desired selectivity while the more porous structure provides the necessary mechanical support. This asymmetric structure is primarily developed from the gradients in temperature and/or composition produced in the polymer casting solutions during membrane formation. Regenerated cellulose membranes are more hydrophilic, reducing both protein adsorption and fouling. New composite regenerated cellulose membranes have excellent mechanical strength and cause little fouling, providing higher flux and better retention characteristics than other UF membranes [4].

Despite the widespread use of polymeric membranes, they suffer from several disadvantages. They have limited mechanical integrity which can lead to deformation during operation and adversely affect membrane performance. Polymeric membranes often have limited temperature and chemical stability, prohibiting their use in applications involving aggressive organic solvents or elevated temperature. Ceramic UF membranes offer greater chemical, mechanical and thermal stability; however, the available pore size range is more limited. The efficiency of ceramic membranes can be significantly reduced by protein fouling and concentration polarization, leading to decrease of product yield and flux. Efforts have focused on developing modules with improved mass-transfer characteristics in order to make them more competitive to polymeric membranes, for example by inclusion of static mixers as turbulence promoters [73], and modification of the ceramic membrane surface [74].

New UF membranes include carbon membranes prepared from spray deposition and pyrolysis of poly(furfuryl alcohol)/poly(ethylene glycol) mixtures on macroporous stainless steel supports [75]. The poly(ethylene glycol), employed as a carbonization template, created a mesoporosity that lead to pores in the UF range. Shah et al. [76] have made nanoporous carbon membranes by first slip-casting silica particles within the macropores of a stainless steel support, with the nanoporous carbon formed by pyrolysis of polyfurfuryl alcohol. A small poly(ethylene glycol) was used as a pore-forming agent. The resulting membrane had dextran sieving curves similar to those of a commercial 100 kDa poly(ether sulfone) UF membrane. In addition, the nanoporous carbon membrane was very stable to aggressive cleaning agents, including very concentrated (3 N) sodium hydroxide.

2.2.3 Surface modification

One potentially attractive approach for producing new UF membrane surfaces is to modify the surface of commercial membranes without substantially changing their filtration characteristics. Several modification methods have been investigated including coating [77], low temperature plasma-induced polymerization [78–80] and photo-oxidation including UV irradiation and UV-assisted graft polymerization [81–84].

Plasma-induced polymerization is one of the techniques that has been successfully used for modifying polymeric membrane surface chemistry [78–80]. Simple treatment with an inert gas such as nitrogen or oxygen plasma, followed by exposure to air, can create peroxides that are suitable reactive sites for subsequent monomer grafting and polymerization. The capability of plasma to alter the physical and chemical properties of polymeric surfaces without affecting the membrane properties (especially mechanical properties) is advantageous for the design and development of surface-modified polymer membranes.

Photochemical modification of UF membranes by the UV-assisted graft polymerization of hydrophilic monomers has been shown to be successful in increasing surface hydrophilicity and decreasing membrane fouling during protein filtration. This has been accomplished with the use of free radical photoinitiators [85–87] or without free radical photoinitiators [81–83,88,89]. A two-step UV process using photoinitiators has been used to modify poly(acrylonitrile) and poly(sulfone) UF membranes [85–87]. Yamagishi and co-workers [81,82] have achieved grafting on the surface of poly(sulfone) and poly(ether sulfone) UF membranes in a single step without the use of a photoinitiator due to the intrinsic photoactivity of poly(sulfone) and poly(ethersulfone). Recently, a one-step technique was developed to render the modification process more easily adaptable to continuous membrane manufacturing processes [83,89]. Using this technique, very hydrophilic poly(ether sulfone) modified membranes were produced that exhibited as low protein fouling as regenerated cellulose membranes during the filtration of bovine serum albumin (BSA) without loss of observed BSA rejection. However, membrane permeability was decreased after modification. This loss of membrane permeability has been observed for other grafted membranes and has been linked to the blockage of membrane pores by the grafted polymer chains [84,85,88].

Other methods for modification of polymeric UF membrane surfaces include surface-initiated atom transfer radical polymerization (ATRP) [90]. ATRP is a catalyst-activated, controlled radical process in which a radical initiator is anchored onto the membrane surface. The catalyst initiates polymerization by reversible abstraction of a halogen atom from the surface, and polymer chains grow by monomer addition from the surface. The technique was used to modify regenerated cellulose UF membranes with poly(PEGMA). The authors concluded that ATRP was a controllable chain growth technique that allowed to control the membrane pore size.

2.2.4 *Environment responsive membranes*

Environment responsive membranes (also called stimuli-responsive membranes) have stimuli-responsive moieties on their surface or embedded within their structure, which allow changing their structural, charge or affinity characteristics in response to changes in their environment, such as pH, temperature, ionic strength, light, electric and magnetic fields and chemicals (Figure 2.8). Wandera et al. [91] recently reviewed the design of stimuli-responsive membranes, including membrane structure and surface characteristics, and their range of use.

Eisenberng and Grodzinsky [92] were among the first to demonstrate the principle of environment-responsive membrane. The permeability of a bovine hide corium collagen membrane towards sucrose was controlled by an electric field applied across the membrane. The change in sucrose flux by varying the electric field strength and the rate of response to stimuli changes were examined. Iwata and Matsuda [93] studied the effects of environment-responsive polymers poly-acrylic acid and poly-acrylamide grafted on the surface of PVDF membranes. By varying the length of the polymer chains, they were able to control the rejection of different macromolecules at different solution conditions. The first comprehensive study of the change in molecular weight cut-off values of such membranes at different conditions was conducted by Masawaki et al. [94]. pH-responsive poly(4-VP-co-AN) membranes were operated at two pH values, 4 and 10. It was found that the molecular weight

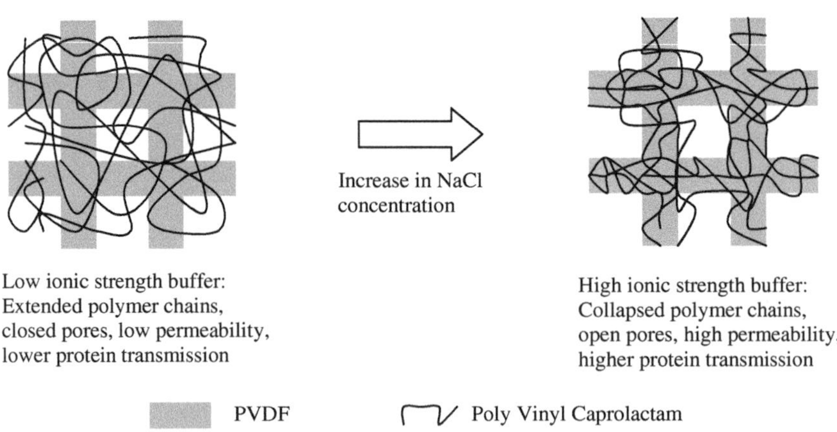

Figure 2.8 Environment-responsive membrane: proposed mechanism for change in solute permeability in response to change in salt concentration (from [104]).

cut-off of the membrane increased as the pH of the solution was increased. Many other environment-responsive membranes have been developed to enable responsiveness to various environmental stimuli including pH/ionic strength [95–99], specific ions [96], electrical field [100] and temperature [101,102].

Membranes that are responsive to environmental stimuli are generally prepared by two distinct methods [103]. The first method combines the responsive polymer component with standard membrane polymers to form a copolymer blend which is then processed through a typical membrane fabrication technique. The product is a membrane with directly embedded functionality within the polymer matrix [94,97,99]. The second method is a two step process where the responsive polymer component is attached to a previously prepared porous matrix. Two different types of membranes can be prepared by this attachment technique. The first type is designated as pore filled-wall grafted and involves covalent bonding of an active polymer to the walls of the porous matrix to create a pseudo-brush structure that extends away from the membrane surface towards the centre of the pore [93,95,96,98]. The second type, designated as pore filled-cross-linked, is prepared by completely filling the porous support with a polymer solution followed by a cross-linking step which locks the active polymer network within the porous matrix [101–103]. The matrix constrains the swelling properties of the gel and gives mechanical strength to the membrane.

When applied to UF, environment responsive membranes allow protein separation by manipulation of environmental and operating conditions. Huang et al. [104] prepared environment-responsive membranes through modification of a commercial microporous PVDF membrane support with a salt-responsive hydrogel composed of poly-N-vinyl-lactams and bisacrylamide cross-linker. The membrane permeability increased significantly with increase in salt concentration due to the collapse of the hydrogel onto the surface of the supporting PVDF membrane. The change in sieving coefficients of model proteins in response to change in salt concentration was studied by parameter scanning UF experiments [105]. A stirred cell UF module fitted with an environment-responsive membrane was integrated with a liquid chromatography system. The membrane allowed total transmission of BSA, moderate transmission of HIgG and low transmission of ferritin and thyroglobulin. The authors concluded on the potential for using environment-responsive membranes for three-component separations.

2.3 Fouling control

Various methods have been proposed to reduce the negative effects of concentration polarization and fouling by improving the hydrodynamics of the cross-flow over the membrane surface. These various techniques to enhance performance of membrane processes were summarized recently in detail [106]. They include chemical, hydrodynamic and physical methods. A direct chemical approach to control membrane fouling is the modification of the surface properties of the membranes, such as surface and hydrophilicity, as described previously. The chemical methods also include pH adjustment to the feed which can alter solute–membrane interactions and reduce fouling. Most of hydrodynamics techniques used for flux enhancement focused on increasing the particle back transport from the membrane–liquid interface by increasing the shear rate and the flow instability in the boundary layer. They often result in an increase in the energy consumption and there is a trade-off between capital cost (related to flux enhancement) and operating cost (related to energy consumption) [106]. The methods involve a cost that must provide a benefit and return on investment.

Dynamic filtration creates the membrane shear rate necessary to maintain the filtration by a rotating disk, or by rotating or vibrating the membrane. Backflushing and backpulsing are *in situ* methods of membrane cleaning by periodically reversing the TMP or permeate flow. Dean vortices, generated by a coiled geometry, can be used to modify the hydrodynamics near the membrane surface. Gas sparging is a hydrodynamic method to enhance membrane filtration by generating bubble-induced secondary flow. The use of corrugated surfaces is another potential approach to enhance mass transfer rates. Finally, electro-UF utilizes an electric field that drags charged proteins away from the membrane surface.

A number of other hydrodynamic techniques have been proposed during these last years. They include pulsed flow, solid–liquid two-phase flow by particle addition, and ultrasound-enhanced filtration [106]. In general, the commercial implementation of these methods has been restricted by several factors, including concerns about energy consumption, complexity of module manufacture and of process development.

2.3.1 Dynamic ultrafiltration

Dynamic filtration, also called shear-enhanced filtration, consists in creating the membrane shear rate necessary to maintain the filtration by a rotating disk, or by rotating or vibrating the membrane. This mode of operation permits to reach very high shear rates, of the order of $1–3 \times 10^5\ s^{-1}$ and to increase both permeate flux and selectivity. Recently, Jaffrin [107] has reviewed various dynamic filtration configurations such as cylindrical rotating membranes, disks or blades rotating near a fixed membrane, and vibrating systems. For biotechnology applications, two systems are reported: rotating membranes (model MSD) commercialized by Westfalia Separator (Germany) and vibrating membranes (model PallSep) sold by Pall Corporation(USA).

The main drawback of rotating disk systems is their complexity and the cost of building industrial scale modules, which require several large diameter disks mounted on the same axis and rotating at high speed. Membrane replacement may be also a complex operation. An original system (PallSep) is commercialized by Pall Corporation (NY, USA). This vibratory shear-enhanced system consists in a stack of circular plane membranes mounted in a cylindrical housing which is spun in torsional oscillations around a vertical axis at a resonant frequency of about 60 Hz. Each compartment contains two membranes separated by a gasket and a permeate collecting system. Since this system is compact, without internal moving parts, with a fluid channel height of about 3.5 mm, it can accommodate up to 140 m^2 of membrane in a single module [108].

When shear-enhanced devices are used in UF, the very high shear rate effectively reduces concentration polarization and it is then advantageous to use high *TMP*, since the permeate flux keeps increasing until higher pressure levels than in conventional cross-flow filtration [109]. For skim milk UF, two different dynamic filtration systems were compared [109]. Tested systems were two home-made rotating disk modules and a commercial pilot with a circular vibrating membrane. The flux was found to be mainly governed by the maximum shear rate and not by details of internal flow. It could be increased to very high levels by increasing rotation speed or vibration amplitude or by equipping the disk with large vanes. Another example was the recovery of trypsin inhibitor and soy milk protein concentration by dynamic filtration [110]. New devices are still tested, for example by the incorporation of vanes of different geometries and arrangements within the rotating disc membrane module [111]. These turbulence promoters were tested with BSA and glucose as model feed solutions. A 45° blade angle vane having holes on the blade surface was shown to produce maximum permeate flux at low *TMP* and membrane speed for even high concentration of the feed solution.

2.3.2 Backflushing and backpulsing

Backflushing is an *in situ* method of membrane cleaning by periodically reversing the TMP or permeate flow [112]. The fouling deposit which accumulates on the membrane surface during forward filtration is lifted off and removed during reverse filtration (Figure 2.9). An alternative strategy, which has been explored within the past 15 years, is backpulsing. The concept of backpulsing is similar to that of backflushing, except that the duration of the reverse flow is much shorter, typically less than 1 s (Figure 2.10). The frequency of the backpulses is more than one every few seconds. High-frequency backpulses can prevent membrane fouling or at least remove the foulants shortly after they are deposited on the membrane.

Backpulsing was first investigated by Rodgers and Sparks [113] for protein UF. These authors demonstrated significant improvements in solute flux using backpulsing frequencies of up to 5 s^{-1}. One proposed mechanism was that TMP pulsing might improve solute flux by removing deposited solute in the pores. The backward flow potentially carried the previously deposited solute out away from the pores,

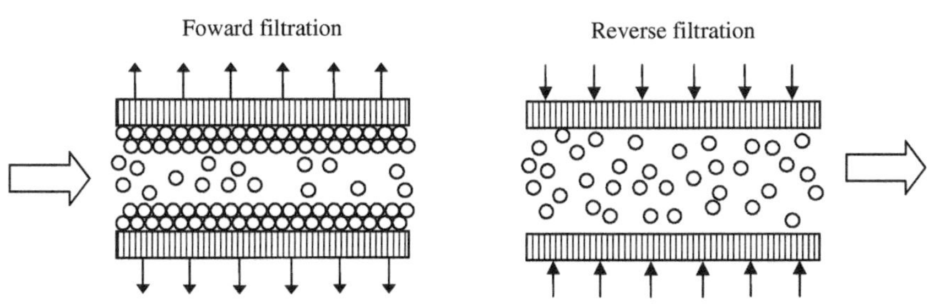

Figure 2.9 Schematic of forward and reverse filtration.

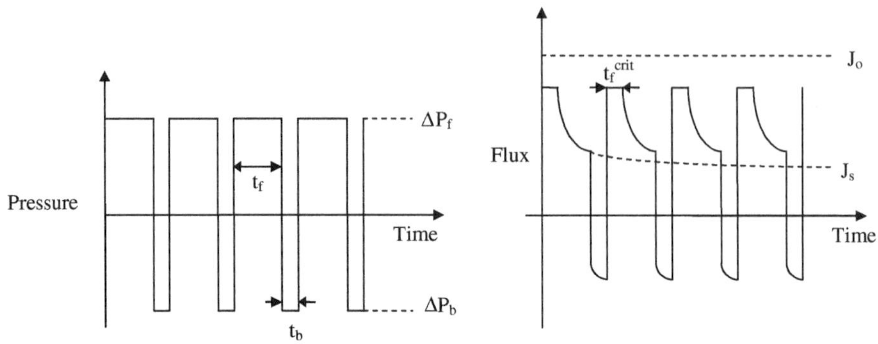

Figure 2.10 TMP and flux profiles during backpulsing.

clearing the pores for solute transport in the next UF cycle [114]. In the purification of conjugated vaccine products, backpulsing was shown to improve flux, reduce fouling and increase protein transmission [115].

2.3.3 Dean vortices

Dean vortices, generated by a coiled geometry, can be used to modify the hydrodynamics near the membrane surface [116]. Dean vortex devices have very high mass-transfer rates, owing to the presence of centrifugal flow instabilities, and are not expected to consume unreasonable amounts of energy. Dean vortices are the secondary flows that occur in the cross section of a curved channel or helically coiled tubes [106]. When fluid flows through the helically curved tube, the faster elements of the fluid in the centre of the tube tend to be moved outward by centrifugal force, while the slower elements of the fluid are forced inward to maintain mass balance. This results in counter rotating vortices in the cross section of the channel. The appearance of the Dean vortices depends on the fluid flow in the tube and the geometric features of the curved channel.

Ghogomu et al. [117] investigated the performance of several curved membrane designs, which provided a high-mixing potential by secondary flows. In all curved channels, the mass transfer was improved compared to classical modules and at the same time, curved channels showed to be more energy efficient. Moulin et al. [118] compared coiled and straight hollow fibre UF modules for two feeds, a colloidal bentonite suspension and a dextran solution. For both solutions it was shown that the formation of dean vortices in the coiled hollow fibres resulted in an effective reduction of fouling and an increase of permeate flux. The limiting flux was enhanced by up to a factor of 2 or 3 for the colloidal suspension and the macromolecular solution, respectively. The flux enhancement increased when the coil diameter was reduced from 11 to 4.1 cm. Kaur and Agarwal [119] evaluated the mass transfer coefficient of the thin spiral channel flow module. The mass transfer coefficient was comparable to rotary modules, about five to ten times higher than in straight modules. These devices show significant increases in protein transmission and capacity, although fouling remains a problem.

2.3.4 Gas sparging

The gas sparging method is a hydrodynamic technique to enhance membrane filtration by generating bubble-induced secondary flow [120,121]. The technique was shown to be effective in reducing concentration polarization in tubular, hollow fibre and spiral wound membrane modules [120,122,123]. The membrane orientation was identified as an important operating parameter, as the flux enhancement was found to be less for horizontally orientated tubular membranes, compared to vertically installed membranes [121]. The effectiveness of the technique was more pronounced at low cross-flow velocities [124]. A three-fold flux increase was found when ultrafiltering dextran solutions in tubular membranes [120], while flux enhancement by gas sparging for hollow fibre [122] and spiral wound [123] membranes was less pronounced, with an observed flux increase from 10 to 60%. Gas sparging was also shown to enhance protein fractionation [125,126]. Under suitable solution conditions, nearly complete separation of BSA and lysozyme was achieved with gas sparged UF. The mechanism of enhancement was explained in terms of disruption of the concentration polarization layer and enhanced mass transfer due to bubble-induced secondary flow. Recently, Taha and Cui [127] proposed a CFD modelling of the gas sparged UF process with the aim of understanding and quantifying the details of the permeate flux enhancement resulting from gas sparging.

2.3.5 Corrugated membrane surfaces

The use of corrugated surfaces is a common approach to enhance heat transfer in heat exchangers. This technique has also been used in membrane applications to enhance mass transfer rates. Corrugated membranes are prepared using several approaches such as pressing originally flat sheet membranes between structured dies [128–130] and casting the polymer solution on or between structured moulds of millimetre or micrometre scale corrugations [131]. Recently, corrugated hollow fibres were obtained by a spinning method [132]. Corrugated membranes enhance flow both by increasing the membrane area per volume and by disrupting the concentration polarization layer. In UF experiments with polysulfone membranes using dextran as solute, van der Waal et al. [130] found that the corrugations resulted in reduced energy consumption or pressure drop compared with flat membranes at a given value of the flux through the membrane.

2.3.6 Electro-ultrafiltration

To limit fouling and concentration polarization in UF of proteins, an electric field can be applied across the membrane. Electro-UF utilizes an electrophoretic force that drags the charged proteins away from the membrane surface. The gel and concentration polarization layer are thereby reduced and the flux increases [133]. Flux improvements in the range of 2–10 times have been reported for BSA [134,135] and gelatin [136]. Another advantage of electro-UF is that it can be carried out at a low cross-flow velocity, since the electric field replaces the need of a high shear rate at the membrane surface.

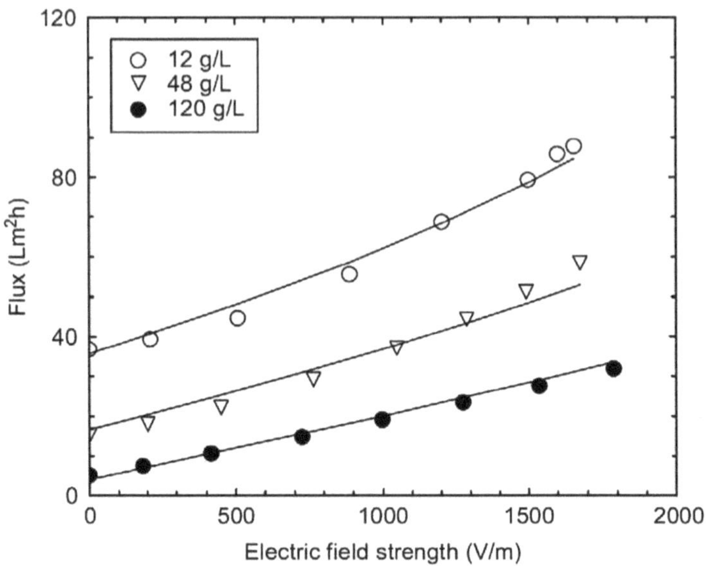

Figure 2.11 Effect of the electric field for different concentrations of an amylase-S solution. The solid lines are an empirical model. *TMP* = 1.5 bar and $v = 0.07$ m s^{-1} [137]. This article was published in Chem. Eng. Sci., Vol. 62, A.D. Enevoldsen, E.B. Hansen, G. Jonsson, Electro-ultrafiltration of amylase enzymes: Process design and economy, pp. 6716–6725, Copyright Elsevier (2007).

Electro-UF of an amylase enzyme solution showed a 3–7 time flux improvement, where the greatest flux enhancement was achieved at high concentrations [137]. An example of the flux enhancement can be seen in Figure 2.11, where the flux was plotted as a function of the applied electric field strength. The economical aspect of electro-UF was found favourable when filtering solutions of high concentration, while it was more favourable for UF when filtering solutions of lower concentration. The authors concluded that electro-UF should be used as a final concentration step in a production line, where the first steps are conventional UF.

Recently, electric field enhanced cross-flow UF has been carried out for fractionation of aqueous solution of two proteins, for example BSA and lysozyme [138]. The study was carried out to investigate the effect of system parameters such as electric field, solution pH, feed concentration ratio, cross-flow velocity and pressure difference on the permeate flux and observed retention of lysozyme. Experimental results showed that the introduction of an electric field of appropriate polarity substantially improved the permeate flux and fractionation. At pH 7.4, by applying electric field 1000 V m^{-1}, permeate flux increases from 23.4 to 36.7 L m^{-2} h^{-1} and observed retention of lysozyme decreases from 73 to 20% compared to zero electric field keeping other operating conditions unchanged.

In addition, a technology named electrodialysis with UF membranes (EDUF) has been developed to fractionate molecules on the basis of their electrical charge and size or molecular weight [139,140]. A conventional electrodialysis cell was used, in which some ion exchange membranes were replaced by UF ones. Compounds of higher molecular weight than the membrane cut-off could be separated so as to extend the field of application of electrodialysis to biologically charged molecules. These studies reported the selective isolation of charged peptides from a β-lactoglobulin (β-LG) hydrolysate. Recently, Firdaous et al. [141] investigated the fractionation of an alfalfa white protein concentrate (AWPC) hydrolysate by electrodialysis with EDUF. Two UF membranes were stacked in an electrodialysis cell to allow a configuration with two recovery compartments (KCl 1 and 2). It was established that EDUF allows the simultaneous separation and concentration of charged peptides with a transport rate up to 7.3 g/(m^2 h). Amongst a total of 70 peptides contained in the AWPC hydrolysate, only 8 peptides were found in the separated adjacent recovery solutions (KCl 1 and 2) at a maximum transmission rate of 18%.

2.4 Configurations

Several operation modes for UF are described in the following. The diafiltration mode is commonly used in many processes to remove or exchange buffers. HPTFF exploits a number of different properties to achieve high resolution, including a proper choice of pH and ionic strength to maximize differences in the hydrodynamic volume of the product and impurity. Affinity UF uses a binding agent in free solution to selectively bind, and thus retain by an UF membrane, the species to be purified. Other modes of operation introduced include carrier phase UF (CPUF), a modification of dead-end UF, and operation at a constant wall concentration, C_w.

2.4.1 Diafiltration

Buffer removal and exchange can be done using a diafiltration mode to wash it out and to replace it with a new buffer (Figure 2.3). The small impurities and buffer components are washed away by the continuous, or discontinuous, addition of new buffer with the desired composition and purity [142]. Diafiltration is used to perform buffer exchange if the product is in the retentate, or to enhance yields if the product is in the filtrate.

The most common approach is to perform the diafiltration using a constant retentate volume in which the buffer is added at the same flow rate as the filtrate flow rate. The impurity concentration in the product solution is then evaluated from a simple mass balance as:

$$\frac{C_i}{C_{io}} = \exp\left[-N_D S_i\right] \tag{2.12}$$

where C_{io} is the initial impurity concentration in the product solution, C_i is the impurity concentration remaining in the retentate at any time and N_D is the number of diavolumes, which is equal to the total collected filtrate volume divided by the constant retentate volume during the diafiltration process. S_i is the solute sieving coefficient, which is the ratio of the solute concentration in the permeate solution to that in the retentate. According to equation (2.12), the diafiltration process with $N_D = 10$ provides more than a 10^4 reduction in the concentration of a completely unretained species.

Many applications involve a combination of UF and diafiltration to achieve both the desired volume reduction and the required impurity or salt removal. If the diafiltration is performed before the species is concentrated, a very large diafiltration volume is needed, although the filtrate flux during the diafiltration process is quite high [143]. The amount of diafiltration buffer decreases as the retentate volume is reduced, but the resulting increase in species concentration causes a reduction in the filtrate flux.

Ng et al. [143] have used the concentration polarization model to evaluate the bulk protein concentration that minimizes the total time for the UF/diafiltration process as:

$$C_b^* = \frac{C_w}{e} \tag{2.13}$$

where $e = 2.718$. If the diafiltration is performed at smaller bulk species concentrations, the process time increases because of the greater volume of diafiltration buffer needed to obtain the desired impurity removal. If the diafiltration is performed at higher bulk species concentrations, the process time increases due to the reduction in the filtrate flux with increasing C_b.

This analysis assumes that the transmission of the impurity is unaffected by the concentration of the retained product. However, many proteins are able to bind a wide range of small molecules, amino acids and ionic species. Shao and Zydney [144] examined the behaviour of an UF/diafiltration process for the removal of an impurity that has a reversible binding interaction with the retained product. Calculations were performed using a Michaelis–Menten binding expression, with the flux evaluated using the classical concentration polarization model. The model was confirmed experimentally using data for the removal of D-tryptophan from BSA. Equations were also developed for the optimal diafiltration conditions to minimize the total process time accounting for the reversible binding of the impurity. Harinarayan et al. [145] recently showed that citrate, a small molecule commonly utilized in downstream processes, had reduced clearance in the UF/diafiltration step of an antigen-binding fragment of a monoclonal antibody. This resulted in high levels of citrate in the final preparation. The authors mentioned potential solutions including increasing the number of diavolumes (taking into account the effect on process time and protein concentration), replacing the small molecule and finding ways to reduce the interaction between the small molecule and the product protein. The authors underlined the importance to understand the clearance of these small molecules, and not assume a theoretical clearance with a sieving coefficient of 1, since protein interactions may exist.

2.4.2 High-performance tangential flow filtration

For optimizing the performance of protein UF, several researchers [14,146–148] have added an additional parameter. The process is conducted at the pI of the transmitted protein while the proteins with significantly different pIs are retained.

The name 'high-performance tangential flow filtration' (HPTFF) was used to describe these very high selectivity membrane processes [5,12,149]. HPTFF exploits a number of different strategies to achieve high-resolution separations, including

(1) proper choice of pH and ionic strength to maximize differences in the hydrodynamic volume of the product and impurity, (2) use of electrically charged membranes to enhance the retention of like-charged proteins, (3) operation in the pressure-dependent regime to maximize the selectivity and (4) use of a diafiltration mode to wash impurities through the membrane.

Several studies demonstrated that HPTFF was able to separate proteins with very similar size, from model binary mixtures. High selectivities ($\geq$50) have been achieved for model solutions containing binary systems like BSA and immunoglobulin G (IgG) [150]. The BSA–IgG separation was obtained by operating at the pI of IgG with a 300 kDa membrane, which allowed IgG to preferentially pass through the membrane. Van Eijndhoven et al. [151] developed a membrane system for the separation of two proteins with similar molecular weight, BSA and haemoglobin, with more than 100-fold purification and nearly 70% yield. Van Reis et al. [12] used HPTFF to separate BSA from an antigen-binding fragment of a monoclonal antibody (Fab), achieving more than 900-fold purification and 90% yield of BSA. More recently, Cheang and Zydney [152] were able to obtain 100-fold purification and greater than 90% recovery of β-LG from a binary mixture with α-lactalbumin (α-LA).

Experimental studies with complex multicomponent feed streams are more limited, and the overall performance of these systems is less impressive. For example, Muller et al. [146] used a combined UF–diafiltration process for the purification of α-LA from liquid whey protein concentrate with the final permeate having an α-LA purity of only 50%. The purification and yield obtained were significantly less than those obtained with model binary mixtures. Cheang and Zydney [153] provided a complete analysis for the separation of two proteins from a complex mixture: the purification of α-LA and β-LG from a commercial whey protein isolate. A two-stage membrane system was used (Figure 2.4), with the performance in each stage optimized to achieve the required selectivity by proper selection of pH, buffer conductivity and filtrate flux. The optimal buffer conditions for the 30 kDa membrane were pH 7 and a conductivity of 1.7 ms/cm, which yielded a selectivity for the separation between α-LA and β-LG of 21. Again, this value was considerably smaller than the selectivity of nearly 60-fold reported previously by Cheang and Zydney [152] for the separation of a binary mixture of α-LA and β-LG. A more optimistic study was reported recently by Lebreton et al. [154]. Using HPTFF with positively charged cellulosic membranes of 100 kDa molecular weight cut-off and operating under a selected range of buffer pH and ionic strength, a 10-fold removal of *Escherichia coli* host cell proteins was obtained with an overall process yield of 98%. The authors stated that when compared with a conventional purification scheme, the proposed process resulted in the elimination of one chromatographic step, a 12% yield improvement and a significant reduction in purification cost of goods.

2.4.3 *Affinity ultrafiltration*

Affinity UF uses a binding agent in free solution to selectively bind, and thus retain by an UF membrane, the species to be purified (Figure 2.12) [6]. The feasibility of affinity UF was demonstrated more than 20 years ago by Mattiasson and Ramstrop

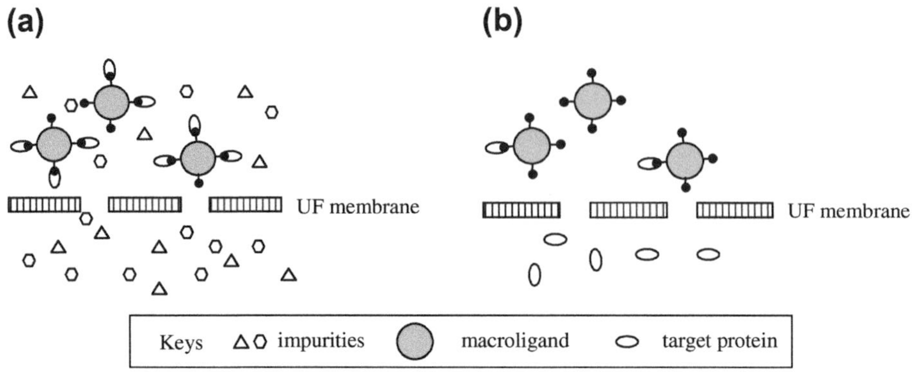

Figure 2.12 Principle of affinity UF: (a) binding of the target protein and washing out impurities; (b) dissociation of the complex target protein-macroligand mediated free ligand and transport of the purified target protein through the membrane [from 6].

[155] for the purification of concanavalin A, using heat killed *Saccharomyces cerevisiae* as the affinity ligand. The separation was performed using a hollow fibre membrane with a molecular weight cut-off of 1000 kDa which was fully retentive to the very large binding complex formed between concanavalin A and the *S. cerevisiae* while allowing unbound species to pass through the membrane.

Subsequent studies of affinity UF have examined: isolation of urokinase using *N*-acryloyl-*m*-aminobenzamide copolymerized with acrylamide as an affinity macroligand [156], purification of avidin using biotinylated liposomes [157], isolation of human serum albumin and lysozyme using Cibacron Blue bound to agarose [158], and recovery of BSA using highly substituted Blue Sepharose [159].

Affinity UF has also been used to separate chiral molecules using stereoselective macroligands. A large stereospecific binding agent in free solution is used to selectively bind, and thus retain by a semi-permeable membrane, one of the stereoisomers [160–166]. Romero and Zydney [164] studied the separation of D- and L-tryptophan by affinity UF using BSA as a stereoselective-binding agent. They were able to obtain 50-fold purification of L-tryptophan, but the product yield was less than 50%. Higher yields could be achieved for D-tryptophan in the permeate solution, but the purification factor for this enantiomer was less than five-fold. Poncet et al. [162] also examined the use of affinity UF for the separation of tryptophan enantiomers, with the optimal process giving 91% purity and 89% recovery of the D-tryptophan.

Although these results were encouraging, the yield and purification remained below the levels obtained with high-performance affinity chromatography. Some approaches were used to enhance the performance of the affinity UF process for protein purification. A multi-stage mode was examined theoretically by Dong et al. [167]. Model calculations clearly demonstrated the potential benefits of the staging, although no experimental results were provided to support the theoretical analysis. Experimental data were obtained for the separation of D- and L-tryptophan using BSA as the affinity macroligand and a two-stage system [166]. The two-stage system gave purification factors of more than 20 at greater than 90% yield. Purification-yield diagrams were used to examine the effects of the ligand concentration, number of stages, and stage volume on the overall separation. The authors concluded that the two-stage system provided better yield and purification factors due to the extra retention of the more highly bound L-tryptophan in the second stage.

Small charged ligands were demonstrated to be suitable for highly selective affinity UF, with the product of interest retained by electrostatic interactions using an electrically charged membrane [168]. A selectivity of more than 30-fold was obtained for the separation of ovalbumin and BSA upon the addition of Cibacron Blue, compared to less than a two-fold selectivity in the absence of the affinity ligand. The very low cost of the affinity dye compared to that of the large affinity macroligands used previously make this process a potentially attractive approach for protein separations.

2.4.4 Other modes of operation

Other modes of operation were introduced recently. A novel mode of operation was termed CPUF [169]. CPUF is a modification of dead-end UF. The principle is to operate the process under optimized conditions throughout an entire operation cycle. A carried phase which is a buffer corresponding to the optimized physicochemical conditions (i.e. buffer type, pH and ionic strength) is pumped through an UF unit in which the permeate flux and system hydrodynamics can be adjusted independently (e.g. stirred cell, or vortex flow UF unit). The flow rate, corresponding to the optimized permeate flux value, is maintained constant by pumping the carrier phase into the module using a high-performance pump. CPUF is carried out at constant flux and hence the pressure can change during the operation. The optimized hydrodynamic condition is maintained using suitable stirring or rotating arrangements (e.g. magnetic, shaft driven stirrer or rotating cylinder). The feed may be introduced into the system using one of the two following modes. In the Pulse Feed Input Mode of operation, the feed is introduced into the UF module in the form of one or several pulses using a suitable sample injector. An alternative way is to introduce the feed in the form of step inputs (Step Feed Input Mode).

Most UF processes are operated at constant *TMP*. An alternative strategy is to operate the process by maintaining a constant wall concentration, C_W, of retained protein [170]. In this study, a constant value of C_W was maintained throughout UF using a control algorithm with either a filtrate flux set point calculated from the polarization concentration model, or a *TMP* set point calculated from combined polarization concentration model and osmotic pressure model. The authors stated that the proper choice of C_W could reduce membrane fouling, minimized yield losses and improved process reproducibility and scale up.

2.5 Applications

In this section, some applications of UF are described. In antibiotic broths, emulsifiers can be removed by UF before solvent extraction to avoid emulsification and to improve extraction efficiency. DNA purification in both dead-end and cross-flow configurations is feasible using UF. Purification of antibodies includes UF in the overall process. UF can be also an alternative to chromatography for whey fractionation. Due to the low content of lysozyme in chicken egg white (CEW), UF is an advantageous process over chromatography and electrophoresis. Finally, some other applications are reported.

2.5.1 Antibiotic production

Most of antibiotics, such as penicillin G, erythromycin and medmycin, are widely used and also serve as raw materials for semisynthetic antibiotics. They are produced by fermentation and recovered from their broths via the conventional steps: filtration (removal of biomass), solvent extraction (isolation and purification) and subsequent crystallization (polish). During solvent extraction, antibiotics are extracted into an immiscible organic phase, *n*-butylacetate, at certain pH values [171]. This extraction step is a successful and established process. However, there exist some components in the culture medium or metabolites which are surface-active such as proteins and polysaccharides, causing a stable emulsion during extraction. De-emulsifiers are then needed, together with high speed and large capacity centrifugal extractors to accelerate the phase separation.

UF was proposed to remove emulsifiers in antibiotic broths before solvent extraction to avoid emulsification and to improve extraction efficiency. An example is the work of Nabais and Cardoso [172], who reported the purification of penicillin G obtained from fermented broths by UF in a diafiltration configuration. This study was done using an UF pilot with tubular membranes, with MWCO of 8 kDa polysulfone, 20 kDa polysulfone and 100 kDa PVDF. Proteins, coloured substances and other impurities were successfully removed giving high penicillin G recovery in the permeate. The authors concluded that UF might be an alternative to the use of flocculants and anti-emulsion agents to obtain good phases separation in penicillin G solvent extraction. Li et al. [173] used bench and pilot scale spiral wound membrane modules with 5, 20 and 50 kDa MWCO to treat benzylpenicillin broth before solvent extraction. The extraction operation was improved in terms of phase separation, elimination of the need for any de-emulsifier or wet agent, and increase extraction recovery and product quality. Similar improvements were observed in the extractions of erythromycin and medmycin (alkaline extraction at pH 10.0 and 8.7, respectively). These results confirmed that UF might be an alternative to the use of de-emulsifier or other wet agent to obtain good phase separation even by gravity without centrifugal extractor in solvent extractions of penicillin G, erythromycin and medmycin.

2.5.2 DNA purification

Over the past 20 years, there has been increasing interest in the use of plasmid DNA for gene therapy and DNA-based vaccines [174,175]. Chromatographic processes are currently used for the large-scale purification of plasmid DNA, although the mass transfer limitations and low-binding capacities associated with these very large biomolecules pose significant technical challenges [176]. Sagar et al. [177] indicated that the chromatography resin accounted for more than 50% of the total raw material cost in manufacturing of plasmid DNA. Membrane processes are an attractive alternative to chromatographic separations since they are only weakly affected by diffusional limitations [178]. Several studies have demonstrated the feasibility of using UF for DNA purification in both dead-end [179] and cross-flow configurations [180,181]. For example, Kahn et al. [181] developed a simple, scalable method for purification of plasmid DNA. The method included the modification of the alkaline-lysis-based plasmid extraction method by extending the solubilization step from less than 30 min to 24 h. The extraction was followed by UF with polyethersulfone 500 kDa and 1000 kDa MCWO membranes for purification of the remaining contaminants. The method yielded 15 to 20 mg of plasmid DNA per litre of bacterial culture and resulted in removal of >99% of RNA and >95% of the protein that remained after the modified alkaline lysis procedure (Figure 2.13). Using this method, plasmids isolated had comparable transfection capability relative to plasmid isolated using a classical, cesium chloride gradient-based method.

Latulippe et al. [178] and Latulippe and Zydney [182] have investigated the factors governing DNA transmission through UF membranes. Experimental data were

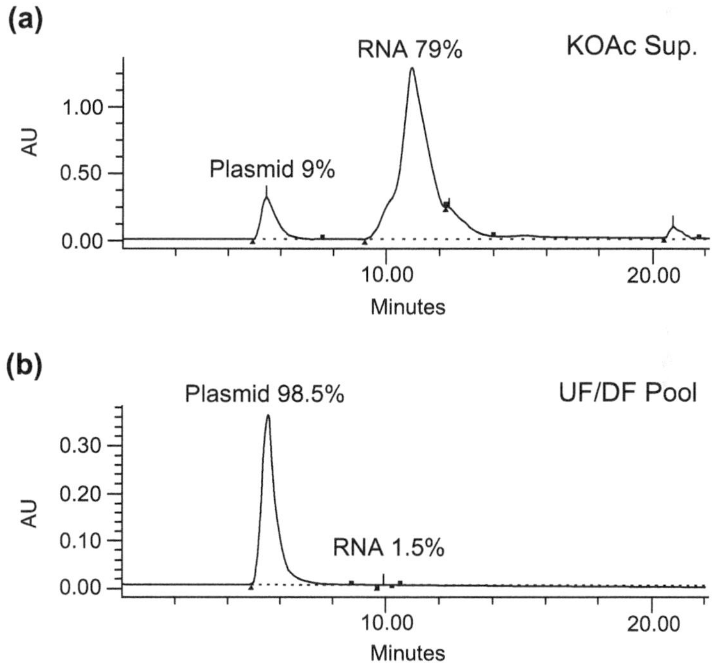

Figure 2.13 Purification of factor VIII plasmid DNA from RNA. Plasmid was prepared using overnight alkaline lysis and UF and then analyzed using size exclusion chromatography. (a) Analysis of the potassium acetate supernatant prior to purification by UF. (b) Analysis of the final ultrafiltred pool. Values indicate the percent of total absorbance at 260 nm [181]. This article was published in Biotechnol. Bioeng., Vol. 69, D.W. Kahn, M.D. Butler, D.L. Cohen, M. Gordon, J.W. Kahn, M.E. Winkler, Purification of plasmid DNA by tangential flow filtration, pp. 101–106, Copyright John Wiley and Sons (2000).

obtained for the transmission of a 3.0 kbp supercoiled plasmid DNA through composite regenerated cellulose UF membranes as a function of solution ionic environment in a stirred UF cell. The dependence on salt concentration was very pronounced, with the sieving coefficient increasing by more than 80-fold as the NaCl concentration increased from 1 to 150 mM at a fixed filtrate flux [182]. In contrast to the behaviour of proteins, plasmid DNA transmission through UF membranes was shown to be governed by the elongation/deformation of the DNA molecules associated with the fluid flow into the membrane pores, leading to a very strong dependence of transmission on the filtrate flux.

2.5.3 Monoclonal antibodies

In recent years, there has been a significant increase in the requirement of purified antibodies due to their potential use in gene therapy and treatment of cancer and autoimmune diseases [183]. By the end of 2002, twelve therapeutic antibodies, antibody fragments, or conjugated antibodies had been licensed by the US Food and Drug Administration (FDA) [184]. To improve patient convenience and compliance, high-concentration liquid mAb formulations for intravenous administration are desirable. MAbs are produced using cell culture supernatant; therefore, the mAb must be purified, concentrated and exchanged into an appropriate formulation. The mAb concentration in the cell culture supernatant is typically in the 0.1–1.0 mg/mL range, although higher yields are also reported [185]. The major impurities present are media proteins such as BSA, bovine transferrin and insulin.

Liquid chromatography is widely used for antibody purification due to its high resolution separation, and is usually associated to membrane techniques such as viral filtration and UF. A typical industrial purification process for monoclonal antibodies purification is shown in Figure 2.14 [186]. The process consists of three chromatographic steps: protein A affinity chromatography and two ion-exchange polishing steps. The sequence also contains viral filtration and UF. The technique has several reported drawbacks including time consuming and price, difficulty to scale-up, possible leaching out of protein A and possible mAb denaturation and dimerization during the elution step of protein A affinity chromatography [185].

Various studies have been proposed focusing on specific applications and/or improvements of the purification performance in terms of recovery and purification factor. The task is particularly difficult due to the very close molecular weight of the species to be separated [187]. Wan et al. [188] reported the fractionation of BSA and mAb using CPUF which is a modified form of batch diafiltration. Using a two-step UF process, high mAb purity was obtained. However, the processing time was high and the concentration of mAb obtained decreased exponentially with time. Recently, Mohanty and Ghosh [185] investigated the purification of Campath-1H, a humanized IgG1 mAb specific for the antigen CD52 from simulated mammalian cell culture supernatant. A novel continuous, three-stage, tangential-flow, cascade UF configuration was used. Purification experiments were carried out with the three-stage cascade system as well as with single-stage and two-stage processes for comparison. The recovery and purification factor were

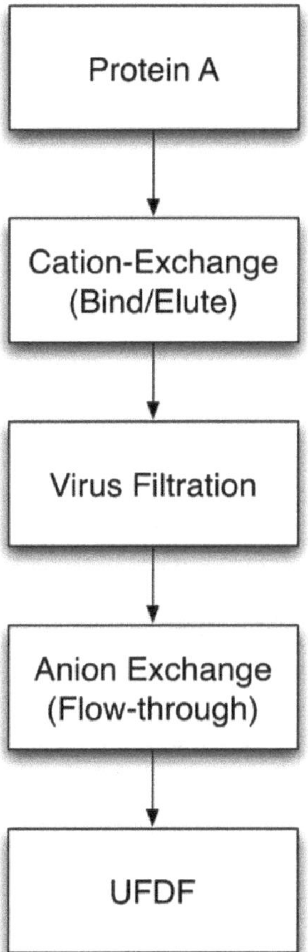

Figure 2.14 Schematic of a typical Genentech monoclonal antibody purification process with three chromatography steps [186].
This article was published in Biotechnol. Prog., Vol. 26, C.A. Teske, B. Lebreton, R. van Reis, Inline ultrafiltration, pp. 1068–1072, Copyright John Wiley and Sons (2010).

both significantly higher with the three-stage system. The highest purity obtained using the three-stage cascade system was 85%.

Teske et al. [186] proposed inline UF in series with the viral filtration step (Figure 2.14) to allow operation in single pass. The outlet of the previous unit operation was connected to the inlet on the UF cassette holder and the retentate sent to the pool tank while the filtrate was sent to drain. Using relatively small commercially available UF cassettes, the authors demonstrated that inline UF can significantly increase recoverable mass in an antibody purification process.

Using UF, Rosenberg et al. [60] prepared concentrated mAb intermediate solutions from a filtered pool of a purified chimeric human Fc (IgG4)/rat Fab antibody in citrate buffer pH 5.5 (IgG A). In addition, two chimeric human Fc (IgG1)/rat Fab antibodies

(IgG B and IgG C) in histidine buffer pH 6.0 were used. The mAbs were provided by Roche Diagnostics GmbH (Penzberg, Germany). An optimized UF concentration method was developed for the production of highly concentrated mAb solutions up to 140 mg/mL by varying the *TMP* and cross-flow conditions systematically depending on the concentration in the retentate. The process showed an improved permeate flux and reduced processing time. Concomitantly, the concentrates showed reduced aggregation compared to other methods operating under constant pressure and flow conditions.

2.5.4 Fractionation of whey proteins

The fractionation of whey proteins is very important in the food, nutraceutical, pharmaceutical and biotechnology industries. The four major whey proteins are β-LG, α- α-LA, BSA and Ig. α-LA can be used in infant formula and as a nutritional additive, β-LG has good emulsifying and foaming properties, and BSA and Ig are both of interest in food and therapeutic applications [189]. Several non-membrane processes have been proposed to fractionate whey proteins, focusing mainly on α-LA and β-LG, the predominant (>60%) whey proteins. These include ion exchange chromatography, affinity chromatography and precipitation [190]. The similar size (14 and 18 kDa) and pI (4.5–4.8 and 5.2) of these proteins complicate the separation process, resulting in inadequate yield and purity.

UF can be an economical alternative to chromatography for whey fractionation. Using model solutions, Cheang and Zydney [152] have obtained selectivity values (defined as the ratio of the observed sieving coefficients of α-LA and β-LG). Few studies have reported on the fractionation of proteins from natural whey. Ehsani et al. [191] have studied the fractionation of natural and model egg-white protein solutions. The difference in the flux and retention characteristics between model and natural suspensions was explained as a result of the interaction with other impurity proteins resulting in aggregation in the natural whey. Thus, industrial whey can pose problems due to the presence of multiple impurities and solutes. This could reduce the efficacy with which separations can be conducted based on pI, due to the propensity to aggregate in the vicinity of the optimally chosen pI.

Baruah et al. [190] developed a two-step membrane isolation and purification process with the goal of recovering heterologous immunoglobulin G (IgG), which comprises 10–15% of the total proteins, from transgenic goat milk. In the first step, MF was used to fractionate the milk proteins and recover >90% of the original IgG at a purity of about 15–20% in the permeate stream. In the second step, UF was used to increase the purity of the target protein to 80%, while maintaining a relatively high IgG yield (80%). UF experiments in diafiltration mode were conducted with 100 kDa cellulose membranes. The approach of conducting the UF process close to the pI of the predominant whey proteins (15–40 kDa, pI 5.2) could not be applied because of precipitation of residual casein at pH values lower than 8.5. Instead, the packing characteristics of the cake layer at the membrane wall was utilized to achieve a selectivity of >15, which was stated to be sufficient to meet the stated goals of purity and yield.

Metsämuuronen and Nyström [192] underlined the challenging task of fractionation of proteins from complex mixtures such as whey. Not only the sizes but also the

Table 2.3 Characteristics of major whey proteins [192]

Protein	Molar mass (kg mol^{-1})	Isoelectric point
β-Lactoglobulin	18.4	5.2
α-Lactoglobulin	14.1	4.5–4.8
Immunoglobulins	150–1000	5.5–8.3
Bovine serum albumin	69	4.7–4.9
Lactoferrin	78	9
Lactoperoxidase	89	9.5
Caseinomacropeptide	7	

isoelectric point of the main whey proteins, α-LA and β-LG, are very close to each other (Table 2.3) and, thus, the exploitation of charge effects is difficult. These authors have shown the role of the UF membrane hydrophobicity and pore size in the enrichment of α-LA to the permeate from diluted whey solutions. The experiments were performed at low TMPs (10–30 kPa) in order to limit fouling. The highest selectivity was obtained with the hydrophilic 30 Da regenerated cellulose membrane. This membrane was also the most resistant to fouling. In all cases, the final α-LA purity was only around 21% due to the high amount of small protein fragments. The α-LA transmission of the most hydrophobic membrane decreased with time and *TMP*, but simultaneously its selectivity increased due to pore narrowing and higher retention of β-LG.

2.5.5 *Purification of lysozyme*

Lysozyme is an enzyme commonly used for different applications, such as food additive in milk products, cell-disrupting agent for extraction of bacterial intracellular products, component of ophthalmologic preparations, and drug for treatment of ulcers and infections. In addition, a large quantity of pure lysozyme is used in research laboratories as model protein. The use of lysozyme in human cancer chemotherapy has also been reported [193,194]. Lysozyme occurs naturally in chicken egg white (CEW). CEW is a mixture of proteins (~54–57% ovalbumin, ~12–15% conalbumin, ~3–4% ovomucin and ~3.4% lysozyme). Commercially available purified lysozyme is manufactured from CEW using a combination of conventional processes such as crystallization, precipitation, centrifugation and adsorption [195].

Due to the low content of lysozyme in CEW, large amount of protein has to be processed to get a small amount of lysozyme. UF is then an advantageous process over chromatography and electrophoresis. If the operation is carried out under sterile conditions, pure lysozyme solution containing no bacteria or particulate matter (suitable for pharmaceutical use) can be obtained. With other types of bioseparation processes, an additional membrane-filtration step would be necessary. Balakrishnan and Agarwal [196] have discussed the fractionation of very dilute simulated mixture of ovalbumin and lysozyme using a vortex flow UF unit and a 100 kDa MWCO polyacrylonitrile membrane. Ehsani et al. [191] have operated the UF of CEW proteins using modified and unmodified 50 kDa MWCO polysulfone membranes. The effects of pH, salt concentration and membrane modification on permeate flux, total protein retention and membrane zeta potential were studied. The highest enrichment of ovalbumin was obtained at pH 4.8 when no salt was added to the feed. Retention of ovalbumin and transmission of lysozyme increased in UF of egg-white solutions in the presence of salt at pH 4.8. Ghosh and Cui [197] have examined the purification of lysozyme from CEW using 25 kDa and 50 kDa MWCO polysulfone membranes. When using a 25 kDa MWCO membrane, the two major components of CEW, ovalbumin and conalbumin were expected to be largely retained by the membrane. However, with the 50 kDa MWCO membrane, significant amounts of ovalbumin and other egg-white proteins were also expected to cross the membrane. In addition to ensuring high transmission of lysozyme, minimization of transmission of ovalbumin and other proteins was therefore also important.

2.5.6 Other applications

Several applications of UF in biotechnologies and pharmaceutics are reported such as the purification and/or concentration of various proteins, and marine microalgae processing.

A number of proteins other than monoclonal antibodies, whey proteins and lysozyme are purified and/or concentrated in a scheme including an UF step. Two recent examples are the purification of recombinant proteins from aqueous corn endosperm and the concentration of the apolipoprotein A1 Milano (ApoA-1 M), a highly self-associating protein. Corn seed consists largely of germ (11% dry weight) and endosperm (83%). The germ contains most of the seed oil and water-soluble protein, and the endosperm mainly starch and water-insoluble proteins [198]. Several techniques are applied for the recovery of recombinant proteins from corn germ and endosperm extracts, such as bed chromatography and aqueous two-phase partitioning. UF is also a suitable technique for the purification of recombinant proteins from aqueous corn endosperm and germ extracts. Aspelund and Glatz [198] investigated the effectiveness of UF by using model proteins of two different sizes: recombinant type I human collagen (rCollagen, 265 kDa) and green fluorescent protein (GFP, 27 kDa). UF with a 300 kDa MWCO membrane resulted in a significant loss of rCollagen, whereas a 100 kDa MWCO membrane completely retained rCollagen. rCollagen purity of 89% was achieved with only diafiltration of endosperm extracts and, when preceded by precipitation, a purity of >99% was obtained. The authors concluded that UF is a suitable method to separate and purify corn-hosted recombinant proteins >100 kDa, particularly when the expression is targeted to the endosperm.

Annathur et al. [199] studied the UF of apolipoprotein A1 Milano (ApoA-1 M), a highly self-associating protein. Apolipoprotein A1 Milano (ApoA-1 M) is a naturally occurring variant of apolipoprotein A-1 (ApoA-1) and is the protein component of ETC-216, a high-density lipoprotein (HDL) mimic with promising potential for reduction of atherosclerotic plaque. The overall manufacturing process for ApoA-1 M dimer includes microbial fermentation and purification that consists of multiple chromatography columns and one reaction step. Urea is used throughout the process to minimize both ApoA-1 M self-association and its cross-association with impurities. The final UF/DF operation is used to concentrate the purified ApoA-1 M and to reduce the molar quantity of urea in the elution pool of the preceding column. The authors found several operating constraints on the filtrate flux even at moderate protein concentrations because of the propensity for gel formation at the membrane surface. Purified ApoA-1 M could not be concentrated by UF beyond 25 g/L in a low-salt buffer without a drastic drop in permeate flux and significant yield loss.

UF, as well as MF and NF, have found applications in marine microalgae processing with several small aquaculture farms using membrane separation techniques for specific applications [200]. Several studies are reported in the literature. Rossi et al. [201] studied the harvesting of *Arthrospira platensis* with ceramic and organic membranes and Rossignol et al. [200], the continuous recovery of *Haslea ostrearia* and *Skeletonema costatum*. UF was found satisfactory in terms of quality of the product (composition of the main cell macro-components, such as proteins and exopolysaccharides) and permeate fluxes. The experimental conditions were low *TMP* (less than 1 bar) and low fluid velocity to reduce shear rate to less than 40,000 s^{-1} in order to preserve micro-organisms and avoid exopolysaccharide synthesis causing severe fouling [202]. Membrane processes are also successful for concentration and/or pre-purification of metabolites issued from microalgae. Jaouen et al. [203] have clarified suspension containing a phycobiliprotein, the C-phycocyanin (C-PC), a natural blue pigment, from disintegrated *Spirulina platensis* cells with 0.2 μm ceramic MF membrane. Permeate fluxes were close to 80 L h^{-1} m^{-2} and C-PC retention ranged from 96 up to 100%. UF and NF membranes were used to concentrate the pigment. Frappart et al. [204] compared two UF systems: a cross-flow UF unit equipped with a flat sheet membrane and a dynamic filtration module, consisting of a disk rotating close to a stationary membrane in order to reduce fouling. *Cylindrotheca fusiformis* and *S. costatum* microalgal suspensions were ultrafiltered at 1 bar with a 40,000 Da PAN membrane with a shear rate equal to 16,000 s^{-1}. The results showed that the dynamic filtration module yielded permeate flux almost twice higher than the cross-flow filtration system for the two microalgae species. Denis et al. [205] have used UF membranes in order to concentrate R-phycoerythrin (another phycobiliprotein) extracted from algae. Best results were obtained with a 30,000 Da UF PES membrane. On the basis of these results, an economical estimation was carried out on a pilot plant allowing the daily production of 31.6 L of an R-phycoerythrin (R-PE) concentrate. The pilot plant would operate 300 days a year and 5 h 30 min per day (Table 2.4) including 1 h 30 min per day for membrane cleaning. It was shown that it was possible to produce, for €1, 1 L of purified R-PE extract having an R-PE concentration equal to 0.245 $g_{R\text{-}PE}$ L^{-1}.

2.6 Theory

This section presents classical models of UF: the concentration polarization model and limiting flux models, including the gel model and the osmotic pressure model. Recent advances in numerical simulations of UF are also presented.

Table 2.4 Standard specifications for economic evaluation of industrial pilot plant [205]

Plant capacity (L d^{-1})	31.6
Annual operating time (d y^{-1})	300
Daily operating time for filtration (h d^{-1})	4
Daily operating time for membrane cleaning (h d^{-1})	1.5
Surface area module (m^2)	0.9
R-PE initial concentration (g L^{-1})	0.049
R-PE recovery rate (%)	100
VRF	5

2.6.1 Concentration polarization

During UF, the TMP is exerted on the solution in contact with the membrane. The concentration of retained solute increases at the membrane surface due to the removal of solvent. This results in a concentration gradient with the maximum solute concentration at the membrane surface (Figure 2.15a). This phenomenon is known as 'concentration polarization.' The model of concentration polarization, also termed 'simple stagnant film model', was originally presented by Michaels [206].

As a result of the increased concentration at the membrane surface, there is a tendency for the solute to diffuse away from it. Under steady state conditions, the

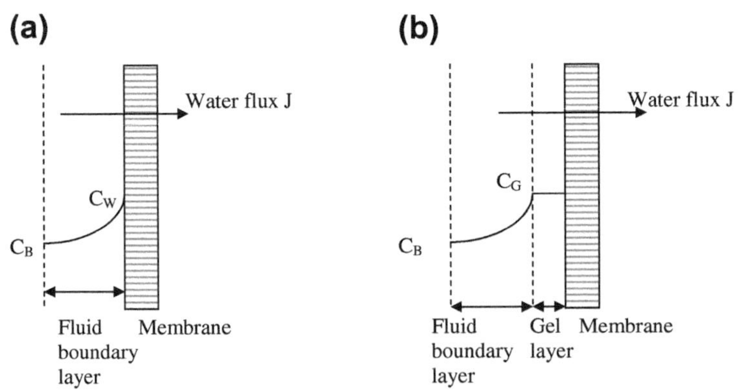

Figure 2.15 Mass transfer during UF: (a) concentration polarization and (b) gel polarization [15].

convective mass transfer due to filtration is balanced by the diffusive movement in the opposite direction. This condition can be expressed as follows:

$$JC - D\frac{dC}{dx} = 0 \tag{2.14}$$

where JC and $D(dC/dx)$ are the convective and diffusive terms, respectively. J is the permeate flux, C is the solute concentration at the distance x from the membrane surface and D is the solute diffusion coefficient. Equation (2.14) can be integrated across the concentration boundary layer to give:

$$J = \left(\frac{D}{\delta}\right)\ln\left(\frac{C_w}{C_b}\right) \tag{2.15}$$

where C_w is the solute concentration at the membrane wall, C_b is the bulk solute concentration, and δ is the boundary layer thickness. Since this term is typically unknown, a mass transfer coefficient, k, usually replaces D/δ:

$$J = k\ln\left(\frac{C_w}{C_b}\right) \tag{2.16}$$

Mass transfer coefficient

Experimental correlations for the mass transfer coefficient are expressed using dimensionless correlations based on heat-mass transfer analogy for the Sherwood number, Sh, in terms of the Reynolds number, Re, and the Schmidt number, Sc:

$$\mathrm{Sh} = \beta \mathrm{Re}^a \mathrm{Sc}^b \left(\frac{d_h}{L}\right)^c \tag{2.17}$$

where

$$\mathrm{Sh} = \frac{k d_h}{D} \tag{2.18}$$

$$\mathrm{Re} = \frac{\rho u d_h}{\mu} \tag{2.19}$$

$$\mathrm{Sc} = \frac{\mu}{\rho D} \tag{2.20}$$

L is the channel length, u the tangential velocity, ρ the solution density and μ the solution viscosity. The parameters β, a, b and c are functions of the specific system geometry and flow conditions. The hydraulic diameter, d_h, is defined as:

$$d_h = 4\frac{\text{cross-sectional area available for flow}}{\text{wetted perimeter of the channel}} \tag{2.21}$$

which is equal to the tube diameter, d, for a tubular or hollow fibre membrane. For a slit-shaped channel formed between two parallel plates, d_h is equal to twice the channel height, 2 h.

The theoretical results for the parameters β, a, b and c are given for fully developed laminar flow, developing laminar flow (i.e. under conditions where the concentration and velocity boundary layers develop simultaneously) and turbulent flow [3]. For the mass transfer coefficient in turbulent flow, Gekas and Hallström [207] have reviewed empirical and theoretical correlations.

In the case of fully developed laminar flow in membrane modules, the Graetz–Leveque correlation can be applied [208]:

$$\mathrm{Sh} = 1.62\,\mathrm{Re}^{0.33}\mathrm{Sc}^{0.33}\left(\frac{d_\mathrm{h}}{L}\right)^{0.33} \tag{2.22}$$

For turbulent flow conditions, the Chilton–Colburn correlation can be used [207]:

$$\mathrm{Sh} = 0.023\mathrm{Re}^{0.8}\mathrm{Sc}^{0.33} \tag{2.23}$$

Other correlations for the mass transfer coefficient have been developed, for example for spacer-filled channel [209] and rotating cylinder module [210].

Transmission and selectivity

In case of partially transmitted species through a membrane, the concentration polarization model is expressed as:

$$J = k\ln\frac{C_\mathrm{w} - C_\mathrm{p}}{C_\mathrm{b} - C_\mathrm{p}} \tag{2.24}$$

where, C_p is the solute concentration in the permeate. This equation can be written as:

$$J = k\ln\frac{(C_\mathrm{w}/C_\mathrm{p}) - 1}{(C_\mathrm{b}/C_\mathrm{p}) - 1} \tag{2.25}$$

The following relation between the intrinsic sieving coefficient, S_i, and the apparent sieving coefficient, S_a, is then obtained:

$$\left(\frac{S_\mathrm{a}}{1 - S_\mathrm{a}}\right) = \left(\frac{S_\mathrm{i}}{1 - S_\mathrm{i}}\right)\exp\left(\frac{J}{k}\right) \tag{2.26}$$

The intrinsic sieving coefficient can be rewritten [150]:

$$S_\mathrm{i} = \frac{S_\infty \exp\left((S_\infty J\delta)/D\right)}{S_\infty + \exp\left((S_\infty J\delta)/D\right) - 1} \tag{2.27}$$

where S_∞ is an asymptotic intrinsic sieving coefficient which is approached at very high J.

From equations (2.26 and 2.27), the following expression for S_a is obtained:

$$S_\mathrm{a} = \frac{S_\infty \exp\left[((S_\infty J\delta)/D) + (J/k)\right]}{(S_\infty - 1)(1 - \exp\left[(S_\infty J\delta)/D\right]) + S_\infty \exp\left[((S_\infty J\delta)/D) + (J/k)\right]} \tag{2.28}$$

The model was used to evaluate the impact of permeate flux and cross-flow velocity on the selectivity of a binary mixture (Figure 2.16) [211]. However, it was not intended to predict experimental data.

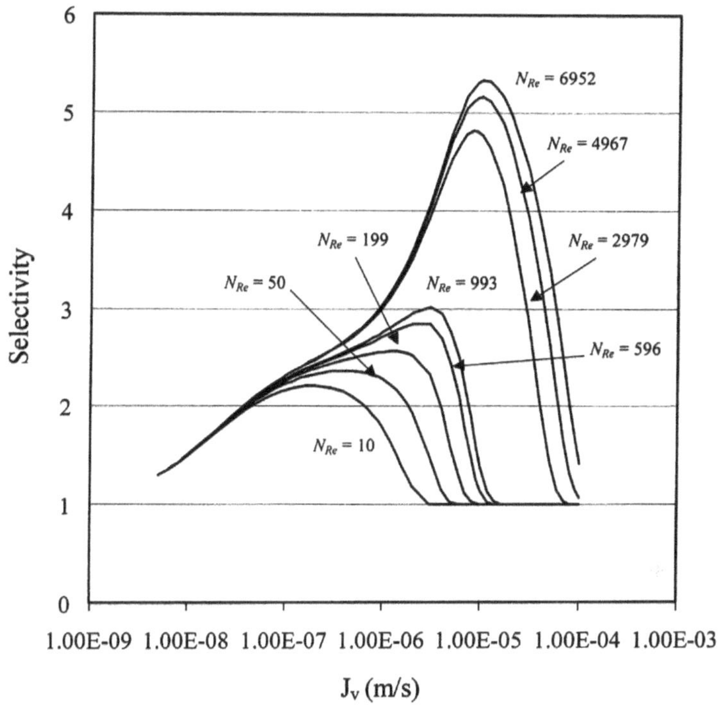

Figure 2.16 Effect of permeate flux on selectivity [211].
This article was published in J. Membr. Sci., Vol. 180, R. Ghosh, Z.F. Cui, Simulation study of the fractionation of proteins using ultrafiltration, pp. 29–36, Copyright Elsevier (2000).

2.6.2 Limiting flux models

Gel model

The gel model was introduced [206,212] to explain the flux plateau obtained in UF, where an increase in the applied pressure yields no flux increase after a certain threshold (Figure 2.5). As the concentration at the membrane surface increases, the solute reaches its solubility limit and precipitates on the membrane surface. The layer of precipitated solute is commonly referred to as the gel layer and the solubility limit is known as the gel concentration. According to this model, an increase in pressure produces a temporary increase in flux which brings more solute to the gel layer and increases its thickness, thereby reducing the flux to its original level (Figure 2.15b).

The flux may be calculated from equation (2.24), which holds for perfect rejection ($R = 1$):

$$J = k \ln \left(\frac{C_g}{C_b} \right) \tag{2.29}$$

where C_g is the 'gel concentration' and may be determined from the flux, J, versus $\ln C_b$ plot as an extrapolation to the $J = 0$ value.

Osmotic pressure model

An important assumption of the gel model is that the osmotic pressure is negligible compared with the applied pressure. However, the wall concentration is usually one or two orders higher than the bulk concentration when the limiting flux is observed. For a lot of macromolecules, the second and third virial coefficients may dominate the magnitude of the osmotic pressure, which may be of the same order of magnitude as the applied pressure. Therefore, the effective *TMP* may not increase with the applied pressure if the wall concentration is high enough. Kedem-Katchalsky [213] derived the osmotic pressure model which describes the permeate flux in relation to the osmotic pressure difference $\Delta\pi$ due to the concentration difference between the two sides of the membrane:

$$J = \frac{\Delta P_{\mathrm{tm}} - \Delta\pi}{R_{\mathrm{m}}} \tag{2.30}$$

where R_{m} is the membrane resistance, $\Delta\pi$ the osmotic pressure difference between both sides of the membrane:

$$\Delta\pi = \pi(C_{\mathrm{w}}) - \pi(C_{\mathrm{P}}) \tag{2.31}$$

The species concentration at the membrane surface, C_{w}, is usually obtained from the concentration polarization model. Since C_{w} is usually high, the corresponding osmotic pressure is calculated from the viral equation:

$$\pi = a_0 + a_1 C + a_2 C^2 + \cdots \tag{2.32}$$

Discussion of the two models

Some proposed improvements for the gel model and the osmotic pressure model have resulted in better fittings between calculated and experimental data. For example, Probstein et al. [214] incorporated the concentration-dependent diffusivity into the gel model and reported excellent agreement with their experimental results of UF of BSA solutions. Trettin and Doshi [215] proposed a closed form integral method for the solute transport equation. Aimar and Sanchez [216] and Aimar and Field [217] demonstrated that the impeded mass transfer in the concentrated layer corresponding to the increased viscosity might lead to a limiting flux. De and DasGupta [218] incorporated the effects of suction and osmotic pressure variations with concentration into the limiting flux analysis together with a modified Sherwood number relationship. This model predicted the limiting flux well, for BSA and dextran UF data, obtained from the literature [219,220].

By fitting the gel model to experimental results, several authors have demonstrated that the gel concentration, C_{g}, was not constant and may depend on other parameters, such as the bulk concentration and cross-flow velocity. In addition, this fitted gel concentration was sometimes physically unreasonable, for example exceeding the solubility. Although the gel layer may occur for some macromolecules under certain conditions, it was most often postulated rather than being directly observed. Many authors concluded that the osmotic pressure model was the most probable one, for UF of BSA and dextran solutions. From experiments with concentrated solutions of dextran and whey, Jonsson [219] concluded that the osmotic pressure of the concentrated boundary layer was responsible for the pressure independence of the permeate flux. Clifton et al. [221] used a model assuming a complete osmotic limitation to flux and the experimental work was performed using dextran and polyvinylpyrrolidone under conditions in which the osmotic effects were likely to predominate. From experiments of unstirred UF of dextran, Nakao et al. [222] calculated the boundary layer concentration and concluded that no gel layer was found. For example, Zaidi and Kumar [223] stated that UF of dextran was osmotically limited and that no gel layer was formed even in the case of a pressure-independent regime for the concentration range studied. Several recent studies also point out the significance of osmotic pressure on the limiting flux phenomena. Wang and Rodgers [224] used the free-solvent-based flux model (FSB) developed from the Kedem–Katchalsky model, film theory and the free solvent representation for osmotic pressure. The FSB model illustrated that osmotic pressure was the primary factor in limiting flux during UF. It provided very good agreement for the experimental results of ovalbumin, BSA, and IgG in moderate NaCl buffered solutions at various pH.

2.6.3 Numerical simulations

Computational fluid dynamics (CFD) is an important tool to model flow and concentration polarization in membrane systems or thin channels with permeable walls. Approaches based on CFD have been developed nearly as 20 years ago to solve the limitations of simple models [225–227]. The basic transport equations that govern the flow of the viscous fluid are mathematical representations of conservation principles under laminar or turbulent conditions [228]. The principle of mass conservation applied to a flowing fluid result in the equation:

$$\frac{\partial u_x}{\partial x} + \frac{\partial u_y}{\partial y} = 0 \tag{2.33}$$

the momentum conservation in the x-direction:

$$\frac{\partial u_x}{\partial t} = -\frac{\partial u_x u_x}{\partial x} - \frac{\partial u_x u_y}{\partial y} - \frac{1}{\rho}\frac{\partial p}{\partial x} + \frac{\partial}{\partial x}\left(\frac{\mu}{\rho}\left[\frac{\partial u_x}{\partial x} + \frac{\partial u_x}{\partial x}\right]\right) + \frac{\partial}{\partial y}\left(\frac{\mu}{\rho}\left[\frac{\partial u_y}{\partial x} + \frac{\partial u_x}{\partial y}\right]\right) \tag{2.34}$$

the momentum conservation in the y-direction:

$$\frac{\partial u_y}{\partial t} = -\frac{\partial u_y u_y}{\partial y} - \frac{\partial u_x u_y}{\partial x} - \frac{1}{\rho}\frac{\partial p}{\partial y} + \frac{\partial}{\partial y}\left(\frac{\mu}{\rho}\left[\frac{\partial u_y}{\partial y} + \frac{\partial u_y}{\partial y}\right]\right) + \frac{\partial}{\partial x}\left(\frac{\mu}{\rho}\left[\frac{\partial u_x}{\partial y} + \frac{\partial u_y}{\partial x}\right]\right) \tag{2.35}$$

finally, the species conservation:

$$\frac{\partial c}{\partial t} = -\frac{\partial u_x c}{\partial x} - \frac{\partial u_y c}{\partial y} + \frac{\partial}{\partial x}\left(D\frac{\partial c}{\partial x}\right) + \frac{\partial}{\partial y}\left(D\frac{\partial c}{\partial y}\right) \tag{2.36}$$

where x, y are the spatial coordinates, p is the pressure, c the species concentration, u the velocity, D the diffusion coefficient, μ the dynamic viscosity and ρ the mass density.

A number of recent numerical simulations have been proposed for UF. Ma et al. [229] used a 2D numerical model to simulate the limiting flux phenomenon observed during the UF of dextran, assuming neither constant permeate velocity nor constant wall concentrations. It was found that the calculated permeate velocity agreed well with the permeate flux of dextran experiments from the literature [219]. These authors acknowledged, however, that their numerical model assumed no cake formation or gel layer. Fernández-Sempere et al. [230] proposed a model of dead-end UF, based on the convection–diffusion mechanism and the osmotic pressure theory, combined with an empirical equation for the adsorption process. The model was validated by a reasonable agreement between experimental PEG data and calculated results (concentration profiles and permeate flux). Moreover, real-time holographic interferometry was used to visualize the appearance, evolution and disappearance of the polarization layer during dead-end UF. For UF of BSA solutions in a parallel cell, Afonso et al. [231] solved simultaneously the Navier–Stokes equations and the mass transport equation. Numerical results were compared with experimental data of the permeate flux in the limiting flux regime published in the literature [220]. This comparison showed that the concentration dependence of the viscosity, diffusivity and osmotic pressure were all necessary to explain experimental results.

Marcos et al. [232] developed a CFD based model for the prediction of the transient concentration process of soy proteins using a hollow fibre UF system. The CFD model was developed by solving the 2D Navier–Stokes equation and the mass conservation equation for transient conditions. A resistance model was used to link the retained protein concentration, the feed and permeate velocity and the pressure at the membrane surface. The resistance model considered the formation of a polarization layer and a cake layer. With these assumptions, the model was able to predict the transient permeate velocity and the pressure profiles with a good fit with the experimental data. The authors also underlined that *TMP* was an important element on the polarization concentration profile and that a constant *TMP* yielded erroneous conclusion on the concentration polarization.

Other recent CFD studies describe and optimize the complex hydrodynamics generated by pulsatile flow and gas sparging, spacers, and Dean and Taylor vortices [228,233,234]. For example, CFD simulation has been a powerful tool to study the flow patterns in spacer-filled channels. The simulations have confirmed that the geometric parameters of spacers are major variables for flow resistance and mass transfer [235]. Other CFD simulations have taken into account the various physical properties of the species retained and the fouling by surface adsorption [236]. Next steps towards better prediction will include replacing the convection–diffusion equation by suitable multi-component transport equations, and spatial resolution of the membrane including the corresponding conservation balances.

References

[1] W. Ho, K. Sirkar, Membrane Handbook, van Mostrand Reinhold, New York, 1992. 597–610.

[2] M. Cheryan, Ultrafiltration Handbook, Technomic Pub. Co., Lancaster, USA, 1986.

[3] L.J. Zeman, A.L. Zydney, Microfiltration and Ultrafiltration, Principles and Applications, Marcel Dekker, New York, 1996.
[4] R. van Reis, A.L. Zydney, Protein ultrafiltration, in: M.C. Flickinger, S.W. Drew (Eds.), Encyclopedia of Bioprocess Technology: Fermentation, Biocatalysis, and Bioseparation, John Wiley & Sons, New York, 1999, pp. 2197–2214.
[5] A.L. Zydney, R. van Reis, High performance tangential flow filtration, in: W.K. Wang (Ed.), Membrane Separations in Biotechnology, Marcel Dekker, New York, 2001, pp. 277–298.
[6] I.Y. Galaev, B. Mattiasson, Protein purification by affinity ultrafiltration, in: W.K. Wang (Ed.), Membrane Separations in Biotechnology, Marcel Dekker, New York, 2001, pp. 243–263.
[7] R. Ghosh, Protein Bioseparation Using Ultrafiltration, Theory, Applications and New Developments, Imperial College Press, London, 2003.
[8] R. van Reis, A.L. Zydney, Protein ultrafiltration, in: M.C. Flickinger (Ed.), Encyclopedia of Industrial Biotechnology: Bioprocess, Bioseparation, and Cell Technology, John Wiley & Sons, New York, 2010, pp. 1–20.
[9] A.B. Koltuniewicz, E. Drioli, Membranes in Clean Technologies, Theory and Practice, Wiley-VCH, Weinheim, 2008.
[10] R. Ghosh, Biopharmaceutical separations by ultrafiltration, in: N.N. Li, A.G. Fane, W.S.W. Ho, T. Matsuura (Eds.), Advanced Membrane Technology and Applications, Wiley, Hoboken, NJ, 2008, pp. 435–450.
[11] R. Ghosh, Ultrafiltration-based protein bioseparation, in: A.K. Pabby, S.S.H. Rizvi, A.M. Sastre (Eds.), Handbook of Membrane Separations, Chemical, Pharmaceutical, Food, and Biotechnological Applications, CRC Press, Taylor & Francis Group, 2009, pp. 497–511.
[12] R. van Reis, J.M. Brake, J. Charkoudian, D.B. Burns, A.L. Zydney, High performance tangential flow filtration using charged membranes, J. Memb. Sci. 159 (1999) 133–142.
[13] A. Allegrezza, T. Ireland, W. Kools, M. Phillips, B. Raghunath, R. Wilkins, A. Xenopoulos, Membranes in the biopharmaceutical industry, in: K.V. Peinemann, S. Pereira-Nunes (Eds.), Membranes for Life Science, Wiley-VCH, Germany, 2008, pp. 91–154.
[14] R. van Reis, E.M. Goodrich, C.L. Yson, L.N. Frautschy, S. Dzengeleski, H. Lutz, Linear scale ultrafiltration, Biotechnol. Bioeng. 55 (1997) 737–746.
[15] C. Charcosset, Downstream processing and product recovery/membrane systems and technology, in: M.Y. Murray (Ed.), Comprehensive Biotechnology, second ed., vol. 2, Elsevier, 2011, pp. 603–618.
[16] D.M. Kanani, W.H. Fissell, S. Roy, A. Dubnisheva, A. Fleischman, A.L. Zydney, Permeability–selectivity analysis for ultrafiltration: effect of pore geometry, J. Memb. Sci. 349 (2010) 405–410.
[17] D.B. Burns, A.L. Zydney, Buffer effects on the zeta potential of ultrafiltration membranes, J. Memb. Sci. 172 (2000) 39–48.
[18] S. Nakao, H. Osada, H. Kurata, T. Tsuru, S. Kimura, Separation of proteins by charged ultrafiltration membranes, Desalination 70 (1988) 191–205.
[19] H. Miyama, H. Yoshida, Y. Nosaka, Negatively charged polyacrylonitrile graft copolymer membrane for permeation and separation of plasma proteins, Makromol. Chem. Rapid Commun. 9 (1988) 57–61.
[20] L. Millesime, C. Amiel, B. Chaufer, Ultrafiltration of lysozyme and bovine serum albumin with polysulfone membranes modified with quaternized polyvinylimidazole, J. Memb. Sci. 89 (1994) 223–234.

[21] J. Howell, R. Field, D. Wu, Ultrafiltration of high-viscosity solutions: theoretical developments and experimental findings, Chem. Eng. Sci. 51 (1996) 1405–1415.
[22] P. Aimar, Ultrafiltration of pseudoplastic fluids, in: M.S. Verrall, M.J. Hudson (Eds.), Separations for Biotechnology, Ellis Horwood, Chichester, UK, 1987, pp. 360–372.
[23] M. Pritchard, The influence of rheology upon mass transfer in cross-flow membrane filtration, Ph.D. Thesis, University of Bath, UK, 1990.
[24] C. Charcosset, L. Choplin, Ultrafiltration of non-Newtonian fluids, J. Memb. Sci. 115 (1996) 147–160.
[25] N.S. Pujar, A.L. Zydney, Electrostatic and electrokinetic interactions during protein transport through narrow pore membranes, Ind. Eng. Chem. Res. 33 (1994) 2473–2482.
[26] D.B. Burns, A.L. Zydney, Effect of solution pH on protein transport through ultrafiltration membranes, Biotechnol. Bioeng. 64 (1999) 27–37.
[27] F.G. Smith, W.M. Deen, Electrostatic double-layer interactions for spherical colloids in cylindrical pores, J. Colloid Interface Sci. 78 (1980) 444–465.
[28] A. Mehta, A.L. Zydney, Effect of membrane charge on flow and protein transport during ultrafiltration, Biotechnol. Prog. 22 (2006) 484–492.
[29] D.B. Burns, A.L. Zydney, Contributions to electrostatic interactions on protein transport in membrane systems, AIChE J. 47 (2001) 1101–1114.
[30] J. Molek, K. Ruanjaikaen, A.L. Zydney, Effect of electrostatic interactions on transmission of PEGylated proteins through charged ultrafiltration membranes, J. Memb. Sci. 353 (2010) 60–69.
[31] S.T. Loh, U. Beuscher, T.K. Poddar, A.G. Porter, J.M. Wingard, S.M. Husson, S.R. Wickramasinghe, Interplay among membrane properties, protein properties and operating conditions on protein fouling during normal-flow microfiltration, J. Memb. Sci. 332 (2009) 93–103.
[32] R. Chan, V. Chen, Characterization of protein fouling on membranes: opportunities and challenges, J. Memb. Sci. 242 (2004) 169–188.
[33] K.J. Kim, A.G. Fane, C.J.D. Fell, D.C. Joy, Fouling mechanisms of membranes during protein ultrafiltration, J. Memb. Sci. 68 (1992) 79–91.
[34] D.N. Lee, R.L. Merson, Examination of cottage cheese whey proteins by scanning electron microscopy: relationship to membrane fouling during ultrafiltration, J. Dairy Sci. 58 (1975) 1423–1432.
[35] G. Gesan, G. Daufin, U. Merin, J.-P. Labbe, A. Quemerais, Fouling during constant flux crossflow microfiltration of pretreated whey. Influence of transmembrane pressure gradient, J. Memb. Sci. 80 (1993) 131–145.
[36] F. Martinez, A. Martin, P. Pradanos, J.I. Calvo, L. Palacio, A. Hernandez, Protein adsorption and deposition onto microfiltration membranes: the role of solute–solid interactions, J. Colloid Interface Sci. 221 (2000) 254–261.
[37] M. Rabiller-Baudry, M. Le Maux, B. Chaufer, L. Begoin, Characterisation of clean and fouled membrane by ATR-FTIR and EDX analysis coupled with SEM: application to UF of skimmed milk with a PES membrane, Desalination 146 (2002) 123–128.
[38] U. Reichert, T. Linden, G. Belfort, M.-R. Kula, J. Thommes, Visualising protein adsorption to ion-exchange membranes by confocal microscopy, J. Memb. Sci. 199 (2002) 161–166.
[39] J.-P. Labbe, A. Quemerais, F. Michel, G. Daufin, Fouling of inorganic membranes during whey ultrafiltration: analytical methodology, J. Memb. Sci. 51 (1990) 293–307.
[40] B.C. Robertson, A.L. Zydney, Protein adsorption in asymmetric ultrafiltration membranes with highly constricted pores, J. Colloid Interface Sci. 134 (1990) 563–575.

[41] M. Nystrom, A. Pihlajamaki, N. Ehsani, Characterization of ultrafiltration membranes by simultaneous streaming potential and flux measurements, J. Memb. Sci. 87 (1994) 245–256.
[42] K.J. Kim, V. Chen, A.G. Fane, Characterization of clean and fouled membranes using metal colloids, J. Memb. Sci. 88 (1994) 93–101.
[43] S.F. Oppenheim, G.R. Buettner, J. Dordick, V.G.J. Rodgers, Applying electron paramagnetic resonance spectroscopy to the study of fouling in protein ultrafiltration, J. Memb. Sci. 96 (1994) 289–297.
[44] K.J. Kim, A.G. Fane, M. Nyström, A. Pihlajamaki, Chemical and electrical characterization of virgin and protein-fouled polycarbonate track-etched membranes by FTIR and streaming-potential measurements, J. Memb. Sci. 134 (1997) 199–208.
[45] T. Maruyama, S. Katoh, M. Nakajima, H. Nabetani, Mechanism of bovine serum albumin aggregation during ultrafiltration, Biotechnol. Bioeng. 75 (2001) 233–238.
[46] T.J. Su, J.R. Lu, Z.F. Cui, R.K. Thomas, Fouling of ceramic membranes by albumins under dynamic filtration conditions, J. Memb. Sci. 173 (2000) 167–178.
[47] G. Yin, J.-C. Janson, Z. Liu, Characterization of protein adsorption on membrane surface by enzyme linked immunoassay, J. Memb. Sci. 178 (2000) 99–105.
[48] R. Chan, V. Chen, M.P. Bucknall, Quantitative analysis of membrane fouling by protein mixtures using MALDI-MS, Biotechnol. Bioeng. 85 (2004) 190–201.
[49] K. Nakamura, K. Matsumoto, Properties of protein adsorption onto pore surface during microfiltration: effects of solution environment and membrane hydrophobicity, J. Memb. Sci. 280 (2006) 363–374.
[50] A. Martín, F. Martinez, J.I. Calvo, P. Pradanos, L. Palacio, A. Hernandez, Protein adsorption onto an inorganic microfiltration membrane: solute–solid interactions and surface coverage, J. Memb. Sci. 207 (2002) 199–207.
[51] M. Cromwell, E. Hilario, F. Jacobson, Protein aggregation and bioprocessing, AAPS J. 8 (2006) E572–E579.
[52] R. van Reis, A.L. Zydney, Bioprocess membrane technology, J. Memb. Sci. 297 (2007) 16–50.
[53] J.A. Koehler, M. Ulbricht, G. Belfort, Intermolecular forces between proteins and polymer films with relevance to filtration, Langmuir 13 (1997) 4162–4171.
[54] I.H. Huisman, P. Pradanos, A. Hernandez, The effect of protein–protein and protein–membrane interactions on membrane fouling in ultrafiltration, J. Memb. Sci. 179 (2000) 79–90.
[55] T. Maruyama, S. Katoh, M. Nakajima, H. Nabetani, T.P. Abbott, A. Shono, K. Satoh, FT-IR analysis of BSA fouled on ultrafiltration and microfiltration membranes, J. Memb. Sci. 192 (2001) 201–207.
[56] M. Meireles, P. Aimar, V. Sanchez, Albumin denaturation during ultrafiltration: effects of operating conditions and consequences on membrane fouling, Biotechnol. Bioeng. 38 (1991) 528–534.
[57] K.J. Kim, V. Chen, A.G. Fane, Some factors determining protein aggregation during ultrafiltration, Biotechnol. Bioeng. 42 (1993) 260–265.
[58] J.S. Bee, J.L. Stevenson, B. Mehta, J. Svitel, J. Pollastrini, R. Platz, E. Freund, J.F. Carpenter, T.W. Randolph, Response of a concentrated monoclonal antibody formulation to high shear, Biotechnol. Bioeng. 103 (2009) 936–943.
[59] K. Ahrer, A. Buchacher, G. Iberer, A. Jungbauer, Effects of ultra-/diafiltration conditions on present aggregates in human immunoglobulin G preparations, J. Memb. Sci. 274 (2006) 108–115.

[60] E. Rosenberg, S. Hepbildikler, W. Kuhne, G. Winter, Ultrafiltration concentration of monoclonal antibody solutions: development of an optimized method minimizing aggregation, J. Memb. Sci. 342 (2009) 50–59.
[61] S. Mochizuki, A.L. Zydney, Dextran transport through asymmetric ultrafiltration membranes: comparison with hydrodynamic models, J. Memb. Sci. 68 (1992) 21–41.
[62] S.R. Wickramasinghe, S.E. Bower, Z. Chen, A. Mukherjee, S.M. Husson, Relating the pore size distribution of ultrafiltration membranes to dextran rejection, J. Memb. Sci. 340 (2009) 1–8.
[63] A.L. Zydney, A. Xenopoulos, Improving dextran tests for ultrafiltration membranes: effect of device format, J. Memb. Sci. 291 (2007) 180–190.
[64] P. Aimar, M. Meireles, Calibration of ultrafiltration membranes against size exclusion chromatography columns, J. Memb. Sci. 346 (2010) 233–239.
[65] A.R. Cooper, D.S. van Derveer, Characterization of ultrafiltration membranes by polymer transport measurements, Sep. Sci. Technol. 14 (1979) 551–556.
[66] G. Trägårdh, Characterization methods for ultrafiltration membranes, Desalination 53 (1985) 25–35.
[67] H. Susanto, S. Franzka, M. Ulbricht, Dextran fouling of polyethersulfone ultrafiltration membranes-causes, extent and consequences, J. Memb. Sci. 296 (2007) 147–155.
[68] G. Tkacik, S. Michaels, A rejection profile test for ultrafiltration membranes and devices, Biotechnology 9 (1991) 941–946.
[69] G. Schock, A. Miguel, R. Birkenberger, Characterization of ultrafiltration membranes: cut-off determination by gel permeation chromatography, J. Memb. Sci. 41 (1989) 55–67.
[70] P. Mulherkar, R. van Reis, Flex test: a fluorescent dextran test for UF membrane characterization, J. Memb. Sci. 236 (2004) 171–182.
[71] D. Venturoli, B. Rippe, Ficoll and dextran vs. globular proteins as probes for testing glomerular permselectivity: effects of molecular size, shape, charge, and deformability, Am. J. Physiol. Renal. Physiol. 288 (2005) F605–F613.
[72] T. Urase, K. Yamamoto, S. Ohgaki, Effect of pore structure of membranes and module configuration on virus retention, J. Memb. Sci. 115 (1996) 21–29.
[73] D.M. Krstić, M.G. Antov, D.M. Peričin, W. Höflinger, M.N. Tekić, The possibility for improvement of ceramic membrane ultrafiltration of an enzyme solution, Biochem. Eng. J. 33 (2007) 10–15.
[74] J. Randon, P. Blanc, R. Paterson, Modification of ceramic membrane surfaces using phosphoric-acid and alkyl phosphonic-acids and its effects on ultrafiltration of BSA protein, J. Memb. Sci. 98 (1995) 119–129.
[75] M.S. Strano, A.L. Zydney, H.B. Wooler, H. Agarwal, H.C. Foley, Ultrafiltration membrane synthesis by nanoscale templating of porous carbon, J. Memb. Sci. 198 (2002) 173–186.
[76] T.N. Shah, H.C. Foley, A.L. Zydney, Performance characteristics of nanoporous carbon membranes for protein ultrafiltration, Biotechnol. Prog. 23 (2007) 1157–1162.
[77] M.J. Steuck, US Patent No. 4,618,533 (1986).
[78] M. Ulbricht, G. Belfort, Low temperature surface modifications of polyacrylonitrile ultrafiltration membranes. 1. Plasma treatment effects, J. Appl. Polym. Sci. 56 (1995) 325–343.
[79] M. Ulbricht, G. Belfort, Surface modification of ultrafiltration membranes by low temperature plasma. II. Graft polymerization onto polyacrylonitrile and polysulfone, J. Memb. Sci. 111 (1996) 193–215.

[80] H. Chen, G. Belfort, Surface modification of poly(ether sulfone) ultrafiltration membranes by low-temperature plasma-induced graft polymerization, J. Appl. Polym. Sci. 72 (1999) 1699–1711.
[81] H. Yamagishi, J. Crivello, G. Belfort, Development of a novel photochemical technique for modifying poly(arylsulfone) ultrafiltration membranes, J. Memb. Sci. 105 (1995) 237–247.
[82] H. Yamagishi, J. Crivello, G. Belfort, Evaluation of photochemically modified poly(aryl sulfone) ultrafiltration membranes, J. Memb. Sci. 105 (1995) 249–259.
[83] J. Pieracci, D.W. Wood, J.V. Crivello, G. Belfort, UV-assisted graft polymerization of N-vinyl-2-pyrrolidinone onto poly-(ether sulfone) ultrafiltration membranes: comparison of dip versus immersion modification techniques, Chem. Mater. 12 (2000) 2123–2133.
[84] B. Kaeselevl, J. Pieracci, G. Belfort, Photoinduced grafting of ultrafiltration membranes: comparison of poly(ether sulfone) and poly(sulfone), J. Memb. Sci. 194 (2001) 245–261.
[85] M. Ulbricht, A. Oechel, J. Lehmann, G. Tomaschewski, H. Hicke, Gas-phase photo-induced graft-polymerization of acrylic-acid onto polyacrylonitrile ultrafiltration membranes, J. Appl. Polym. Sci. 55 (1995) 1707–1723.
[86] M. Ulbricht, H. Matuschewski, A. Oechel, H. Hicke, Photo-induced graft polymerization surface modifications for the preparation of hydrophilic and low-protein-adsorbing ultrafiltration membranes, J. Memb. Sci. 115 (1996) 31–47.
[87] M. Ulbricht, K. Richau, H. Kamusewitz, Photomodification of ultrafiltration membranes–part 11–chemically and morphologically defined ultrafiltration membrane surfaces prepared by heterogeneous photo-initiated graft polymerization, Colloids. Surf. 138 (1998) 353–366.
[88] J. Pieracci, J.V. Crivello, G. Belfort, Photochemical modification of 10 kDa polyethersulfone ultrafiltration membranes for reduction of biofouling, J. Memb. Sci. 156 (1999) 223–240.
[89] J. Pieracci, J.V. Crivello, G. Belfort, Increasing membrane permeability of UV-modified poly(ether sulfone) ultrafiltration membranes, J. Memb. Sci. 202 (2002) 1–16.
[90] N. Singh, Z. Chen, N. Tomer, S.R. Wickramasinghe, N. Soice, S.M. Husson, Modification of regenerated cellulose ultrafiltration membranes by surface-initiated atom transfer radical polymerization, J. Memb. Sci. 311 (2008) 225–234.
[91] D. Wandera, S.R. Wickramasinghe, S.M. Husson, Stimuli-responsive membranes, J. Memb. Sci. 357 (2010) 6–35.
[92] S. Eisenberng, A. Grodzinsky, Electrically modulated membrane permeability, J. Memb. Sci. 19 (1984) 173–194.
[93] H. Iwata, T. Matsuda, Preparation and properties of novel environment-sensitive membranes prepared by graft-polymerization onto a porous membrane, J. Memb. Sci. 38 (1988) 185–199.
[94] T. Masawaki, M. Taya, S. Tone, Evaluation of molecular weight cut-off profiles of pH responsive ultrafiltration membranes prepared under various casting conditions, J. Chem. Eng. Jpn. 29 (1996) 180–183.
[95] Y. Ito, Y. Ochiai, Y.S. Park, Y. Imanishi, pH-sensitive gating by conformational change of a polypeptide brush grafted onto a porous polymer membrane, J. Am. Chem. Soc. 119 (1997) 1619–1623.
[96] M.A. Islam, A. Dimov, A.L. Malinova, Environment-sensitive properties of polymethacrylic acid-grafted polyethylene membranes, J. Memb. Sci. 66 (1992) 69–78.

[97] M.S. Oak, T. Kobayashi, H.Y. Wang, T. Fukaya, N. Fujii, pH effect on molecular size exclusion of polyacrylonitrile ultrafiltration membranes having carboxylic acid groups, J. Memb. Sci. 123 (1997) 185–195.
[98] H.J. Zhang, Y. Ito, pH control of transport through a porous membrane self assembled with a poly(acrylic acid) loop brush, Langmuir 17 (2001) 8336–8340.
[99] J.F. Hester, S.C. Olugebefola, A.M. Mayes, Preparation of pH-responsive polymer membranes by self-organization, J. Memb. Sci. 208 (2002) 375–388.
[100] Y. Ly, Y.L. Cheng, Electrically-modulated variable permeability liquid-crystalline polymeric membrane, J. Memb. Sci. 77 (1993) 99–112.
[101] L. Liang, X.D. Feng, L. Peurrung, V. Viswanathan, Temperature-sensitive membranes prepared by UV photopolymerization of N-isopropylacrylamide on a surface of porous hydrophilic polypropylene membranes, J. Memb. Sci. 162 (1999) 235–246.
[102] G.V.R. Rao, M.E. Krug, S. Balamurugan, H.F. Xu, Q. Xu, G.P. Lopez, Synthesis and characterization of silica-poly(N-isopropylacrylamide) hybrid membranes: switchable molecular filters, Chem. Mater. 14 (2002) 5075–5080.
[103] D.R. Latulippe, A.M. Mika, R.F. Childs, R. Ghosh, C.D.M. Filipe, Flux performance and macrosolute sieving behavior of environment responsive formed-in-place ultrafiltration membranes, J. Memb. Sci. 342 (2009) 227–235.
[104] R. Huang, L.K. Kostanski, C.D.M. Filipe, R. Ghosh, Environment-responsive hydrogel-based ultrafiltration membranes for protein bioseparation, J. Memb. Sci. 336 (2009) 42–49.
[105] R. Ghosh, Y. Wan, Z.F. Cui, G. Hale, Parameter scanning ultrafiltration: rapid optimization of protein separation, Biotechnol. Bioeng. 81 (2003) 673–682.
[106] A.G. Fane, S. Chang, Techniques to enhance performance of membrane processes, in: A.K. Pabby, S.S.H. Rizvi, A.M. Sastre (Eds.), Handbook of Membrane Separations, Chemical, Pharmaceutical, Food, and Biotechnological Applications, CRC Press, Taylor & Francis Group, 2009, pp. 193–232.
[107] M.Y. Jaffrin, Dynamic shear-enhanced membrane filtration: a review of rotating disks, rotating membranes and vibrating systems, J. Memb. Sci. 324 (2008) 7–25.
[108] O. Al-Akoum, M.Y. Jaffrin, L.H. Ding, P. Paullier, C. Vanhoutte, An hydrodynamic investigation of microfiltration and ultrafiltration in a vibrating membrane module, J. Memb. Sci. 197 (2002) 37–52.
[109] M.Y. Jaffrin, L.H. Ding, O. Akoum, A. Brou, A hydrodynamic comparison between rotating disk and vibratory dynamic filtration systems, J. Memb. Sci. 242 (2004) 155–167.
[110] O. Al-Akoum, D. Richfield, M.Y. Jaffrin, L.H. Ding, P. Swart, Recovery of trypsin inhibitor and soy milk protein concentration by dynamic filtration, J. Memb. Sci. 279 (2006) 291–300.
[111] D. Sen, W. Roy, L. Das, S. Sadhu, C. Bhattacharjee, Ultrafiltration of macromolecules using rotating disc membrane module (RDMM) equipped with vanes: effects of turbulence promoter, J. Memb. Sci. 360 (2010) 40–47.
[112] R.H. Davis, Crossflow microfiltration with backpulsing, in: W.K. Wang (Ed.), Membrane Separations in Biotechnology, Marcel Dekker, New York, 2001, pp. 277–299.
[113] V.G.J. Rodgers, R.E. Sparks, Reduction of membrane fouling in the ultrafiltration of binary protein mixtures, AIChE J. 37 (1991) 1517–1528.
[114] C. Wilharm, V.G.J. Rodgers, Significance of duration and amplitude in transmembrane pressure pulsed ultrafiltration of binary protein mixtures, J. Memb. Sci. 121 (1996) 217–228.

[115] S.M. Finnigan, J.A. Howell, The effect of pulsatile flow on ultrafiltration fluxes in a baffled tubular membrane system, Chem. Eng. Res. Des. 67 (1989) 278–282.
[116] C.H. Müller, G.P. Agarwal, T. Melin, T. Wintgens, Study of ultrafiltration of a single and binary protein solution in a thin spiral channel module, J. Memb. Sci. 227 (2003) 51–69.
[117] J.N. Ghogomu, C. Guigui, J.C. Rouch, M.J. Clifton, P. Aptel, Hollow-fibre membrane module design: comparison of different curved geometries with Dean vortices, J. Memb. Sci. 181 (2001) 71–80.
[118] P. Moulin, P. Manno, J.C. Rouch, C. Serra, M.J. Clifton, P. Aptel, Flux improvement by Dean vortices: ultrafiltration of colloidal suspensions and macromolecular solutions, J. Memb. Sci. 156 (1999) 109–130.
[119] J. Kaur, G.P. Agarwal, Studies on protein transmission in thin channel flow module: the role of dean vortices for improving mass transfer, J. Memb. Sci. 196 (2002) 1–11.
[120] Z.F. Cui, K.I.T. Wright, Gas–liquid two-phase cross-flow ultrafiltration of BSA and dextran solutions, J. Memb. Sci. 90 (1994) 183–189.
[121] Z.F. Cui, K.I.T. Wright, Flux enhancement with gas sparging in downwards crossflow ultrafiltration: performance and mechanism, J. Memb. Sci. 117 (1996) 109–116.
[122] S.R. Bellara, Z.F. Cui, D.S. Pepper, Gas sparging to enhance permeate flux in ultrafiltration using hollow fibre membranes, J. Memb. Sci. 121 (1996) 175–184.
[123] Z.F. Cui, K.I.T. Wright, D.H. Glass, Experimental evidence on the mechanism of enhancement of ultrafiltration with gas sparging, in: Proc. IChemE Research Event, Leeds, UK, 1996, p. 304.
[124] Q.Y. Li, R. Ghosh, S.R. Bellara, Z.F. Cui, D.S. Pepper, Enhancement of ultrafiltration by gas sparging with flat sheet membrane modules, Sep. Purif. Technol. 14 (1998) 79–83.
[125] Q.Y. Li, Z.F. Cui, D.S. Pepper, Fractionation of HSA/IgG by gas sparged ultrafiltration, J. Memb. Sci. 136 (1997) 181–190.
[126] R. Ghosh, Q.Y. Li, Z.F. Cui, Fractionation of BSA and lysozyme using ultrafiltration: effect of gas sparging, AIChE J. 44 (1998) 61–67.
[127] T. Taha, Z.F. Cui, CFD modelling of gas-sparged ultrafiltration in tubular membranes, J. Memb. Sci. 210 (2002) 13–27.
[128] K. Scott, A.J. Mahmood, R.J. Jachuck, B. Hu, Intensified membrane filtration with corrugated membranes, J. Memb. Sci. 173 (2000) 1–16.
[129] M.J. van der Waal, I.G. Racz, Mass transfer in corrugated-plate membrane modules. I. Hyperfiltration experiments, J. Memb. Sci. 40 (1989) 243–260.
[130] M.J. van der Waal, S. Stevanovic, I.G. Racz, Mass transfer in corrugated-plate membrane modules. II. Ultrafiltration experiments, J. Memb. Sci. 40 (1989) 261–275.
[131] A.M. Gronda, S. Buechel, E.L. Cussler, Mass transfer in corrugated membranes, J. Memb. Sci. 165 (2000) 177–187.
[132] P.Z. Çulfaz, E. Rolevink, C. van Rijn, R.G.H. Lammertink, M. Wessling, Microstructured hollow fibers for ultrafiltration, J. Memb. Sci. 347 (2010) 32–41.
[133] H.M. Huotari, G. Trägårdh, I.H. Huisman, Crossflow membrane filtration enhanced by an external dc electric field: a review, Trans. IChemE 77 (1999) 461–468.
[134] S. Oussedik, D. Belhocine, H. Grib, H. Lounici, D.L. Piron, N. Mameri, Enhanced ultrafiltration of bovine serum albumin with pulsed electric field and fluidized activated alumina, Desalination 127 (2000) 59–68.
[135] P.V. Zumbusch, W. Kulcke, G. Brunner, Use of alternating electrical fields as anti fouling strategy in ultrafiltration of biological suspensions – introduction of a new experimental procedure for crossflow filtration, J. Memb. Sci. 142 (1998) 75–86.

[136] H. Yukawa, K. Shimura, A. Suda, A. Maniwa, Characteristics of cross flow electro-ultrafiltration for colloidal solution of protein, J. Chem. Eng. Jpn. 16 (1983) 246–248.
[137] A.D. Enevoldsen, E.B. Hansen, G. Jonsson, Electro-ultrafiltration of amylase enzymes: process design and economy, Chem. Eng. Sci. 62 (2007) 6716–6725.
[138] B. Sarkar, S. DasGupta, S. De, Electric field enhanced fractionation of protein mixture using ultrafiltration, J. Memb. Sci. 341 (2009) 11–20.
[139] J.F. Poulin, J. Amiot, L. Bazinet, Simultaneous separation of acid and basic bioactive peptides by electrodialysis with ultrafiltration membrane, J. Biotechnol. 123 (2006) 314–328.
[140] J.-F. Poulin, J. Amiot, L. Bazinet, Improved peptide fractionation by electrodialysis with ultrafiltration membrane: influence of ultrafiltration membrane stacking and electrical field strength, J. Memb. Sci. 299 (2007) 83–90.
[141] L. Firdaous, P. Dhulster, J. Amiot, A. Gaudreau, D. Lecouturier, R. Kapel, Concentration and selective separation of bioactive peptides from an alfalfa white protein hydrolysate by electrodialysis with ultrafiltration membranes, J. Memb. Sci. 329 (2009) 60–67.
[142] R.T. Kurnick, A.W. Yu, G.S. Blank, A.R. Burton, D. Smith, A.M. Athalye, R. van Reis, Buffer exchange using size exclusion chromatography, countercurrent dialysis, and tangential flow filtration: models, development, and industrial application, Biotechnol. Bioeng. 45 (1995) 149–157.
[143] P. Ng, J. Lundblad, G. Mitra, Optimization of solute separation by diafiltration, Sep. Sci. 11 (1976) 499–502.
[144] J.H. Shao, A.L. Zydney, Optimization of ultrafiltration/diafiltration processes for partially bound impurities, Biotechnol. Bioeng. 87 (2004) 286–292.
[145] C. Harinarayan, K. Skidmore, Y. Kao, A.L. Zydney, R. van Reis, Small molecule clearance in ultrafiltration/diafiltration in relation to protein interactions: study of citrate binding to a Fab, Biotechnol. Bioeng. 102 (2009) 1718–1722.
[146] A. Muller, G. Daufin, B. Chaufer, Ultrafiltration modes of operation for the separation of α-lactalbumin from acid casein whey, J. Memb. Sci. 153 (1999) 9–21.
[147] M. Nyström, P. Aimar, S. Luque, M. Kulovarra, S. Metsamuuronen, Fractionation of model proteins using their physicochemical properties, Colloids Surf. 138 (1998) 185–205.
[148] M. Rabiller-Baudry, B. Chaufer, P. Aimar, B. Bariou, D. Lucas, Application of a convection-diffusion-electrophoretic migration model to ultrafiltration of lysozyme at different pH values and ionic strengths, J. Memb. Sci. 179 (2000) 163–174.
[149] R. van Reis, S. Gadam, L.N. Frautschy, S. Orlando, E.M. Goodrich, S. Saksena, R. Kuriyel, C.M. Simpson, S. Pearl, A.L. Zydney, High performance tangential flow filtration, Biotechnol. Bioeng. 56 (1997) 71–82.
[150] S. Saksena, A.L. Zydney, Effect of solution pH and ionic strength on the separation of albumin and immunoglobulin (IgG) by selective filtration, Biotechnol. Bioeng. 43 (1994) 960–968.
[151] R.H.C.M. van Eijndhoven, S. Saksena, A.L. Zydney, Protein fractionation using electrostatic interactions in membrane filtration, Biotechnol. Bioeng. 48 (1995) 406–414.
[152] B. Cheang, A.L. Zydney, Separation of α-lactalbumin and β-lactoglobulin using membrane ultrafiltration, Biotechnol. Bioeng. 83 (2003) 201–209.
[153] B. Cheang, A.L. Zydney, A two-stage ultrafiltration process for fractionation of whey protein isolate, J. Memb. Sci. 231 (2004) 159–167.

[154] B. Lebreton, A. Brown, R. van Reis, Application of high-performance tangential flow filtration (HPTFF) to the purification of a human pharmaceutical antibody fragment expressed in *Escherichia coli*, Biotechnol. Bioeng. 100 (2008) 964–974.

[155] B. Mattiasson, M. Ramstrop, Ultrafiltration affinity purification: isolation of concanavalin A from seeds of Canavalia ensiformis, J. Chromatogr. 283 (1984) 323–330.

[156] K.B. Male, A.L. Nyugen, J.H.T. Luong, Isolation of urokinase by affinity ultrafiltration, Biotechnol. Bioeng. 35 (1990) 87–93.

[157] J.D. Powers, P.K. Kilpartik, R.G. Carbonell, Trypsin purification by affinity binding to small unilamellar liposomes, Biotechnol. Bioeng. 36 (1990) 506–519.

[158] D.C. Herrak, E.W. Merrill, Affinity cross-flow filtration: experimental and modelling work using the system of HSA and Cibacron Blue agarose, Biotechnol. Prog. 5 (1989) 9–17.

[159] L.-Z. He, X.-Y. Dong, Y. Sun, Modeling and analysis of the affinity filtration process, including broth feeding, washing, and elution steps, Biotechnol. Prog. 14 (1998) 594–600.

[160] A. Higuchi, Y. Ishada, T. Nakagawa, Surface modified polysulfone membranes: separation of mixed proteins and optical resolution of tryptophan, Desalination 90 (1993) 127–136.

[161] B. Mattiason, M. Ramstorp, Ultrafiltration affinity purification, Ann. N. Y. Acad. Sci. 413 (1983) 307–309.

[162] S. Poncet, J. Randon, J. Rocca, Enantiomeric separation of tryptophan by ultrafiltration using the BSA solution system, Sep. Sci. Technol. 32 (1997) 2029–2038.

[163] Y. Itoh, Y. Saura, Y. Tsuda, H. Yamada, Stereoselectivity and enantiomer–enantiomer interactions in the binding of ibuprofen to human serum albumin, Chirality 9 (1997) 643–649.

[164] J. Romero, A.L. Zydney, Chiral separations using ultrafiltration with a stereoselective binding agent, Sep. Sci. Technol. 36 (2001) 1575–1594.

[165] J. Romero, A.L. Zydney, Affinity ultrafiltration: effects of ligand binding on process optimization, Biotechnol. Bioeng. 77 (2002) 256–265.

[166] J. Romero, A.L. Zydney, Staging of affinity ultrafiltration processes for chiral separations, J. Memb. Sci. 209 (2002) 107–119.

[167] X. Dong, L. Li, Y. Sun, Multi-stage affinity cross flow filtration: process optimization, Bioprocess Eng. 16 (1997) 229–235.

[168] S. Rao, A.L. Zydney, High resolution protein separations using affinity ultrafiltration with small charged ligands, J. Memb. Sci. 280 (2006) 781–789.

[169] R. Ghosh, Fractionation of biological macromolecules using carrier phase ultrafiltration, Biotechnol. Bioeng. 74 (2001) 1–1.

[170] R. van Reis, E.M. Goodrich, C.L. Yson, L.N. Frautschy, R. Whiteley, A.L. Zydney, Constant C_{wall} ultrafiltration process control, J. Memb. Sci. 130 (1997) 123–140.

[171] T.A. Hatton, Liquid–liquid extraction of antibiotics, in: M. Moo-Young (Ed.), Comprehensive Biotechnology, vol. 2, Pergaman Press, Oxford, 1985, p. 439.

[172] A.M.A. Nabais, J.P. Cardoso, Purification of benzylpenicillin filtered broths by ultrafiltration and effect on solvent extraction, Bioprocess Eng. 21 (1999) 157–163.

[173] S.Z. Li, X.Y. Li, Z.F. Cui, D.Z. Wang, Application of ultrafiltration to improve the extraction of antibiotics, Sep. Purif. Technol. 34 (2004) 115–123.

[174] G. Brooks, An introduction to DNA and its use in gene therapy, in: G. Brooks (Ed.), Gene Therapy – The Use of DNA as a Drug, Pharmaceutical Press, London, 2002, pp. 1–22.

[175] M. Giese, DNA-antiviral vaccines: new developments and approaches – a review, Virus Genes 17 (1998) 219–232.
[176] M.M. Diogo, J.A. Queiroz, D.M.F. Prazeres, Chromatography of plasmid DNA, J. Chromatogr. A 1069 (2005) 3–22.
[177] S. Sagar, M.P. Watson, A. Lee, Y.G. Chau, Capacity challenges in chromatography based purification of plasmidDNA, in: A.S. Rathore, A. Velayudhan (Eds.), Scaleup and Optimization in Preparative Chromatography, Marcel Dekker, New York, 2003, pp. 251–272.
[178] D.R. Latulippe, K. Ager, A.L. Zydney, Flux-dependent transmission of supercoiled plasmid DNA through ultrafiltration membranes, J. Memb. Sci. 294 (2007) 169–177.
[179] A. Higuchi, T. Yoshimura, K. Kato, M. Hara, M. Higuchi, N. Minoura, T. Hirasaki, T. Sato, G. Ishikawa, H. Nakano, S. Satoh, S. Manabe, Sieving study of chromatin and histone-DNA complex by porous hollow fiber membranes, J. Memb. Sci. 126 (1997) 7–17.
[180] A. Eon-Duval, R.H. Macduff, C.A. Fisher, M.J. Harris, C. Brook, Removal of RNA impurities by tangential flow filtration in an RNase-free plasmid DNA purification process, Anal. Biochem. 316 (2003) 66–73.
[181] D.W. Kahn, M.D. Butler, D.L. Cohen, M. Gordon, J.W. Kahn, M.E. Winkler, Purification of plasmid DNA by tangential flow filtration, Biotechnol. Bioeng. 69 (2000) 101–106.
[182] D.R. Latulippe, A.L. Zydney, Salt-induced changes in plasmid DNA transmission through ultrafiltration membranes, Biotechnol. Bioeng. 99 (2008) 390–398.
[183] M. Mayani, K. Mohanty, C. Filipe, R. Ghosh, Continuous fractionation of plasma proteins HSA and HIgG using cascade ultrafiltration systems, Sep. Purif. Technol. 70 (2009) 231–241.
[184] R.J. Harris, S.J. Shire, C. Winter, Commercial manufacturing scale formulation and analytical characterization of therapeutic recombinant antibodies, Drug Dev. Res. 61 (2004) 137–154.
[185] K. Mohanty, R. Ghosh, Novel tangential-flow countercurrent cascade ultrafiltration configuration for continuous purification of humanized monoclonal antibody, J. Memb. Sci. 307 (2008) 117–125.
[186] C.A. Teske, B. Lebreton, R. van Reis, Inline ultrafiltration, Biotechnol. Prog. 26 (2010) 1068–1072.
[187] C. Charcosset, M.Y. Jaffrin, L.H. Ding, Effect of permeate to feed ratio on protein recovery in membrane plasma fractionation, J. Memb. Sci. 60 (1991) 87–101.
[188] Y. Wan, R. Ghosh, G. Hale, Z. Cui, Fractionation of bovine serum albumin and monoclonal antibody alemtuzumab using carrier phase ultrafiltration, Biotechnol. Bioeng. 90 (2005) 303–315.
[189] A.L. Zydney, Protein separations using membrane filtration: new opportunities for whey fractionation, Int. Dairy J. 8 (1998) 243–250.
[190] G.L. Baruah, A. Nayak, E. Winkelman, G. Belfort, Purification of monoclonal antibodies derived from transgenic goat milk by ultrafiltration, Biotechnol. Bioeng. 93 (2006) 747–754.
[191] N. Ehsani, S. Parkinnen, M. Nyström, Fractionation of natural and model egg-white protein solutions with modified and unmodified polysulfone ultrafiltration membranes, J. Memb. Sci. 123 (1997) 105–119.
[192] S. Metsämuuronen, M. Nyström, Enrichment of α-lactalbumin from diluted whey with polymeric ultrafiltration membranes, J. Memb. Sci. 337 (2009) 248–256.

[193] A. Vacca, N. Campobasso, G. Iodice, M. Ronco, F. Dammaco, Cyclic lysozyme administration as a tool for immunopotentiation in patients with multiple myeloma, Chemotherapy 19 (1985) 145–155.
[194] F. Cartei, G. Cartei, V. Ceschia, S. Pacor, G. Sava, Recovery of lymphocyte CD4+–CD8+ ratio in patients treated with lysozyme, Drug Invest 4 (1992) 51–57.
[195] T. Imoto, L.N. Johnson, A.C.T. North, D.C. Phillips, J.A. Rupley, Vertebrate lysozyme, in: P.D. Boyer (Ed.), third ed., The Enzymes, vol. VII, Academic Press, New York, 1972, pp. 665–868.
[196] M. Balakrishnan, G.P. Agarwal, Protein fractionation in a vortex flow filter. II: separation of simulated mixtures, J. Memb. Sci. 112 (1996) 75–84.
[197] R. Ghosh, Z.F. Cui, Purification of lysozyme using ultrafiltration, Biotechnol. Bioeng. 68 (2000a) 191–203.
[198] M.T. Aspelund, C.E. Glatz, Purification of recombinant plant-made proteins from corn extracts by ultrafiltration, J. Memb. Sci. 353 (2010) 103–110.
[199] G.V. Annathur, S. Kawas, T.K. Das, S.V. Ho, Ultrafiltration of a highly self-associating protein, J. Memb. Sci. 353 (2010) 41–50.
[200] N. Rossignol, L. Vandanjon, P. Jaouen, Membrane technology for the continuous separation microalgae/culture medium: compared performances of cross-flow microfiltration and ultra-filtration, Aquacul. Eng. 20 (1999) 191–208.
[201] N. Rossi, I. Petit, P. Jaouen, P. Legentilhomme, M. Derouiniot, Harvesting of cyanobacterium Arthrospira platensis using inorganic filtration membranes, Sep. Sci. Technol. 40 (2005) 3033–3050.
[202] O. Morineau-Thomas, P. Jaouen, P. Legentilhomme, The role of exopolysaccharides in fouling phenomenon during ultrafiltration of microalgae (Chlorella sp. and Porphyridium purpureum): advantage of a swirling decaying flow, Bioprocess Biosyst. Eng. 25 (2002) 35–42.
[203] P. Jaouen, B. Lépine, N. Rossignol, R. Royer, F. Quéméneur, Clarification and concentration with membrane technology of phycocyanin solution extracted from Spirulina platensis, Biotechnol. Tech. 13 (1999) 877–881.
[204] M. Frappart, A. Massé, M.Y. Jaffrin, J. Pruvost, P. Jaouen, Influence of hydrodynamics in tangential and dynamic ultrafiltration systems for microalgae separation, Desalination 265 (2011) 279–283.
[205] C. Denis, A. Massé, J. Fleurence, P. Jaouen, Concentration and pre-purification with ultrafiltration of a R-phycoerythrin solution extracted from macro-algae Grateloupia turuturu: process definition and up-scaling, Sep. Purif. Technol. 69 (2009) 37–42.
[206] A.S. Michaels, New separation technique for the CPI, Chem. Eng. Prog. 64 (1968) 31–43.
[207] V. Gekas, B. Hallström, Mass transfer in the membrane concentration polarization layer under turbulent cross flow. 1. Critical literature review and adaptation of existing Sherwood correlations to membrane operations, J. Memb. Sci. 30 (1987) 153–170.
[208] M.D. Leveque, Les lois de transmission de chaleur par convection, Ann. Mine. 13 (1928) 201–299.
[209] A.R. Da Costa, A.G. Fane, D.E. Wiley, Spacer characterization and pressure drop modelling in spacer-filled channels for ultrafiltration, J. Memb. Sci. 87 (1994) 79–98.
[210] U.B. Holeschovsky, C.L. Cooney, Quantitative description of ultrafiltration in a rotating filtration device, AIChE J. 37 (1991) 1219–1226.
[211] R. Ghosh, Z.F. Cui, Simulation study of the fractionation of proteins using ultrafiltration, J. Memb. Sci. 180 (2000b) 29–36.

[212] W.F. Blatt, A. Dravid, A.S. Michaels, L. Nelson, Solute polarization and cake formation on membrane ultrafiltration: causes, consequences, and control techniques, in: J.E. Flinn (Ed.), Membrane Science and Technology, Plenum Press, New York, 1970, pp. 47–97.
[213] O. Kedem, A. Katchalsky, Thermodynamic analysis of the permeability of biological membranes to non-electrolytes, Biochim. Biophys. Acta. 27 (1958) 229–246.
[214] R.F. Probstein, J.S. Shen, W.F. Leung, Ultrafiltration of macromolecular solutions at high polarization in laminar channel flow, Desalination 24 (1978) 1–16.
[215] D.R. Trettin, M.R. Doshi, Limiting flux in ultrafiltration of macromolecular solutions, Chem. Eng. Commun. 4 (1980) 507–522.
[216] P. Aimar, V. Sanchez, A novel approach to limiting phenomena during ultrafiltration of macromolecules, Ind. Eng. Chem. Fund. 25 (1986) 789–798.
[217] P. Aimar, R. Field, Limiting flux in membrane separations: a model based on the viscosity dependency of the mass transfer coefficient, Chem. Eng. Sci. 47 (1992) 579–586.
[218] S. De, S. DasGupta, Role of mass transfer coefficient with suction including property variations to predict limiting phenomena during ultrafiltration, J. Memb. Sci. 161 (1999) 41–53.
[219] G. Jonsson, Boundary layer phenomena during ultrafiltration of dextran and whey protein solutions, Desalination 51 (1984) 61–77.
[220] W. Leung, R.F. Probstein, Low polarization in laminar ultrafiltration of macromolecular solutions, Ind. Eng. Chem. Fund. 18 (1979) 274–278.
[221] M.J. Clifton, N. Abidine, P. Aptel, V. Sanchez, Growth of the polarization layer in ultrafiltration with hollow fibre membranes, J. Memb. Sci. 21 (1984) 233–246.
[222] S. Nakao, J.G. Wijmans, C.A. Smolders, Resistance to the permeate flux in unstirred ultrafiltration of dissolved macromolecular solutions, J. Memb. Sci. 26 (1986) 165–178.
[223] S.K. Zaidi, A. Kumar, Experimental studies in the dead-end ultrafiltration of dextran: analysis of concentration polarization, Sep. Purif. Technol. 36 (2004) 115–130.
[224] Y. Wang, V.G.J. Rodgers, Free-solvent model shows osmotic pressure is the dominant factor in limiting flux during protein ultrafiltration, J. Memb. Sci. 320 (2008) 335–343.
[225] C. Rosen, C. Trägårdh, Computer simulations of mass transfer in the concentration boundary layer over ultrafiltration membranes, J. Memb. Sci. 85 (1993) 139–156.
[226] V. Gekas, P. Aimar, J.P. Lafaille, V. Sanchez, A simulation study of the adsorption–concentration polarisation interplay in protein ultrafiltration, Chem. Eng. Sci. 48 (1993) 2753–2765.
[227] F. Ruiz-Beviá, V. Gomis-Yagües, J. Fernández-Sempere, M.J. Fernández-Torres, An improved model with time-dependent adsorption for simulating protein ultrafiltration, Chem. Eng. Sci. 52 (1997) 2343–2352.
[228] R. Ghidossi, D. Veyret, P. Moulin, Computational fluid dynamics applied to membranes: state of the art and opportunities, Chem. Eng. Process. 45 (2006) 437–454.
[229] S. Ma, S.C. Kassinos, D. Kassinos, Direct simulation of the limiting flux: I. Interpretation of the experimental results, J. Memb. Sci. 337 (2009) 81–91.
[230] J. Fernández-Sempere, F. Ruiz-Beviá, P. García-Algado, R. Salcedo-Díaz, Visualization and modelling of the polarization layer and a reversible adsorption process in PEG-10000 dead-end ultrafiltration, J. Memb. Sci. 342 (2009) 279–290.
[231] A. Afonso, J.M. Miranda, J.B.L.M. Campos, Numerical study of BSA ultrafiltration in the limiting flux regime – effect of variable physical properties, Desalination 249 (2009) 1139–1150.

[232] B. Marcos, C. Moresoli, J. Skorepova, B. Vaughan, CFD modeling of a transient hollow fiber ultrafiltration system for protein concentration, J. Memb. Sci. 337 (2009) 136–144.

[233] V. Magueijo, M.N. de Pinho, V.M. Geraldes, Numerical and experimental study of mass transfer in lysozyme ultrafiltration, Desalination 145 (2002) 193–199.

[234] Z. Cao, D.E. Wiley, A.G. Fane, CFD simulations of net-type turbulence promoters in a narrow channel, J. Memb. Sci. 185 (2001) 157–176.

[235] F. Li, W. Meindersma, A.B. de Haan, T. Reith, Optimization of commercial net spacers in spiral wound membrane modules, J. Memb. Sci. 208 (2002) 289–302.

[236] P. Schausberger, N. Norazman, H. Li, V. Chen, A. Friedl, Simulation of protein ultrafiltration using CFD: comparison of concentration polarisation and fouling effects with filtration and protein adsorption experiments, J. Memb. Sci. 337 (2009) 1–8.

3 Microfiltration

Chapter Outline

Membrane Processes in Biotechnology and Pharmaceutics. DOI: 10.1016/B978-0-444-56334-7.00003-4

3.1 Principles

3.1.1 Introduction

Microfiltration (MF) is a pressure-driven separation process, which is widely used in concentrating, purifying or separating macromolecules, colloids and suspended particles from solution. MF membranes typically have nominal pore sizes on the order of 0.1–1.0 μm. MF processing is widely used in the food industry for applications such as wine, juice and beer clarification, for wastewater treatment, and plasma separation from blood for therapeutic and commercial uses. In biotechnology industries, MF concerns applications such as cell recycle and harvesting, separation of recombinant proteins from cell debris, and purification of process streams.

MF is usually operated at relatively low *TMP*s (<4 bar or 0.4 MPa) in a cross-flow configuration. The feed stream flows tangentially to the membrane surface to prevent cake formation and hence fouling of the membrane. The operation of cross-flow MF is often limited by membrane fouling caused by suspended solids in the feed stream. Permeate flux decreases with time as the retained particles accumulate on and within the membrane. Accumulation of cells, cell debris or other rejected particles on the membrane surface (external fouling or cake formation) is usually reversible, while deposition and adsorption of small particles or macromolecules within the internal pore structure of the membrane (internal fouling) is often irreversible.

In highly fouled membranes, the loss of effective pore area or pore numbers can result in filtrate fluxes that are less than those observed in UF. For example, Gatenholm et al. [1] filtering fermentation broth (*Escherichia coli*) on 100 kD UF and 0.2 μm MF membrane have found that the flux decline was much greater with the MF membrane, and that final MF fluxes were less than those of the UF membrane. To minimize the negative effects of membrane fouling, optimization of the module device is required.

3.1.2 Previous reviews

An important literature is available on the main aspects of MF including membrane materials, configurations, models and the very broad range of applications. Books and book chapters have presented the major principles and applications of this filtration process [2,3]. Other book chapters and review articles have described more specific aspects. Belfort et al. [4] provided a very complete review on fouling of MF membranes by focussing on the formation of cakes, the behaviour of suspension flows and particle transport in simple geometry ducts and on the formation and behaviour of fouling layers including those resulting from macromolecules, colloids and particles. Kuriyel and Zydney [5] have presented several aspects of sterile filtration, performed with MF filters, and virus filtration, which involves specific membranes. Wakeman and Williams [6] have reviewed techniques used to limit or retard fouling during MF processes including feed pre-treatment, choice of membrane material, flow modification, the use of high shear, gas sparging and additional force fields. Russotti and Göklen [7] have provided a review of cross-flow membrane filtration of fermentation broth, including single-cell organisms/extracellular product, single-cell organisms/cell-associated product, multicellular organism/extracellular product and finally, multicellular organisms/cell-associated product. Examples of each type of processes were given.

3.1.3 Basic definitions

Parameters involve in MF processes are typically the same as those defined for UF. One major difference is that MF usually operates with suspensions of cells or particles. The cells or particles are characterized by their size defined as the average cell diameter for single-cell organisms [7]. Filamentous organisms are typically characterized by diameter of hyphae. The cell or particle density or suspension concentration in the feed is expressed either in concentration units (particles per unit volume) or as a volume fraction. The suspension can also be characterized by its maximum or close-packed density which can be achieved for a given cell or particle. Two properties of cells are particularly important in MF processes: cell disruption and cell deformability.

Cell disruption

In a number of MF applications, cell disruption has to be realized before the filtration step [8]. The cytoplasm of bacteria and yeast cell is a rich source of bio-products (proteins, cytoplasmic enzymes, polysaccharides, etc.) valuable in biotechnology, pharmaceutical and food industries. The efficient breakage of the cell walls preserving cell compounds and the effective separation of cell debris are essential for the good recovery of the intracellular products. The nature of the bacteria or yeast cell wall is greatly dependent on the type of organism, as well as on external growth conditions including temperature of cultivation, presence of chemical, and composition of the growth medium, growth phase or specific growth rate [9].

The composition of a bacterial cell surface is different for Gram-negative or Gram-positive bacteria [9]. In Gram-positive bacteria, including strains of lactic acid bacteria (*Lactobacillus*, *Streptococcus*, *Lactococcus* or *Leuconostoc* sp.), and those of the *Bacillus* sp., a thick layer of mucopeptide (peptidoglycan network) confers a high rigidity to the cell wall. In general, the walls of Gram-positive organisms are composed predominantly of a mucopeptide, teichoic acids and other polysaccharides. In Gram-negative bacteria (e.g. *E. coli*), the cell wall is thinner than that of the Gram-positive bacteria, and its structure is more complex, both in physical configuration and in the number of macromolecular components. It is formed of two distinct layers. Lipopolysaccharides, lipoproteins and phospholipids compose the outer membrane, while the thin inner layer consists of mucopeptides without teichoic acids. The major resistance to disruption of bacterial cells is the peptidoglycan network. Because of their larger size and different cell-wall structure, disruption of yeasts is generally easier than bacteria. The basic structural components of the yeast cell wall are glucans, mannans and proteins. Glucan fibrils constitute the inner most part of the cell wall and give the cell its shape. The overall structure is thicker than in Gram-positive bacteria.

High-pressure homogenization is one of the most effective ways for the mechanical disruption of microbial cell walls and is employed for the large-scale industrial recovery of intracellular products [9]. Disruption in a high-pressure homogenizer is achieved by passing a cell suspension under high pressure through an adjustable, restricted orifice discharge valve. The main parameters determining efficiency are operating pressure and number of passes through the valve, suspension temperature and homogenizer valve design. This method results in breakage of cells and high recovery of products. However, high-pressure homogenization requires multiple product passages with supplementary cooling, and the final products contain a large quantity of cell debris, which complicates the downstream processes of purification. Other recent techniques were introduced recently such as underwater high-voltage electrical discharges (40 kV) of microsecond duration. This method creates an electric arc in liquid, powerful light radiation, pressure shock waves, bubble cavitations and other physical phenomena, which provoke the mechanical disintegration of cell walls and cell membranes [10].

The method chosen for the cell disruption step has a final effect on the MF results. Liu et al. [8] used for yeast cells disruption both methods: high-pressure homogenization and high-voltage electrical discharges. These methods enabled to obtain complex suspensions with different content of disrupted and undisrupted cells, cell debris and extracted biomolecules. The high-pressure homogenization treatment was found to be more effective for the extraction of intracellular compounds (proteins and nucleic acids) than the high-voltage electrical discharge treatment. However, the high-pressure homogenization treated suspensions showed higher fouling of the MF membrane.

Cell deformability

The deformability of biological particles (red blood cells, yeast, bacteria) during filtration experiments has to be considered to describe the performance of UF and MF. This aspect is of importance in MF because the size of the particles is close to the membrane pore size; therefore, they can pass differently through the membrane.

Many studies consider the deformability properties of red blood cells on the filtration performance of MF membranes. This phenomenon is of particular importance in cross-flow MF plasmapheresis, the separation of plasma (including all the blood proteins and smaller molecular weight solutes) from the cellular components of blood. This process is used in the collection of blood components and in the treatment of a variety of diseases. The cross-flow configuration, in which blood flow is parallel to the membrane, minimizes the accumulation of retained cells near the membrane surface, resulting in significantly larger filtration rates than that can be obtained in a dead-end system. Zydney et al. [11] studied the filtration of red blood cells through MF membranes pores and the properties of the resulting cake. Red blood cells are highly deformable and the cell bed is compressible. Relationships were developed between hydraulic permeability, porosity and compressive pressure based on experimental data obtained by filtration and centrifugation and including pore blockage effect. These relationships were incorporated into a model to predict the cross-flow permeation flux, assuming that the hydraulic resistance provided by cross-flow MF of red blood cells was equivalent to that observed in an unstirred filtration system. Recently, scanning electron microscopy observations of the transfer of bovine red blood cells [12] through MF membranes of average pore diameter of 0.3 and 0.4 μm showed a considerable deformation of the cells enabling them to penetrate in smaller pores. This phenomenon was attributed to the high flexibility of the bilipidic membrane of the red blood cells.

A number of studies investigated the hydraulic resistance of deposits of baker's yeast (*Saccharomyces cerevisiae*). Variations in results were attributed to the differences in particle size, state of agglomeration, age, pH of liquors and concentration of extracellular compounds. Meireles et al. [13] compared some of these experimental data for the specific resistance of baker's yeast filtered in the pressure range 0–400 kPa [14–16] with values calculated using the Carman–Kozeny equation assuming Stokes flow through a granular bed of 5 μm diameter particles (close to the mean diameter of yeast particles) and a void fraction of 0.27 corresponding to the maximum packing of a face-centred cubic array. The Carman–Kozeny equation represents the hydraulic resistivity of a porous medium:

$$r_{\mathrm{h}} = K_0(\phi)\frac{S^2}{(1-\phi)^3} \tag{3.1}$$

where r_{h} is the hydraulic resistivity, K_0 is the Kozeny coefficient, ϕ is the volume fraction of solid and S is the surface area per unit volume. The experimental cake resistances of yeast were found more than a hundred times higher than the values predicted. Experimental values also showed a pressure dependence which was not accounted for by the equation. Most authors explained the discrepancy between Carman–Kozeny predictions and experimental data by the compressible nature of biological cells. Meireles et al. [13] explained the experimental values observed for the hydraulic resistance of cell beds by the first layer of cells in contact with the membrane which partially blocked the membrane area open to flow.

Bacteria present a complex cell-wall structure, composed not only of a lipidic bilayer but also of a cross-linked polymer (peptidoglycan) network responsible for mechanical strength and stiffness of their cell wall. Lebleu et al. [17] studied the transfer of bacteria larger than the nominal pore size through MF track-etched membranes for several kinds of bacterial strains (Figure 3.1). The authors concluded that the kind of bacteria (Gram-positive or Gram-negative) was the most important parameter as a consequence of the cell-wall flexibility. Gram-negative and Gram-positive bacteria may be distinguished according to their cell-wall structure. The bacterial cell-wall is composed of a specific layer of a cross-linked polymer, the peptidoglycan. In Gram-negative bacteria, this layer thickness is around 2–6 nm, whereas Gram-positive bacteria present a thicker peptidoglycan layer around 20–80 nm. The more the peptidoglycan layer is thin and elastic, the

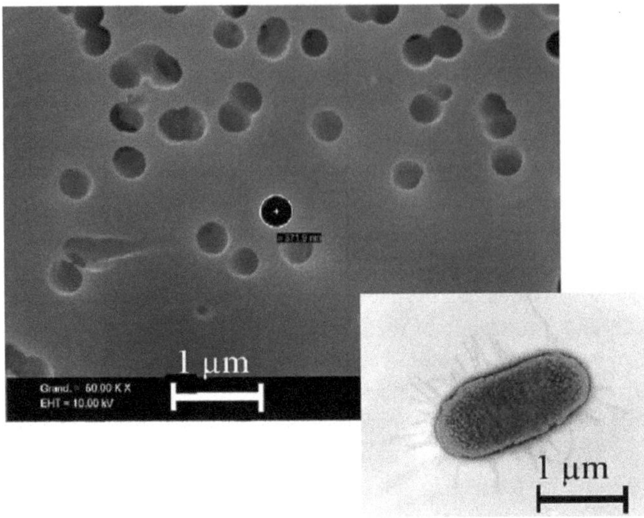

Figure 3.1 Electron microscope images of track-etched membrane pore (0.4 µm) and bacteria *E. coli* (2 µm × 1 µm) at the same scale [17].
This article was published in J. Membr. Sci., Vol. 326, N. Lebleu, C. Roques, P. Aimar, C. Causserand, Role of the cell-wall structure in the retention of bacteria by microfiltration membranes, pp. 178–185, Copyright Elsevier (2009).

more the bacteria is deformable and likely to pass through pores of smaller size than the bacteria size. As a consequence, bacteria of equal size can exhibit different filtration performance: Gram-positive bacteria which present a thicker peptidoglycan layer are less deformable and so better rejected than Gram-negative ones.

The effect of particle deformability on MF performance was also underlined in a recent study [18]. Three kinds of dextran–MnO_2 soft gel particles, with nearly the same particle size distributions, behaved differently due to their softness and compression properties. The filtration flux decreased with increasing particle softness because softer particles constructed a more compact and higher resistance cake. The specific cake filtration resistance was estimated using the Kozeny–Carman equation. In the early filtration period, the cake porosity suddenly decreased and the specific cake filtration resistance drastically increased due to the deformability of the particles.

3.1.4 Membranes

Original microporous membranes were solvent cast from cellulose nitrate and cellulose acetate [4]. Depending on the application, a large range of polymers and inorganic materials with wide pH and temperature ranges are now available. They include pure cellulose acetate, poly(vinylidene fluoride), regenerated cellulose, acrylic polymers, polyamide and polysulfone polypropylene, polyethylene, polycarbonate, ceramic, zirconium oxide, borosilicate glass, stainless steel, silver and poly(tetrafluoroethylene).The first materials are solvent cast. The others are produced by a radiation track-etched process (polycarbonate), by hot and cold stretching and heat setting (polypropylene), by an electrochemical deposition process (some ceramic membranes), by a thermal quenching procedure (polypropylene), by controlled stretching (poly(tetrafluoroethylene), by a sintering process (stainless steel, silver and ceramic) and by sol-gel processing (some ceramic membranes).

A very large range of developments have been proposed to increase membrane selectivity, and/or decrease fouling and its negative effects [19]. Surface modification is one potential method. In particular, polypropylene MF membranes have been intensively investigated with respect to introducing functional groups or layers into the polymer surface in order to overcome limitations, due to their low surface energy. Surface modifications aiming at increasing the hydrophilicity and antifouling properties or the biocompatibility of polypropylene membrane surfaces can be classified into physical (non-covalent) and chemical (covalent) methods [20]. Wetting of the pores of polypropylene membranes with polar solvents, for example isopropanol, can temporarily increase hydrophilicity. Adsorption of various substances, especially amphiphilic polymers, from solutions in the pores has also been used, but the hydrophilization was not stable [21].

In contrast, chemical grafting is more popular because of the high efficiency and the stability of the modification. One group of methods is based on the pre-treatment of polypropylene membrane with high-energy irradiation, for example plasma or UV, to generate reactive groups, and then chemically graft functional moieties [22,23]. The resulting covalent functionalization is reported to be stable, but undesired degradations of chemical or pore structure were observed. An alternative strategy is to adsorb a photo-initiator on or into the membrane surface, and perform a subsequent UV-initiated graft copolymerization to yield tethered polymer chains on the polypropylene membrane surface [24–27]. It had been demonstrated that such 'grafting-from' functionalization can be better controlled by entrapping the photo-initiator in the surface layer of the polypropylene membrane pores (via swelling the polypropylene in an initiator solution in an apolar solvent and then immobilizing it by deswelling the polypropylene in a polar solvent), rather than by an adsorption process [28,29]. Other modified surface membrane included an Anopore™ alumina membrane [30]. The nanocomposite membrane was made hydrophilic by two methods: adsorption of Pluronic F108 onto the membrane with an octadecylsilane overlayer and the grafting of poly(ethylene glycol) (PEG) to either amine-functionalized surfaces or dendrons with amines on their outer periphery. It was shown that both approaches led to a significant reduction in membrane fouling with the PEG-grafted materials displaying higher performance.

Poly(vinylidene fluoride) (PVDF) is another widely used polymer material for MF membranes due to its excellent chemical resistance, well-controlled porosity, and good thermal and mechanical properties. However, the highly hydrophobic property of PVDF leads to rather low permeability and high fouling. Many studies have been carried out to enhance the performance of PVDF membranes by introducing hydrophilic functional groups to the surfaces by chemical modification, plasma treatment and surface grafting. For example, negatively charged PVDF membranes were prepared using direct sulfonation with chlorosulfonic acid [31]. Fouling of charged poly(styrene sulfonic acid) molecules on the surface-modified membrane decreased and rejection values increased with increasing degree of sulfonation due to the electrostatic repulsion between the negatively charged molecules and the negatively charged membrane. PVDF microporous membranes with surface-immobilized poly(ethylene glycol) (PEG) were prepared by argon plasma-induced grafting of PEG and were shown to exhibit good antifouling properties [32].

Mixing PVDF with another membrane material is an effective method for improving membrane performance. For example, Zhao et al. [33] blended three kinds of amphiphilic polymers, including the tri-block copolymer of (polyethylene oxide)–(polypropylene oxide) (polyethylene oxide), the comb-like copolymer of polysiloxane with polyethylene oxide and polypropylene oxide side chains and the hyperbranched star copolymer of polyester-*g*-methoxyl polyethylene glycol, with PVDF to fabricate porous membranes via a phase inversion process. The three polymers investigated showed better stability and protein resistance than those of the membrane using PEG as modifier. Recently, Mu et al. [34] used cellulose acetate as an additive to improve the permeability and antifouling properties of the PVDF membranes. X-ray photoelectron spectroscopy analysis confirmed the enrichment of CA segments on the PVDF membrane surfaces. The MF membranes prepared had higher wettability, lower protein adsorption and higher water flux in comparison with the initial PVDF membrane.

Several recent studies have described the fabrication of specifically patterned MF membranes with well-defined pore geometry and spacing, with the goal of increasing the flux and/or selectivity compared to conventional membranes. For example, Kuiper et al. [35] developed a new class of membranes using a combination of micromachining and micro-lithographic processes. Stork Veco and Micropore Technologies Ltd have produced metallic membranes with well-defined pores and a very narrow pore-size distribution having either slot-shaped or circular pore geometries [36] (Figure 3.2). Wang et al. [37] designed a membrane casting cell for self-assembly of silica spheres into a colloidal crystal template. The resulting close-packed colloidal crystal was infiltrated with a reactive monomer solution. After polymerization, the silica spheres were etched away, resulting in an inverse colloidal crystal membrane with high porosity and uniform pores that were highly interconnected. The membranes were fabricated with different pore sizes and thicknesses. The authors underlined that this membrane fabrication process may be limited for scale up.

Although metal membranes have not received much attention in bioprocessing, they should have high chemical and thermal resistance, providing the possibility of steam sterilization, enhanced cleaning and reuse, and longer lifetime. In addition, the rigidity and durability of these metal membranes would be highly desirable for applications employing very high-shear rates, oscillatory flow, vibratory membranes and/or backpulsing.

3.2 Fouling control

In MF, external fouling of the membrane is the result of cell, cell fragments and rejected particles accumulation on the membrane surface. Deposition of macromolecules and

(a)

(b)

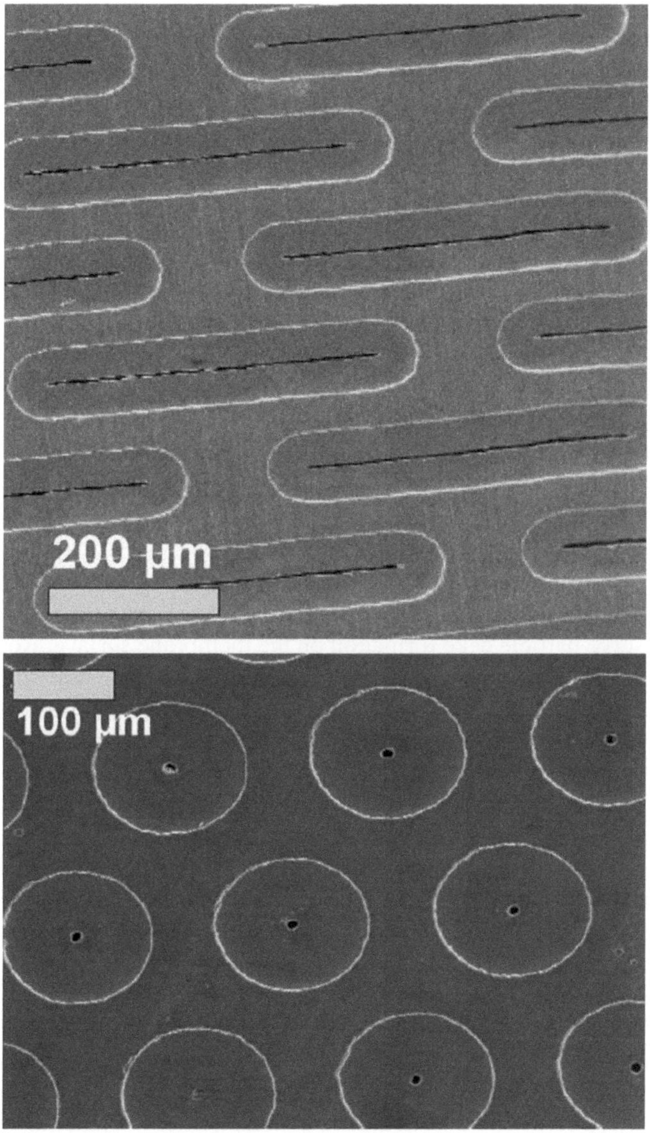

Figure 3.2 SEM images of (a) the Micropore™ slotted pore membrane and (b) the Micropore™ circular pore membrane [36].
This article was published in J. Membr. Sci., Vol. 285, M. Chandler, A. Zydney, Effects of membrane pore geometry on behavior during yeast cell microfiltration, pp. 334–342, Copyright Elsevier (2006).

small particles inside the internal membrane structure results in internal fouling, which is often irreversible contrary to usually reversible external fouling. Proteins can bind to membrane surfaces by a variety of mechanisms including electrostatic attraction, hydrophobic interactions, hydrogen bonding and van der Waals forces, which in turn result in internal fouling (see Chapter 2).

In MF, the size of the retained particles allows direct observation using microscopy techniques. Results in this field are first presented. Similarly to UF, various methods have been proposed in MF to reduce the negative effects of concentration polarization and fouling by improving the hydrodynamics of the cross-flow over the membrane surface. These methods are presented in the following and include dynamic filtration, which uses a rotating disk, or by rotating or vibrating the membrane, vortex systems, backflushing and backpulsing which periodically reverse the *TMP* or permeate flow, and an electric field that drags charged species away from the membrane surface. Other techniques were proposed to reduce the effects of concentration polarization and fouling such as constant filtrate flux operation, flocculation, gas sparging, secondary membranes and turbulence promoters.

3.2.1 Direct visual observation of deposition

In situ studies of fouling and cleaning processes have been developed to characterize MF. To observe particle cake formation on the membrane surface, Romero and Davis [38] used a cross-flow microfilter with glass sidewalls and a cathetometer to measure the thickness of the stagnant and flowing cake layers at different points along the filter. Using channel heights (1.0 and 0.5 cm) and particle diameters (50–150 μm) that are much larger than those typically used in MF, they showed that the cake layer thickness increased with increased particle concentration, increased permeate flux and decreased shear rate. Mackley and Sherman [39] used a video camera to observe cake growth and particle motions on the cake surface, again with particle sizes (125–180 μm) that were well above those typically used in MF. Similarly, Wakeman [40] used a high-speed camera but with a high-magnification zoom lens to observe the dynamics of smaller particles (2.6 μm) on a woven surface, as well as the growth of the cake layer along the centre axis of the filter. The immediate formation of a cake layer prevented direct observations of particle deposition and particle–particle interactions near the surface.

Hodgson et al. [41] developed a technique, termed direct observation through the membrane (DOTM), to investigate membrane–foulant interactions. Their studies involved mounting a microscope on the permeate side of a transparent membrane (Anopore anodized-alumina) and observing the fouling as it occurred on the membrane surface. They used 0.1 dry wt% baker's yeast and 0.1 wt% sulfonated styrene-based ion-exchange particles to foul the membranes with varying cross-flow velocity and *TMP*. They observed that the cross-flow broke up cell aggregates, diminishing the mean size from 32 to 6 μm. They also observed that the foulant cake was made of smaller particles rather than larger particles, because of a shear classification mechanism. The cross-flow velocity had little effect on flux, though increasing the *TMP* led to irreversible fouling of the membrane. Hodgson et al. [41]

also observed the apparent slowing of particles adjacent to the cake, suggesting a trajectory mechanism for particle/cake contact in laminar flow. Li et al. [42,43] used the DOTM technique, by keeping the permeate flux constant. They used yeast (5 μm), algae (3 μm) and latex beads (3, 6.4 and 12 μm) for filtration through Anopore anodized-alumina membranes. The fouling was observed through a microscope placed on the permeate side of the module. The critical flux, measured by observing when the particles formed a cake on the membrane surface, was shown to increase with particle size and cross-flow velocity.

In addition, Mores and Davis [44] performed direct visual observation (DVO) of fouling and cleaning of MF membranes during forward filtration and after back-washing or single backpulses. A microscope mounted on the feed side of the membrane was used to make direct observations (Figure 3.3). Experiments involved first depositing a foulant layer on the membrane surface. Foulant removal was subsequently observed during the cleaning period. The experimental results were compared with model predictions of the recovered flux based on the fraction of the membrane surface cleaned by reverse filtration. Later, Kang et al. [45] developed a technique enabling quantitative determination of the mechanisms governing initial deposition and release of biofoulants in cross-flow MF processes. They used a laboratory-scale, cross-flow membrane filtration flow cell and direct microscopic observation apparatus similar to that of Mores and Davis [44]. Yeast cells, bacteria cells, and latex particles served as model microbial particles. A polysulfone membrane and two modified polyacrylonitrile membranes were tested as model MF membranes. In addition, an image analysis procedure enabled rapid quantification of net deposition rates, which were interpreted with an interaction force balance model.

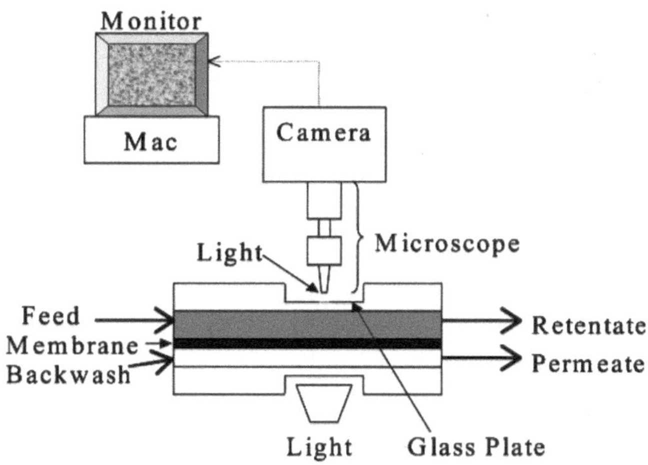

Figure 3.3 DVO apparatus [44].
This article was published in J. Membr. Sci., Vol. 189, W.D. Mores, R.H. Davis, Direct visual observation of yeast deposition and removal during microfiltration, pp. 217–230, Copyright Elsevier (2001).

3.2.2 Dynamic microfiltration

Shear-enhanced or dynamic filtration consists in creating the shear rate at the membrane by a relative motion between the membrane and a moving part such as a rotating disk or an impeller. This method has shown to be very effective in MF of biological suspensions [46–50], especially for macromolecule recovery [51–53]. The reason for this performance is that very high-shear rates are produced with a low inlet flow, and therefore, a low pressure drop in the module [54]. Consequently, the *TMP* can be kept very low, and the combination of high-shear rates and low *TMP* facilitates macromolecules transmission through the membrane.

Jaffrin et al. [54] compared the effects of various hydrodynamic parameters on the permeate flux provided by two different dynamic filtration systems using the same membrane and the same fluids. Tested systems were two home-made rotating disk modules and a vibrating membrane filtration (VMF) system (PallSep) commercialized by Pall Corp. (USA). The PallSep system consisted in a stack of circular plane membranes mounted in a cylindrical housing which was spun in torsional oscillations around a vertical axis at a resonant frequency of about 60 Hz. Each compartment contained two membranes separated by a gasket and a permeate collecting system. The flux was found to be mainly governed by the maximum shear rate and not by details of internal flow and can be increased to very high levels by increasing rotation speed or vibration amplitude. Postlethwaite et al. [55] reported the performance of the PallSep VMF system to process a model biological feed stream consisting of yeast (200–500 g l^{-1} wet weight) and BSA (0.75 g l^{-1}) and two types of real industrial fermentation broth. The use of the vibrating membrane system allowed the processing of feed streams with very high solids loading whilst maintaining permeate flux and protein transmission at acceptable levels.

These systems are, however, more complex and expensive to build than cross-flow filtration units, and may also consume more energy [50]. Novel dynamic MF systems have been proposed including a rotating disk equipped with straight vanes of various heights which was tested with baker's yeast suspensions [50] and a vibrating hollow fibre membrane module [56].

3.2.3 Backflushing and backpulsing

Backflushing and backpulsing studies have shown promise for a variety of MF applications [57–61]. The basic principles of these processes are detailed in Chapter 2. Applications in the biotechnological field include Kroner et al. [57]'s study, who achieved 50% enhancement in the net flux during the removal of *E. coli* bacteria from a fermentation broth using cross-flow MF with a 0.3 μm membrane which was backflushed by reversing the *TMP* for 5 s every 5 min. Matsumoto et al. [59,60] reported up to 10-fold flux increase with backflushing (for 5 s every 3 min) for yeast suspensions. These authors also compared different backflushing techniques and found backwashing with permeate to be the most effective. Nipkow et al. [61] reported initial improvements of 42% in the permeate flux with backflushing of a MF cell–recycle pilot-scale system for continuous cultivation of *Clostridium thermosulfurogenes*.

Redkar and Davis [62] studied the backpulsing process for MF of washed yeast suspensions. Under optimum conditions, flux enhancement of approximately 40-fold was observed. The authors also proposed a model which assumed completely reversible fouling and instantaneous cake removal. Mallubhotla and Belfort [63] modified this model by characterizing the cake removal during reverse filtration as gradual rather than instantaneous. Kuberkar et al. [64] and Mores et al. [65] further refined the model to include partially irreversible fouling. Parnham and Davis [66] studied MF of bacterial lysates using backpulsing. More than 10-fold increases in the net flux and improved protein transmission from 60 to 100% were obtained with backpulsing under optimum operating conditions. Mores and Davis [67] found that multiple short backpulses were more effective in cleaning the membrane surface than fewer, longer backpulses at high-shear rate and backpulse pressure. However, the reverse was true at low shear rate or backpulse pressure. In addition, longer and weaker backpulses led to higher recovered fluxes than did shorter, stronger backpulses.

Crossflushing is a much weaker cleaning technique than backpulsing or backflushing. It involves periodic stoppage of permeate flow while maintaining cross-flow over the membrane surface. In the absence of permeate flow, the cake layer may be partially removed by the cross-flow. This removal increases the permeate flux. Kuruzovich and Piergiovanni [68] studied crossflushing of yeast suspensions using crossflushing intervals of 5–90 s every 30–360 s. The authors observed more than 100% flux increase under optimal conditions. Redkar and Davis [62] also studied crossflushing with washed yeast suspensions. The authors observed nearly 10-fold flux increase due to crossflushing of 2 s every 9 s. However, this flux increase was much smaller than the nearly 30-fold flux increase observed when backpulsing for 2 s was done every 9 s. In addition, the net flux was nearly steady with backpulsing, but it continuously declined with crossflushing.

Recently, Kuberkar and Davis [69] performed crossflushing and backflushing experiments with yeast suspensions, BSA solutions and yeast–BSA mixtures. Backflushing was found highly effective, while crossflushing was partially effective, in removing the external cake formed during filtration of yeast suspensions. Crossflushing was completely ineffective and backflushing was only partially effective for removal of internal foulants during filtration of BSA solutions. During backflushing or crossflushing of yeast–BSA mixtures, complete or partial removal of the yeast cake reduced the hydraulic resistance to permeate flow. However, this removal also exposed the primary membrane to internal fouling by BSA, against which neither crossflushing nor backflushing was very effective.

3.2.4 Vortex systems

Belfort et al. have developed tubular and slit membrane modules in which the retentate is forced to twist as it flows across the cartridge in curved membrane channels [70–73]. The induced secondary flow produces counter rotating vortices that are able to re-entrain deposited particles or cells from the membrane surface to the bulk. Such flows are used to enhance membrane filtration performance, over that

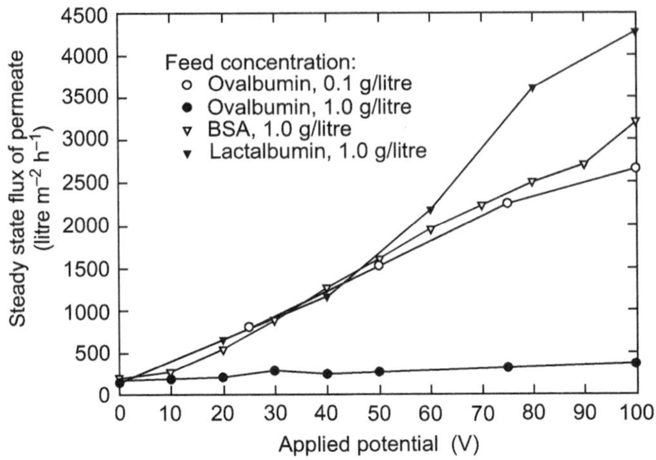

Figure 3.4 Variation of steady state permeate flux with applied potential for BSA, ovalbumin and lactalbumin dispersions (0.2 μm cellulose nitrate membranes; 204 kPa *TMP*; 1.5 ms^{-1} cross-flow velocity; 25 °C for the BSA, 30 °C for the other experiments) [79].
This article was published in Trans IChemE Part C, Vol. 76, R.J. Wakeman, Electrically enhanced microfiltration of albumin suspensions, pp. 53–59, Copyright Elsevier (1998).

obtained with conventional MF modules [74,75]. The additional cost for these systems is associated to the additional pumping energy [74].

For example, Schutyser et al. [76] used cross-flow MF and diafiltration to concentrate and purify recombinant brain-derived neutrophic factor (rBDNF) inclusion bodies from an *E. coli* cell suspension and a homogenized *E. coli* cell suspension (homogenate/lysate). Two hollow fibre membrane modules (helical and linear) with 0.1 μm pore size were used. The linear module consisted of straight parallel fibres simulating current commercial modules, whereas the helical module contained hollow fibres wound in a single-wrap helix around a rod, as shown in Figure 3.4. The purpose of the rod was to stabilize the membrane helix and establish the desired curvature. With the feeds containing cell and homogenate suspensions, improvements in flux of about 50 and 70%, respectively, were obtained with the helical module as compared with that obtained with the linear module. For diafiltration with the homogenate suspension as feed, solute transport was from 100 to 40% higher after 40 and 100 min, respectively, with the helical module as compared with that obtained with the linear module. The authors concluded that improved filtration performance, and a purer and more concentrated product could be expected with these Dean vortex filters.

3.2.5 Electric field

In cross-flow MF, the convective flow of the solutes or particles towards the membrane can be counterbalanced by giving an electrophoretic velocity to the particles through the application of a d.c. electric field gradient [6,77,78].

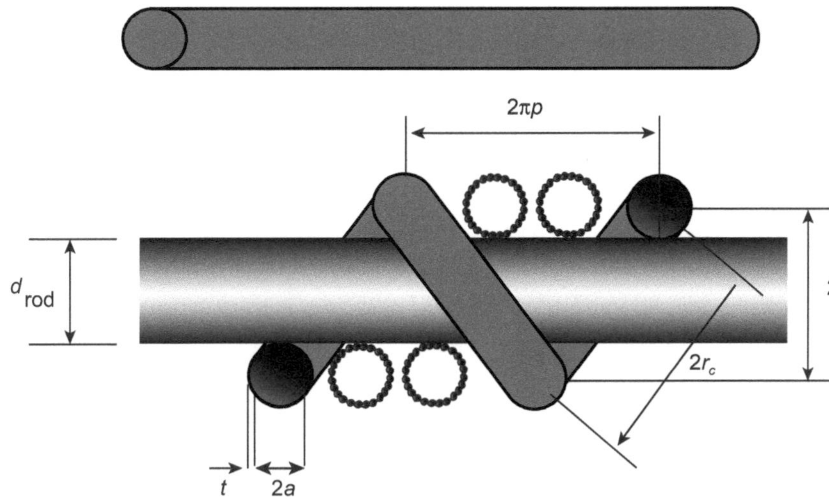

Figure 3.5 Configurations of the linear and helical modules [125].
This article was published in Biotechnol. Prog., Vol. 18, M. Schutyser, R. Rupp, J. Wideman, G. Belfort, Dean vortex membrane microfiltration and diafiltration of rBDNF E. coli inclusion bodies, pp. 322–329, Copyright John Wiley and Sons (2002).

The contribution of fine particles and colloids to membrane fouling can be reduced, thereby enhancing the filtrate flux. The critical voltage gradient defines the voltage gradient at which the net particle migration towards the membrane is zero. That is, the electrical (and diffusive and radial when they exist) migration away from the membrane exactly balances the convective transport of fine particles and colloids towards the membrane.

The extent of flux improvement is dependent primarily on solute molecule or particle size and surface charge and on the magnitude of the imposed field gradient [6]. Greater flux enhancements are possible for finer particles carrying higher surface charges, and when using steeper field gradients. The field gradient is dependent on the applied voltage and the inter-electrode distance. Electric power consumption can be reduced by designing the separation distance to be as small as possible. A potential disadvantage associated with using d.c. electric fields is gas evolution by electrolysis [6].

Figure 3.5 shows the effects of ovalbumin feed concentration and applied voltage on steady state flux; the electric field had greater impact on permeate flux enhancement when the feed concentration was low [79]. Permeate flux was increased by up to an order of magnitude and protein rejection was similarly increased. The electric field allowed choice of larger pore-sized membranes than would normally be acceptable, thus enabling further increase in flux to be obtained. However, the potential application of the technique to protein solutions was limited by degradation effects. In higher conductivity solutions, protein degradation occurred due to the combined effects of prolonged shear and localized heating acting on protein in the adsorbed layer formed on the membrane surface.

3.2.6 Constant filtrate flux operation

One approach that is used to minimize the adverse effects of fouling is to operate at constant filtrate flux, instead of at constant *TMP*, below a critical flux. This avoids the very high flux, and the associated rapid fouling, that can occur at the start of a constant pressure filtration. The concept of critical flux was first proposed by Field et al. [80]. It states that on start-up of the cross-flow MF process, there exists a flux below which a decline of flux with time does not occur. Above this value, fouling is observed. This flux is the critical flux and its value depends on the hydrodynamics and other variables. The critical flux values can be determined by *TMP* monitoring, direct observation and/or mass balance [81]. *TMP* monitoring is the most commonly used method since it can be done simply by conducting a 'flux-stepping' test and monitoring the *TMP* increase. With *TMP* monitoring, the critical flux is defined as the highest flux value that does not cause a steady increase in *TMP*. For example, Chen et al. [82] studied the profiles of flux versus *TMP* of colloidal silica in cross-flow MF. The flux was first increased and then decreased in the same manner for 30 min in each step. The *TMP*s for the decreasing flux steps were found to exhibit hysteresis and did not follow the values for the respective increasing flux steps even when there was no *TMP* increase detected. Other methods were developed, for example, combining mass balance and image analysis [81].

Quantitative studies of the pressure rise during constant flux filtration are almost all based on the classical pore blockage, pore constriction and cake filtration models, which were originally developed for constant pressure filtration [83]. For example, Hlavacek and Bouchet [84] analysed data for the *TMP* as a function of time during the constant flux MF of BSA solutions and found best agreement using the linear form of the intermediate blocking law. This model assumes that the pores are fully blocked when the protein deposits on the membrane surface, but it allows for superposition of the fouling 'particles' on the membrane. Ho and Zydney [85] recently developed a model that accounts for both pore blockage and cake formation as previously presented by Ho and Zydney [86] to describe constant pressure filtration. The objective of the study was to quantitatively examine the increase in *TMP* during constant flux MF. The model was shown to be in very good agreement with experimental data obtained during constant flux MF of BSA solutions through track-etched membranes. Model simulations were used to examine the underlying phenomena governing the rate of pressure increase during constant flux filtration.

3.2.7 Other techniques

Flocculation

Flocculation has found extensive use in the wastewater and biotechnology industries to aid in separation techniques, such as filtration, flotation, sedimentation and centrifugation [87]. Several studies have shown that the addition of a polyelectrolyte flocculant results in increased permeate flux rates for cross-flow MF of yeast cells [88–90], Chinese Hamster Ovary cell suspensions [89] and a complex media broth [87]. However, flocculation of cell suspensions by cationic polyelectrolytes also leads to a widening of the particle size distribution that can reduce benefits of increased size. This effect is likely due to selective deposition of smaller particles because of greater shear-induced back diffusion of larger particles [87].

For example, Kim et al. [90] investigated the flocculation of yeast suspensions prior to MF using commercially available synthetic polymeric flocculants. Cationic polymers were shown successful in flocculating the yeast suspensions. Flocculation prior to MF lead to an increase in the permeate flux, and an optimum polymer concentration, which maximized the degree of flocculation, was found. Flocculation of the yeast cells by aluminium salts instead of cationic polymers was also found to increase the flux rate during the cross-flow filtration through zirconia and alumina membranes [91]. In this study, the major role of electrostatic interactions between particles were underlined. Karim et al. [92] showed that polymeric flocculants could also enhance MF of disrupted cells (homogenized lysate of *E. coli*). Recently, Aspelund et al. [87] tested several polyelectrolytes for the MF of a bacterial cell suspension (Figure 3.6). Larger flocculated particles were formed by flocculation with polyelectrolytes of longer-chain length. In cross-flow MF, long-term permeate flux improvements occurred despite complete disruption of the floc in the recycled feed and the extent of the improvements were dependent on both the polyelectrolyte amount and cross-flow rate. The authors concluded that flocculation with cationic

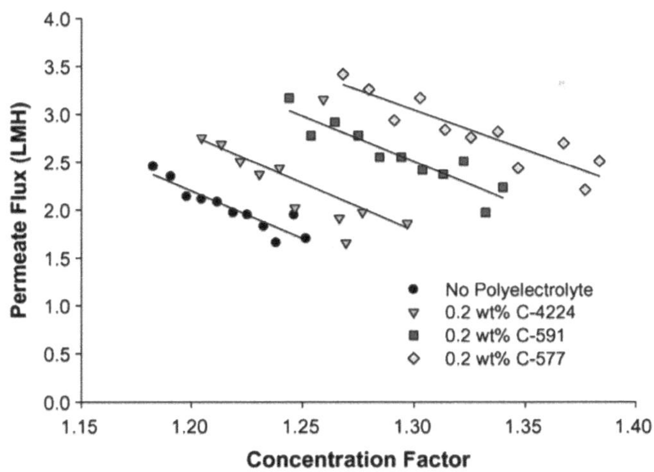

Figure 3.6 Comparison of different polyelectrolytes for their enhancement of permeate flux rates for unstirred dead-end filtrations [87].
This article was published in J. Membr. Sci., Vol. 324, M.T. Aspelund, G. Rozeboom, M. Heng, C.E. Glatz, Improving permeate flux and product transmission in the microfiltration of a bacterial cell suspension by flocculation with cationic polyelectrolytes, pp. 198–208, Copyright Elsevier (2008).

polyelectrolytes increased permeate flux and reduced product rejection in the MF of a complex industrial fermentation broth.

Gas sparging

Injecting gas into the feed stream to create a gas–liquid two-phase cross-flow operation proved to be effective in enhancing MF flux. Cui and Wright [93] observed up to a 175% in permeate flux increase in yeast filtration using a polymeric membrane module. Mercier et al. [94] also applied slug flow to achieve a significant increase in MF application in a continuous alcoholic fermentation. Sur and Cui [95] studied gas-sparged MF with baker's yeast suspensions using a 0.15 μm PCI multi-tubular membrane module. The authors demonstrated that injecting air to create a gas–liquid two-phase cross-flow operation can significantly increase the permeate flux. The observed enhancement, resulting from injecting a relatively small amount of gas, ranged from 10% when fouling was insignificant to 135% when fouling tendency was severe. The same team investigated the effect of gas sparging on cross-flow MF of 2,3-butanediol from blackstrap molasses by *Klebsiella oxytoca* [96]. A plate-type cross-flow MF unit with a 0.2 μm cellulose acetate membrane was employed. Results showed that the permeate flux declined rapidly due to fouling caused by the natural impurities of blackstrap molasses. The permeate flux was enhanced by air-sparging. The average permeate flux increased by 39 and 54% for an air-sparging rate of 0.5 L min^{-1} and 1.0 L min^{-1} respectively, in the case of an initial biomass concentration of 4.66 g L^{-1}. For an initial biomass concentration of 14.2 g L^{-1}, the flux increased by 105 and 146% for the gas rate of 0.5 and 1.0 L min^{-1}, respectively. The authors concluded that gas sparging could be beneficial in cross-flow MF of thick suspensions like a fermentation broth.

Secondary membranes

When filtering a feed solution containing fine particles, a dynamic or secondary layer of rejected particles is formed on the underlying primary filter surface. The dynamic or secondary layer often controls the separation performance [97,98]. The basic operation of rotary vacuum filters exploits this concept by using a filter aid which forms a dynamic layer on a cloth screen. In contrast to traditional filter operations, membrane-based separations typically employ a membrane whose pore size and properties are chosen to provide the desired separation performance. Then, the formation of a fouling layer of rejected particles, which can alter the performance, is undesired. However, in some applications, the fouling layer serves as a secondary membrane which can aid the separation. Secondary membranes are produced when filtering feed solutions containing membrane-forming materials such as inorganic hydrous oxides and macromolecular materials through membranes or other porous supports. The advantages of a secondary membrane include its high permeability and its ease of removal and regeneration.

Davis and coworkers [99–101] demonstrated that protein fouling of MF membranes could be significantly reduced by forming a yeast cake on the membrane surface, either during or before protein filtration. They observed higher protein

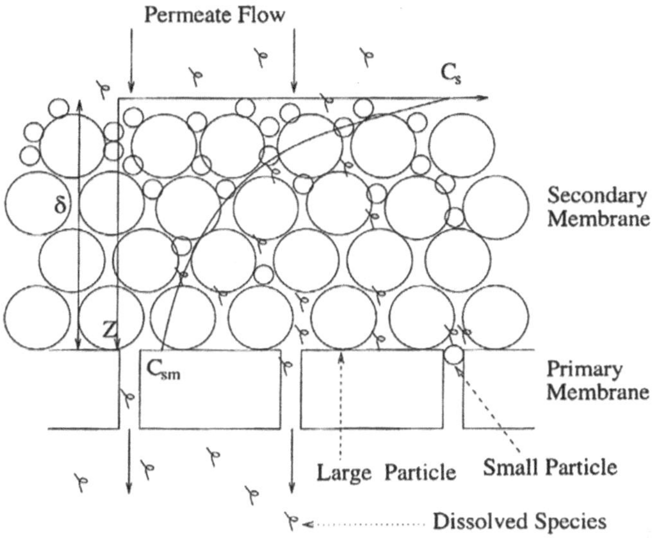

Figure 3.7 Schematic of an external cake of large particles preventing small particles from fouling the membrane, but allowing a dissolved species to pass through [97].
This article was published in J. Membr. Sci., Vol. 168, V.T. Kuberkar, Davis R.H., Modeling of fouling reduction by secondary membranes, pp. 243–258, Copyright Elsevier (2000).

transmission and higher flux in the presence of the yeast cake than in its absence. They proposed that the yeast cake acted as a pre-filter for the protein solution, sieving out protein aggregates which would otherwise initiate the protein fouling of the membrane. Kuberkar and Davis [97] presented a mechanistic model of the formation and function of secondary membranes during filtration of a multi-component mixture (Figure 3.7). The larger particles were rejected by the membrane and formed an external cake. The cake acted as a secondary membrane or depth filter which captured small particles and prevents them from fouling the primary membrane. The model predicted an optimum concentration of the large particles which maximized the permeate volume for a given duration of filtration. To test the model, cross-flow MF experiments were done with yeast suspensions, protein solutions, and yeast–protein mixtures using cellulose acetate membranes. During filtration of yeast–protein mixtures, protein aggregates were captured or screened by the yeast cake, which reduced fouling of the primary membrane by the protein aggregates. Under optimum conditions, the presence of yeast increased the total permeate volume by two-fold.

Turbulence promoters

The use of turbulence promoters or inserts in the device is another technique applying hydrodynamic methods in controlling membrane fouling [102]. In cross-flow MF with tubular ceramic membranes, turbulence promoters are inserted into the membrane channel where they generate turbulence which subsequently reduces membrane fouling by producing a helical flow pattern and generating secondary flow to hinder the formation of a particle layer above the membrane surface.

Turbulence promoters or inserts have many shapes and sizes. There are static rods, Kenics static mixers, metal grills, spiral wire, cone-shaped inserts, disc- and doughnut-shaped inserts. A number of studies has been conducted in order to investigate the influence of turbulence promoters on the permeate flux increase during filtration process. Gupta et al. [103] used helical baffles in membrane filtration of baker's yeast and dodecane–water emulsion through ceramic membranes. The helically wound baffle allowed to increase the permeate flux at the same hydraulic dissipated power and without any additional equipment such as pulsating pump or any backwashing system. Recently, Jokić et al. [102] studied the influence of operating factors on the MF of baker's yeast (*S. cerevisiae*) in presence of static mixer as turbulence promoter. The flux improvement was in the range between 89 and 258%, depending on experimental conditions. The best results considering flux improvement as well as reduction of specific energy consumption were obtained for moderate feed flow rates.

3.3 Applications

Bacterial removal is the most common application of MF and is also known as sterile filtration. The concentration and washing of cultures of single-cell organisms where the product is intracellular or cell associated is another common application. The third example is the recovery of intracellular molecules produced from fermentation broth including proteins, antibiotics, lactic acid, and polysaccharides. Finally, MF is used for the purification of nanoparticle suspensions to eliminate solvent and surfactant which are inherent to the preparation process.

3.3.1 Bacterial removal (sterile filtration)

The most common application of MF is sterile filtration (bacterial removal) prior to final formulation of many products, and in the initial clarification of fermentation broths to remove the suspended cell mass and other particulate debris. Sterile filtration is performed in a dead-end configuration using 0.2 μm pore size membranes that have been validated for the absolute removal of *Brevundimonas diminuta* [5]. However, sterilizing filters can pass very small microorganisms under some process conditions. Therefore, some users employ 0.1 μm pore size membranes to provide enhanced sterility assurance in pharmaceutical processes [104].

Some users employ two filters in series (redundant filtration) to minimize the risk of expensive product loss and batch rejection if one filter fails during sterile filtration. The drawback is to increase the design complexity of the process. Rajniak et al. [105] developed a methodology for filter sizing in redundant, constant pressure sterile filtration. Models of pore fouling based on the standard blocking or complete blocking mechanisms were derived to simulate filtration of solutions using filters of different area and construction. The authors concluded that scale up of sterile filtration required understanding and correct prediction of (1) the fouling characteristics (resulting mainly from properties and interactions between solution and membrane) and (2) the flow characteristics (resulting from the filter construction and the filtration system configuration).

Integrity testing is critical for all sterile filters to ensure that the system operates at the required level of performance [5]. The real test would use *B. diminuta*; however, such a test would be destructive, that is to say that the filter could not be used after the test. The most common integrity tests used are the bubble point and the liquid intrusion techniques (see Chapter 1).

Allegrezza et al. [106] covers in detail sterile filtration including filter selection (compatibility, retention, membrane system sizing) and device selection (sterilization, implementation). Useful data can be found on membrane system sizing including flow rate requirements and capacity. Methods to measure capacity at constant flow or at constant pressure are detailed.

3.3.2 Concentration and clarification of cells

Another common application of MF is the concentration and washing of cultures of single-cell organisms where the product is intracellular or cell associated [7]. Common cases include recombinant yeast cultures producing proteins and antigen particles, and recombinant *E. coli* producing proteins in the form of solid inclusion bodies. The next step is usually cell lysis and recovery of the product from cell debris. Separation of product from cell debris can often be performed by MF, as detailed in the next section.

Russotti et al. [107] have used cross-flow MF for the harvest of recombinant yeast in a short period of time to minimize the risk for product degradation. MF studies with flat sheet membranes showed high throughput with initial fluxes on the order of water fluxes (>1000 l m^{-2} h^{-1}, regime I, <2 min), followed by a rapid decay towards a low pseudo-steady state flux (20 l m^{-2} h^{-1}, regime II, >2 min). Large pore membranes (0.65 μm) were found to be more suitable for harvesting yeast (10 μm size) without cell leakage than smaller pore ones (0.22 μm and 0.45 μm). Among operating parameters, feed flow rate (i.e. shear rate) had a significant impact on average flux, whereas change in *TMP* afforded little improvement. In another recent example, Lee [108] concentrated recombinant yeast cells using a cross-flow MF unit containing a 0.2 μm membrane device. The concentrated cells were homogenized by several passes through a high-pressure homogenizer. The homogenate was then clarified using cross-flow MF. The clarified material was concentrated by UF and buffer-exchanged before delivering the material to down-stream for further purification.

3.3.3 Separation of products from fermentation broth

Microorganisms are sources of valuable enzymes, proteins and other bio-products. They produce two basic types of biological molecules: extracellular, which are excreted into a growth medium, and intracellular, which are retained inside the cytoplasm of the cells [9]. A variety of host microorganisms have been studied. The most often used organisms are *E. coli*, *S. cerevisiae* and *Bacillus subtilis*. Several other microbial strains have been used for production of microbial enzymes, such as *Aspergillus niger* and *Kluyveromyces fragilis* (for production of catalase), *Saccharomyces lactis* and *Kluyveromyces lactis* (β-galactosidase), *Bacillus coagulans* and

Streptomyces sp. (glucose isomerase) and *Penicilium notatum* (glucose oxidase). A large proportion of potentially useful microbial products is retained within the cells of the microorganisms. The isolation of intracellular molecules requires either the cell to be genetically engineered (so that intracellular molecules can be excreted into the growth medium) or the cells must be disintegrated by physical, chemical or enzymatic means to release their cytoplasmic content.

A major application of MF in biotechnologies is the recovery of intracellular molecules produced from fermentation broth [66]. In typical processes, fermentation is followed by cell harvesting, where the cells are concentrated into a paste by centrifugation or MF. The paste is then diluted in a buffer and homogenized by shear impingement or milling to rupture the cell walls and release the product. The over-expressed molecule needs to be separated from the cell debris in the resulting lysate before it is introduced to the down-stream purification process. Changes in fermentation media affect not only the performance of the fermentation itself (with regard to the kinetics of biomass and product formation and the yields obtained) but also the initial product recovery operations downstream of the fermentor. In the case of cross-flow MF, this may occur due to a reduction in permeate flux or reduced transmission of the target molecule through the membrane. In addition, several studies have observed the effects of individual media components on MF operations, for example antifoams [109] and oils [110].

Proteins

Cross-flow MF may be used to recover soluble proteins from the cellular debris and other insoluble components [46,57,111–115]. As reviewed by Belfort et al. [4], however, a rapid flux decline occurs initially due to cake formation, followed by a relatively slow decline due to protein fouling of the cake and membrane. Moreover, the proteins are also retained by the microporous membrane and the overlying cake layer of rejected solids, so that the protein transmission coefficient is typically less than unity. For example, Parnham and Davis [46] investigated MF of bacterial cell debris and obtained low long-term permeate fluxes in the range 10^{-4}–10^{-3} cm s^{-1}, with an average total protein transmission of only 60%. Low protein transmission values are undesirable, as they represent a significant reduction in yield of valuable product. Backflushing is one approach to the problem in which the permeate is periodically forced back through the membrane in the reverse direction to normal permeate flow in order to flush out the accumulated fouling material from the membrane pores and the membrane surface.

Proteins recovered by MF from fermentation broth include heterogeneous IgG from transgenic goat milk [116] and naturally glycosylated therapeutic proteins produced from animal cell cultures [49]. The work of Baruah and Belfort [116] on the optimization of monoclonal antibodies recovery from transgenic goat milk by MF is an interesting example. The optimization involved varying pH, *TMP*, milk feed concentration, membrane module type and axial velocity. Operation in the pressure-dependent regime at low uniform *TMP*s using permeate circulation in co-flow, at the pI of the protein (9 in this case) was shown to increase IgG recovery from

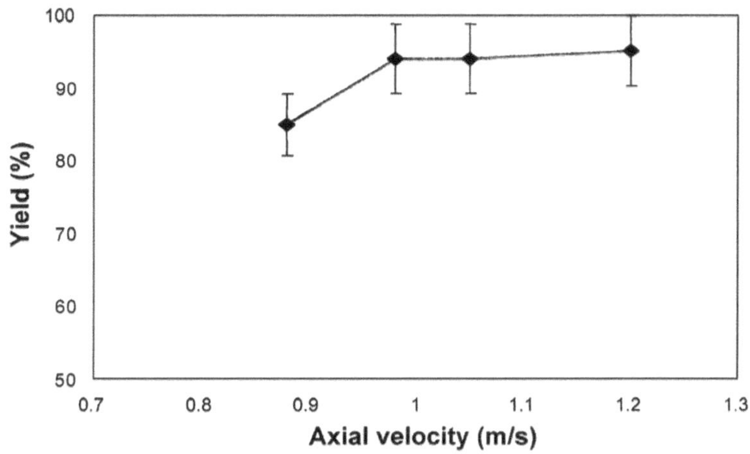

Figure 3.8 Yield versus axial velocity (Re ≈ 830–1170) for the short helical hollow fibre module at pH = 9.0, ≤2 psi uniform *TMP*, 1X milk concentration and $N_d = 4$ [116].
This article was published in Biotechnol. Bioeng., Vol. 87, G.L. Baruah, G. Belfort, Optimized recovery of monoclonal antibodies from transgenic goat milk by microfiltration, pp. 274–285, Copyright John Wiley and Sons (2004).

less than 1% to over 95%. Figure 3.8 shows, for example, the effect of axial velocity evaluated with a short helical module at pH 9.0. A yield of 95% was obtained for velocities ≥0.95 m/s. This method was claimed to be generalizable to the recovery of target proteins found in other complex suspensions of biological origin.

Antibiotics

One of the main advantages of membrane systems for the initial recovery of antibiotics from a fermentation broth is the ability to obtain very high yield using a combined filtration and diafiltration process. Complete retention of the cells and particulate matter can be achieved using membranes with pores sizes up to 0.2–0.45 μm, and essentially complete passage of the antibiotics can be achieved as long as the nominal MWCO is greater than about 20 kDa [117]. Davies et al. [118] showed that a change of medium composition affected the production of an antibiotic, erythromycin, and its transmission and rate of flow through the membrane used to separate the biomass from whole fermentation broths. As a model system, the authors studied the growth of *Saccharopolyspora erythraea* (a spore forming, Gram-positive, hyphae-producing bacterium) on both a soluble complex medium and an oil-based process medium. These systems were then processed using a Minitan II (Millipore) cross-flow filtration module, fitted with a single 60 cm^2 Durapore (Millipore) hydrophilic 0.2 μm membrane, operated in concentration mode. The media composition changed throughout the period of the fermentation as media components were utilized and biomass growth took place. Residual glucose and oil concentrations are shown in Figure 3.9. The variation of permeate

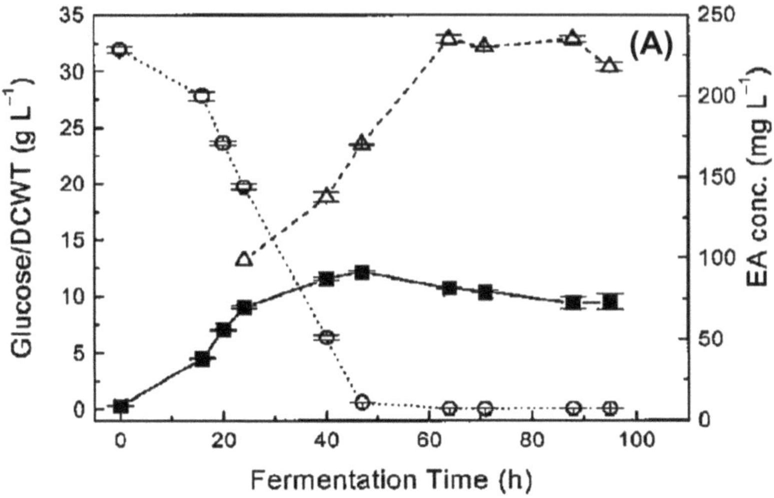

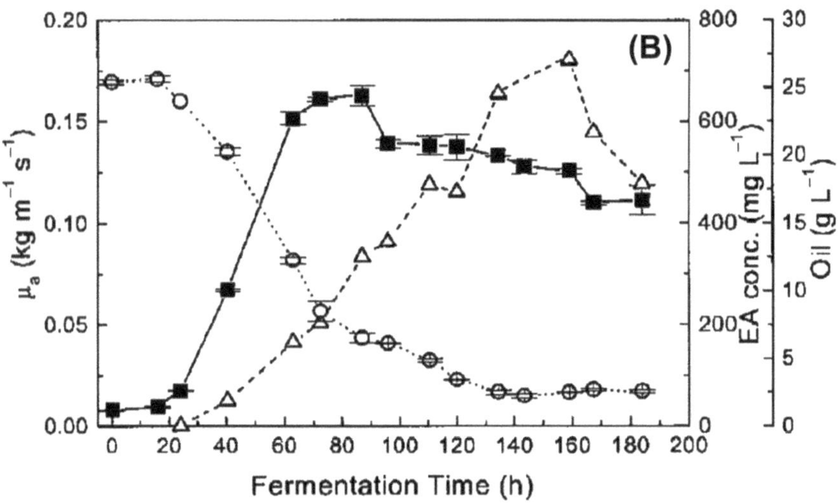

Figure 3.9 Examples of batch fermentation with *S. erythraea* CA340. (A) In soluble complex medium (SCM). Dry cell weight (DCWT: ■), residual glucose (○) and erythromycin A concentrations (EA: △). (B) In oil-based medium (OBM). Apparent viscosity at 71 s^{-1} shear rate (■), residual oil (○) and erythromycin A concentrations (EA: △) [119].

This article was published in Biotechnol. Bioeng., Vol. 69, J.L. Davies, F. Baganz, A.P. Ison, G.J. Lye, Studies on the interaction of fermentation and microfiltration operations: Erythromycin recovery from Saccharopolyspora erythraea fermentation broths, pp. 429–439, Copyright John Wiley and Sons (2000).

flux and erythromycin transmission as a function of fermentation time was also investigated.

Lactic acid

Lactic acid and lactate salts are widely used in the food industry as acidulants, preservatives or flavour enhancers. In the food industry, the natural grade of lactic acid of biological origin makes it more attractive as a food additive than that of chemical origin. However, the production of lactic acid from fermentation requires the use of an efficient and economic downstream process to recover lactic acid and isolate it from various impurities present in the fermentation broth [119,120]. In some studies, UF membranes have been used for separation of microbial cells and protein molecules from lactic acid fermentation broth [121]. However, it was observed that UF membrane in the first step of separation, presented serious fouling problem. The use of MF in the first step was shown to be more effective in subsequent downstream purification [119,120,122].

Polysaccharides

Polysaccharides are increasingly used in many industrial applications, for instance as thickening or gelling agents (agar) in the food industry. Polysaccharides produced by cultures of microorganisms (also called biopolymers) constitute a promising alternative to petrochemical and vegetal polymers because they are produced from renewable resources and their characteristics can be adjusted by controlling cell culture conditions in fermentors. Some authors investigated the extraction of polysaccharides from fermentation broths of *Sinorhizobium meliloti* M5N1CS using cross-flow filtration through ceramic membranes of various pore sizes from 0.1 to 0.8 μm [52,51,123]. The most interesting results were obtained with dynamic filtration (0.2 μm nylon membrane using a rotating disc filter) because it allowed operation at high-shear rates and low *TMP*. Sieving coefficients remained between 90 and 100%.

Other products

Other products recovered from fermentation broth include yeast alcohol dehydrogenease (ADH) from baker's yeast [124], rBDNF inclusion bodies from *E. coli* cell suspensions by cross-flow MF and diafiltration [76] and the separation of hyaluronic acid from fermentation broth by cross-flow MF and UF [125].

3.3.4 Purification of nanoparticle suspensions

Nanoparticles are solid colloidal particles ranging in size from about 10 to 1000 nm. They have received considerable attention in recent years, in particular those prepared with biodegradable polymers, because of their potential use as delivery systems for drugs, contrast agents, proteins, peptides, DNA, vaccines and other biologically active agents. Nanoparticle is a collective term used to describe both nanospheres and nanocapsules. The difference between these two forms lies in the morphology and body architecture. Nanospheres have a matrix-type structure. Drugs or tracers may be adsorbed at their surface, entrapped in the particle or dissolved in it. Nanocapsules have a polymeric shell and an inner liquid core. In this case, the active substances are usually dissolved in the inner core, but may also be adsorbed at their surface. Most of the nanoparticle preparations are designed for the purpose of transporting the diagnostic or therapeutic agent to particular targets in the body following oral or parenteral administration. Therefore, purification is needed to ensure the biological tolerance for the product. The presence of impurities will not only cause potential biological intolerance, but may also alter the physicochemical and release characteristics of nanoparticle systems.

Nanoparticles can be prepared from either natural or synthetic polymeric materials. There are numerous methods for preparation of drug-loaded nanoparticles. Some involve polymerization of monomers, and others form nanoparticles by manipulation of polymers via processes such as emulsification-solvent evaporation, solvent diffusion, multiple emulsion, salting out, phase inversion, ionic gelation and nanoprecipitation [126]. Depending on the method of preparation, the following potentially toxic impurities must be eliminated: organic solvents, stabilizers, residual monomers, polymerization initiators and large polymer aggregates.

Polymer aggregates in the micrometer range are usually eliminated by filtration through sintered glass filters. Organic solvents, especially the chlorinated ones, must absolutely be eliminated. Most of the stabilizers are also potentially toxic. Purification methods involve dialysis [127], gel filtration [128,129], evaporation under reduced pressure [130] and centrifugation or ultracentrifugation [131–133]. Centrifugation or ultracentrifugation, in combination with washing nanoparticles with an appropriate medium such as deionized water, is used to remove large quantities of process impurities. However, the impact of the centrifugation force can cause cake formation and difficulties in redispersing nanoparticles [134]. In addition, when insufficient centrifugation force is applied, a significant loss of nanoparticles to the supernatant can occur, resulting in a low yield of nanoparticles.

Cross-flow MF is a suitable technique for nanoparticle suspension purification which can be scaled-up for industrial applications. For nanoparticle suspensions with very low size distribution, UF is used instead of MF. A high degree of purification is obtained which is usually associated to membrane fouling. MF and UF membranes were used to remove the electrolyte, polyvinyl alcohol (PVA) and the acetone in savoxepine-loaded poly(lactic acid) nanoparticles prepared using a salting-out process [135]. Batches of nanoparticles were purified completely in less than 3 h. On the other hand, Quintanar-Guerrero et al. [136] investigated the purification by cross-flow MF of poly(D, L-lactic acid) (PLA) nanoparticles prepared by an emulsification–diffusion technique using PVA or poloxamer 188 (P-188). The results indicated a difference between the two stabilizers, suggesting that the nature of the coating layer has an important role during cross-flow MF. Nanoparticles prepared with PVA were filtered without fusion. In contrast, aggregation of nanoparticles was observed during cross-flow MF of batches prepared with P-188.

The purification of nanoparticle suspensions prepared by an emulsion–diffusion technique was performed using cross-flow MF [137]. The purification process involved two steps: first a concentration step where the preparation is concentrated by a factor of 1/5, then a diafiltration step, where the feed volume is kept constant by adding pure water continuously. It was shown that PVA and ethyl acetate concentrations decreased to an acceptable level, after 40 min of concentration and 2 h of diafiltration (Figure 3.10). Fouling was reported to be high with the flow rate reaching 15% of its initial value after 50 min of concentration and 2 h of diafiltration (from 33 to 5 ml/min). Fouling was of attributed to: (1) the adsorption and the deposition of surfactant molecules on the membrane surface, (2) the deposition of nanocapsules on the membrane surface, (3) the deposition of nanocapsules of size around 80 nm in the membrane pores.

Different purification processes were evaluated for removing surfactant from nanoparticle dispersions [134]. Nanoparticle formulations prepared from poly(D, L-lactide-co-glycolide) and PVA were purified by a diafiltration centrifugal device (DCD) using 300 K and 100 K molecular weight cut-off membranes and a tangential flow filtration (TFF) system with a 300 K MWCO membrane. These two techniques were compared to a common ultracentrifugation technique. Both DCD and TFF systems with a 300 K MWCO membrane removed maximal percent PVA from nanoparticle suspensions (89.0 and 90.7%, respectively). Purified nanoparticle

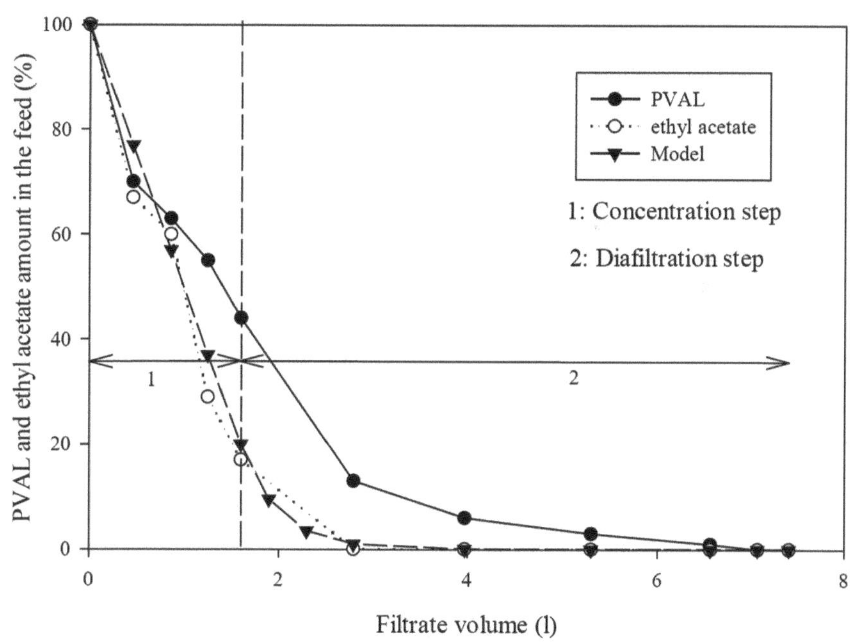

Figure 3.10 PVA and ethyl acetate amount in the feed during cross-flow microfiltration. $T = 25$ °C, $Q_i = 100\ \mathrm{l\ h^{-1}}$, and $V_D = 2$ l [137].
This article was published in Sep. Purif. Technol., Vol. 38, I. Limayem, C. Charcosset, H. Fessi, Purification of nanoparticle suspensions by a concentration/diafiltration process, pp. 1–9, Copyright Elsevier (2004).

suspensions were stable and free of aggregation at ambient conditions over 3 days. This was in contrast to the commonly used ultracentrifugation technique, which showed nanoparticle aggregation and loss of fine particles in the supernatant.

3.3.5 Other applications

A number of other applications of MF in biotechnologies and pharmaceutics are reported. Wickramasinghe et al. [138] used MF and UF flat sheet membranes for concentration of human influenza A virus, which may be used for viral vaccines and gene therapy. Influenza A virus is a lipid-enveloped RNA virus that belongs to the *Orthomyxoviridae* family. Human influenza virus was propagated in Madin Darby Canine Kidney cells in serum-free medium and depth filtered to remove lysed cells and cell debris. The virus-containing feed was then chemically inactivated and filtered using flat sheet MF and UF membranes. Several uses of MF and UF membranes were reported. Membranes with pore sizes larger than the virus particles were used to remove host cell fragments and attached viruses. Membranes with pore sizes similar to the virus particles were applied to the fractionation and hence isolation of the virus particles of interest from damaged virus fragments. Finally, membranes with pore sizes smaller than the virus particles were used to concentrate the virus particles and remove host cell proteins (HCP) and DNA.

Aspelund and Glatz [139] investigated cross-flow MF for the solid/liquid clarification of corn protein extracts from endosperm- and germ-rich dry milling fractions. Corn has been explored as an expression host for several recombinant proteins including monoclonal antibodies, vaccines and, more recently, collagen-related proteins. A major step that is required in the purification process is the solid/liquid clarification of the protein extracts. Currently, the clarification of recombinant proteins from corn extracts uses centrifugation, rotary drum vacuum filtration, or expanded bed chromatography. Cross-flow MF using a poly(vinylidene fluoride) membrane resulted in rather low internal fouling and no measurable rejection of host cell proteins (HCP), green fluorescent protein (GFP), or recombinant human collagen (rCollagen). MF was therefore presented as an effective method for the solid/liquid clarification of corn protein extracts.

Another example of MF is provided by Li et al. [140] for MF of a Chinese herb decoction which is a complex mixture that includes many macromolecules such as tannin, starch and proteins. Li et al. [140] used a ZrO_2 ceramic MF membrane with mean pore size of 0.15 μm and the aqueous extract of a kind of Chinese herb (Radix Rehmannia) at various pH values. The authors showed that fouling was significantly affected by pH. At the IEP (7.2) of ZrO_2, the steady permeate flux was minimum and the rejection of total solids maximum.

3.4 Theory

In this section, some approaches to MF modelling are presented including balances between convectively driven particle deposition and particle removal, and force and torque balance models acting on one particle. Some other approaches to simulate MF phenomena are briefly discussed.

3.4.1 Balances between convectively driven particle deposition and particle removal

In the models of steady state behaviour in cross-flow MF devices, the convectively driven particle deposition onto the membrane surface is balanced by mechanisms of particle removal. The earliest works have proposed a mechanism of particle back-transport from the membrane surface based on the inertial lift of the particles [141–143]. If the conditions are such that the inertial lift velocities are sufficient to offset the permeate velocity, then the particles are not deposited on the membrane [144]. Models have also obtained good agreement with experimental data by using a shear-enhanced particle diffusivity in the classical film theory [145]. Other models have considered the removal of particles from the membrane surface as a tangentially flowing layer of concentrated particles [146,147]. Romero and Davis [148] have included both enhanced diffusion due to the shear of the concentrated-particle layer and the tangential flow of this layer. Sethi and Wiesner [149] presented an extended model for predicting the transient permeate flux in low pressure membrane filtration processes by combining Brownian diffusion and inertial lift with the shear-induced diffusion and flowing boundary layer model formulated by Romero and Davis [148].

3.4.2 Force balance models acting on one particle

Surface transport models of membrane fouling during cross-flow MF use a balance of forces acting on one particle in proximity to the membrane surface. In these models, a critical operating condition is derived under which a particle will deposit on the surface. The main hydrodynamic and intermolecular forces acting on the particle are shown in Figure 3.11. The drag force associated with the filtrate flux and

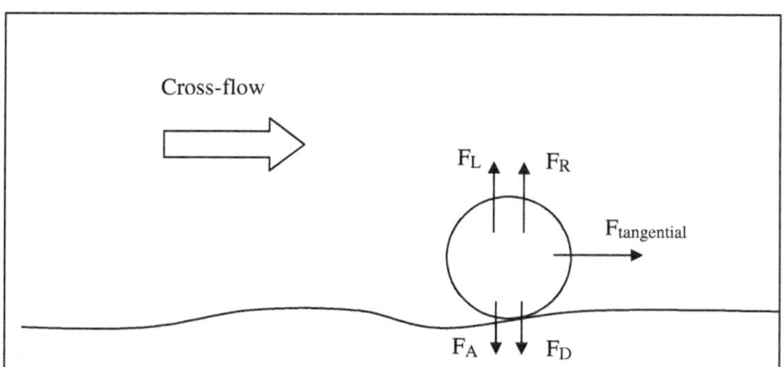

Figure 3.11 Schematic representation of the forces acting on a particle in the boundary and at the membrane surface [from 3].

other attractive forces (e.g. van de Waals interactions) tend to move the particle towards the membrane surface. The hydrodynamic forces (inertial lift) and other repulsive (e.g. electrostatic) interaction tend to transport the particle back into the feed suspension. In addition, particles that actually reach the membrane surface may flow or 'roll' along the membrane. This type of surface flow occurs if the drag force due to the cross-flow velocity is sufficient to overcome the forces acting on the particle in the direction normal to the membrane surface. The force expressions derived by Kang et al. [45] are given below. Other expressions are found in the literature [150,151].

Permeate drag force

The drag force on a particle approaching an impermeable surface becomes infinite at small gaps because large pressures develop in the zone of closest approach due to the entrapped liquid film between the particle and wall [45,152]. The drag force increases as an inverse function of the separation gap and must be modified by an appropriate correction factor. The corrected drag force is usually expressed as:

$$F_D = -6\pi\mu_w a_p v_w \phi_H \tag{3.2}$$

where a_p is the particle/cell radius, μ_w is the dynamic water viscosity and v_w is the permeate water velocity or flux. $\phi_H(= F/F_\infty)$ is the hydrodynamic correction factor to the Stokes drag force, F being the drag force at fluid–membrane interface and F_∞ drag force in the bulk. In the vicinity of the membrane surface, it is assumed that the particle velocity is sufficiently small to be ignored and inertial effects are negligible. Goren [153] derived a theoretical description of the hydrodynamic correction factor for a spherical particle approaching a permeable planar surface that was a complex function of the surface's Darcy permeability, particle size and separation distance $[\phi_H = f(h/a_p,\ R_m a_p)]$, where h is the separation gap and R_m is the membrane hydraulic resistance, or the inverse of the Darcy permeability. Tabulated results for $\phi_H = (h/a_p)$ were provided for a range of $R_m a_p$ values.

Cross-flow lift force

A particle in contact with a planar surface in a channel with laminar cross-flow may experience a lift force due to tangential convection [154]. This force acts normal to and away from the membrane surface and may be described by an equation of the form:

$$F_L = 81.2\left(\rho_w \mu_w \dot{\gamma}_0^3\right)^{1/2} a_p^3 \tag{3.3}$$

where ρ_w is the water density and $\dot{\gamma}_0$ is the wall shear rate. For a fully developed laminar flow field in a thin rectangular channel, the wall shear rate is given by $\dot{\gamma}_0 = 6Q_f/W_c H_c^2$ (Q_f is the volumetric feed flow rate, and H_c and W_c are the channel height and width, respectively) [155].

van der Waals force

The attractive van der Waals force between a spherical particle and flat membrane can be calculated from:

$$F_A = \frac{-A_H a_p}{6h^2(1 + 5.3(h/\lambda))} \tag{3.4}$$

where A_H is the Hamaker constant and λ is a characteristic wavelength ($\approx$100 nm) [156]. This expression accounts for retardation effects due to the finite time of propagation of electromagnetic waves travelling between the interacting bodies. The retardation effect significantly reduces the van der Waals force at separations larger than the characteristic wavelength down to about 10 nm of interfacial separation. The Hamaker constant is usually estimated to 1 kT [45].

Electrostatic double layer force

In case the zeta potentials and separation distances considered are small ($ze\zeta/kT < 1$ and $h/a_p << 1$), the constant surface potential sphere-plate electrical double layer force expression:

$$F_R = 2\pi a_p \varepsilon \varepsilon_0 \kappa \left(\zeta_m^2 + \zeta_p^2\right) \times \left(\frac{2\zeta_m \zeta_p}{\zeta_m^2 + \zeta_p^2} - e^{-\kappa h}\right)\left(\frac{e^{-\kappa h}}{1 - e^{-2\kappa h}}\right) \tag{3.5}$$

of Hogg et al. [157] can be used [45,151]. ε and ε_0 are the relative permittivity of water and that of vacuum, κ is the inverse Debye length (proportional to $I^{1/2}$, where I is the solution ionic strength), ζ_m and ζ_p are the membrane and particle zeta potentials and h is the interfacial separation distance.

Models

Lu and Ju [158] were among the first authors to have quantitatively described particle deposition using a surface transport model. Force and torque balances acting on a deposited particle in contact with a roughened, permeable surface were used to calculate conditions under which a particle will deposit. These conditions were a function of the tangential fluid velocity, permeate flux and angle of repose between the particle and the membrane surface. Inertial lift and buoyancy forces were also considered. Blake et al. [159] developed a similar model for a tubular geometry. Stamatakis and Tien [160] proposed that particle deposition occurred once the particle contacted a surface roughness element that was taller than a critical height. This critical height was calculated using force and torque balances around the particle. Contact forces and normal and tangential hydrodynamic drag forces were also considered. The model predicted that the size of depositing particles decreased with increased cross-flow and increases with increased permeate flux.

Other authors have adapted the force and torque balance models to include the effects of van der Waals and double layer interactions, as well as back-transport mechanisms such as shear-induced and Brownian diffusion [161–169]. Several studies [161,162,166,168] have indirectly quantified particle deposition by measuring correlations between *TMP* and permeate flux, to experimentally validate the models. Chellam and Wiesner [170] quantified particle deposition and translational speed by measuring the residence time distributions of particles as they passed through the feed channel. A more direct method was used by Altmann and Ripperger [167], who employed a laser triangulometer to measure cake thickness. Moreover, Elzo et al. [165] used a microscope to directly observe the release of glass particles from a membrane as the cross-flow velocity was increased. Recently, Knutsen and Davis [151] have extended the surface transport models to create a dynamic model that predicted not only conditions required for particle deposition, but also the average tangential rolling speed of foulant particles interacting with the membrane surface. The model was developed to explain phenomena observed using DVO such as the fractional deposition and rolling speed of foulant particles.

Kang et al. [45] used the force model to predict a set of 'critical operating conditions' at which cells were reversibly deposited. Model equations suggested that cell aggregates were more strongly influenced by tangential shear and lift forces, which tend to reduce deposition. Direct observation experiments were used to verify that most cells deposited under the predicted 'critical' conditions were easily removed when permeation ceased – without the need for backpulsing or chemical cleaning.

3.4.3 Other models

Recent models of MF phenomena include the use of artificial neural network models for membrane fouling induced by polydispersed feed suspensions during cross-flow MF [171]. The artificial neural network model incorporated the integral effects of complex colloidal transport and deposition mechanisms as well as changes in cake morphology and resistance with hydrodynamics. The initial permeate flux was found the most important operational variable influencing membrane fouling. In contrast to cross-flow filtration of monodispersed suspensions, the shear rate was found to only weakly influence permeate flux. A Monte Carlo method was also developed for cross-flow membrane filtration to determine the critical flux of hard sphere suspensions [172] Brownian and shear-induced diffusion were incorporated into an effective hydrodynamic force exerted on the hard spheres in a concentrated shear flow.

Another example is the particle trajectories evaluation in cross-flow filtration by numerical integration of the Langevin equation accounting for the combined effects of electrostatic repulsion, enhanced hydrodynamic drag, Brownian diffusion, inertial lift and van der Waals attraction [150]. The membrane remained completely free of particles below a critical filtration velocity due to the electrostatic repulsion between the charged particles and the charged membrane. This critical flux increased with increasing surface potential and decreasing ionic strength due to the increase in electrostatic repulsion. The critical flux also increased with increasing wall shear rate due to the reduction in residence time over the pore. In addition, particle transmission increased with increasing filtrate flux and ionic strength, and decreased with increasing particle size, wall shear rate and electrostatic potential.

References

[1] P. Gatenholm, S. Paterson, A.G. Fane, C.J.D. Fell, Performance of synthetic membranes during cell harvesting of *Escherichia Coli*, Process Biochem. 23 (1988) 79–81.
[2] W. Ho, K. Sirkar, Membrane Handbook, van Mostrand Reinhold, New York, 1992.
[3] L.J. Zeman, A.L. Zydney, Microfiltration and Ultrafiltration: Principles and Applications, Marcel Dekker, New York, 1996.
[4] G. Belfort, R.H. Davis, A.L. Zydney, The behavior of suspensions and macromolecular solutions in crossflow microfiltration, J. Memb. Sci. 96 (1994) 1–58.
[5] R. Kuriyel, A.L. Zydney, Sterile filtration and virus filtration, in: M. Desai (Ed.), Methods in Biotechnology: Downstream Protein Processing, vol. 9, Humanae Press, Totowa, New Jersey, 2000, pp. 185–194.
[6] R.J. Wakeman, C.J. Williams, Additional techniques to improve microfiltration, Sep. Purif. Technol. 26 (2002) 3–18.
[7] G. Russotti, K.E. Göklen, Crossflow membrane filtration of fermentation broth, in: W.K. Wang (Ed.), Membrane Separations in Biotechnology, Marcel Dekker, New York, 2001, pp. 85–160.
[8] D. Liu, R. Savoire, E. Vorobiev, J.L. Lanoiselle, Effect of disruption methods on the dead-end microfiltration behavior of yeast suspension, Sep. Sci. Technol. 45 (2010) 1042–1050.
[9] J. Geciova, D. Bury, P. Jelen, Methods for disruption of microbial cells for potential use in the dairy industry – a review, Int. Dairy J. 12 (2002) 541–553.
[10] M. Kimeldorf, S. Gleizer, Y.E. Krasik, J. Felsteiner, V. Brumfeld, H. Zuckerman, Effect of underwater high-current discharge on the properties of low-concentration β-lactoglobulin solutions, Innovative Food Sci. Emerg. Technol. 4 (2003) 151–159.
[11] A.L. Zydney, W.M. Saltzman, C.K. Colton, Hydraulic resistance of red cell beds in an unstirred filtration cell, Chem. Eng. Sci. 44 (1) (1989) 147–159.
[12] T. Suchecka, W. Piatkiewicz, T.R. Sosnowski, Is the cell retention by MF membrane absolutely safe—a hypothetical model for cell deformation in a membrane pore, J. Memb. Sci. 250 (2005) 135–140.
[13] M. Meireles, C. Molle, M.J. Clifton, P. Aimar, The origin of high hydraulic resistance for filter cakes of deformable particles: cell-bed deformation or surface-layer effect? Chem. Eng. Sci. 59 (2004) 5819–5829.
[14] E. Piron, F. René, E. Latrille, A cross flow filtration model based on integration of the mass transport equation, J. Memb. Sci. 108 (1995) 57–70.
[15] K. Nakanishi, T. Tadokoro, R. Matsuno, On the specific resistance of cakes of microorganisms, Chem. Eng. Commun. 62 (1987) 187–201.
[16] A. Rushton, E. Khoo, The filtration characteristics of yeast, J. Appl. Chem. Biotechnol. 27 (1977) 99–109.
[17] N. Lebleu, C. Roques, P. Aimar, C. Causserand, Role of the cell-wall structure in the retention of bacteria by microfiltration membranes, J. Memb. Sci. 326 (2009) 178–185.
[18] K.J. Hwang, Y.T. Wang, E. Iritani, N. Katagiri, Effect of gel particle softness on the performance of cross-flow microfiltration, J. Memb. Sci. 365 (2010) 130–137.

[19] M. Ulbricht, Advanced functional polymer membranes, Polymer 47 (2006) 2217–2262.
[20] H. Guo, M. Ulbricht, Surface modification of polypropylene microfiltration membrane via entrapment of an amphiphilic alkyl oligoethyleneglycolether, J. Memb. Sci. 349 (2010) 312–320.
[21] C.G.P.H. Schroen, M.C. Wijers, M.A. Cohen-Stuart, A. Van der Padt, K. Van't Riet, Membrane modification to avoid wettability changes due to protein adsorption in an emulsion/membrane bioreactor, J. Memb. Sci. 80 (1993) 265–274.
[22] H.Y. Yu, Y.J. Xie, M.X. Hu, J.L. Wang, S.Y. Wang, Z.K. Xu, Surface modification of polypropylene microporous membrane to improve its antifouling property in MBR: CO_2 plasma treatment, J. Memb. Sci. 254 (2005) 219–227.
[23] H. Kim, S.S. Kim, Plasma treatment of polypropylene and polysulfone supports for thin film composite reverse osmosis membrane, J. Memb. Sci. 286 (2006) 193–201.
[24] M. Ulbricht, Photograft-polymer-modified microporous membranes with environment sensitive permeabilities, React. Funct. Polym. 31 (1996) 165–177.
[25] S.A. Piletsky, H. Matuschewski, U. Schedler, A. Wilpert, E.V. Piletska, T.A. Thiele, M. Ulbricht, Surface functionalization of porous polypropylene membranes with molecularly imprinted polymers by photograft copolymerization in water, Macromolecules 33 (2000) 3092–3098.
[26] Q. Yang, Z.K. Xu, Z.W. Dai, J.L. Wang, M. Ulbricht, Surface modification of polypropylene microporous membranes with a novel glycopolymer, Chem. Mater. 17 (2005) 3050–3058.
[27] Q. Yang, J. Tian, M.X. Hu, Z.K. Xu, Construction of a comb-like glycosylated membrane surface by a combination of UV-induced graft polymerization and surface-initiated ATRP, Langmuir 23 (2007) 6684–6690.
[28] M. Ulbricht, H. Yang, Porous polypropylene membranes with different carboxyl polymer brush layers for reversible protein binding via surface-initiated graft copolymerization, Chem. Mater. 17 (2005) 2622–2631.
[29] A.H.M. Yusof, M. Ulbricht, Polypropylene-based membrane adsorbers via photo-initiated graft copolymerization: optimizing separation performance by preparation conditions, J. Memb. Sci. 311 (2008) 294–305.
[30] S. Yeu, J.D. Lunn, H.M. Rangel, D.F. Shantz, The effect of surface modifications on protein microfiltration properties of Anopore™ membranes, J. Memb. Sci. 327 (2009) 108–117.
[31] G.N.B. Barona, B.J. Cha, B. Jung, Negatively charged poly(vinylidene fluoride) microfiltration membranes by sulfonation, J. Memb. Sci. 290 (2007) 46–54.
[32] P. Wang, K.L. Tan, E.T. Kang, K.G. Neoh, Plasma-induced immobilization of poly(ethylene glycol) onto poly(vinylidene fluoride) microporous membrane, J. Memb. Sci. 195 (2002) 103–114.
[33] Y. Zhao, Y. Qian, B. Zhu, Y. Xu, Modification of porous poly(vinylidene fluoride) membrane using amphiphilic polymers with different structures in phase inversion process, J. Memb. Sci. 310 (2008) 567–576.
[34] C. Mu, Y. Su, M. Sun, W. Chen, Z. Jiang, Remarkable improvement of the performance of poly(vinylidene fluoride) microfiltration membranes by the additive of cellulose acetate, J. Memb. Sci. 350 (2010) 293–300.
[35] S. Kuiper, C.J.M. van Rijn, W. Nijdam, M.C. Elwenspoek, Development and applications of very high flux microfiltration membranes, J. Memb. Sci. 150 (1998) 1–8.
[36] M. Chandler, A. Zydney, Effects of membrane pore geometry on behavior during yeast cell microfiltration, J. Memb. Sci. 285 (2006) 334–342.

[37] X. Wang, S.M. Husson, X. Qian, S.R. Wickramasinghe, Inverse colloidal crystal microfiltration membranes, J. Memb. Sci. 365 (2010) 302–310.
[38] C.A. Romero, R.H. Davis, Experimental verification of the shear-induced hydrodynamic diffusion model of crossflow microfiltration, J. Memb. Sci. 62 (1991) 249–273.
[39] M.R. Mackley, N.E. Sherman, Cross-flow cake filtration mechanisms and kinetics, Chem. Eng. Sci. 47 (1992) 3067–3084.
[40] R.J. Wakeman, Visualization of cake formation in crossflow microfiltration, Trans. Inst. Chem. Eng. Part A 72 (1994) 530–540.
[41] P.H. Hodgson, V.L. Pillay, A.G. Fane, Visual study of crossflow microfiltration with inorganic membranes: resistance of biomass and particulate cake, in: Proceedings of the Sixth World Filtration Congress, 1993, pp. 607–610. Nagoya.
[42] H. Li, A.G. Fane, H.G.L. Coster, S. Vigneswaran, Direct observation of particle deposition on the membrane surface during crossflow microfiltration, J. Memb. Sci. 149 (1998) 83–97.
[43] H. Li, A.G. Fane, H.G.L. Coster, S. Vigneswaran, An assessment of depolarization models of crossflow microfiltration by direction observation through the membrane, J. Memb. Sci. 172 (2000) 135–147.
[44] W.D. Mores, R.H. Davis, Direct visual observation of yeast deposition and removal during microfiltration, J. Memb. Sci. 189 (2001) 217–230.
[45] S.T. Kang, A. Subramani, E.M.V. Hoek, M.A. Deshusses, M.R. Matsumoto, Direct observation of biofouling in cross-flow microfiltration: mechanisms of deposition and release, J. Memb. Sci. 244 (2004) 151–165.
[46] C.S. Parnham, R.H. Davis, Protein recovery from cell debris using crossflow microfiltration, Biotechnol. Bioeng. 47 (1995) 155–164.
[47] U. Frenander, A.S. Jönsson, Cell harvesting by cross-flow microfiltration using a shear-enhanced module, Biotechnol. Bioeng. 52 (1996) 397–403.
[48] A. Pessoa, M. Vitolo, Evaluation of cross-flow microfiltration membrane using a rotary disc-filter, Process Biochem. 33 (1998) 39–45.
[49] J.H. Vogel, K.H. Kroner, Controlled shear filtration: a novel technique for animal cell separation, Biotechnol. Bioeng. 63 (1999) 663–674.
[50] A. Brou, L.H. Ding, M.Y. Jaffrin, Dynamic microfiltration of yeast suspensions using rotating disks equipped with vanes, J. Memb. Sci. 197 (2002) 269–282.
[51] C. Harscoat, M.Y. Jaffrin, R. Bouzerar, J. Courtois, Influence of fermentation conditions and microfiltration processes on membrane fouling during recovery of Glucuronane fermentation broths, Biotechnol. Bioeng. 65 (1999) 500–511.
[52] A. Brou, M.Y. Jaffrin, L.H. Ding, J. Courtois, Microfiltration and ultrafiltration of polysaccharides produced by fermentation using a rotating disk dynamic filtration system, Biotechnol. Bioeng. 82 (2003) 429–437.
[53] S.A. Lee, A. Burt, G. Russoti, B. Buckland, Microfiltration of recombinant yeast cells using a rotating disk dynamic filtration system, Biotechnol. Bioeng. 48 (1995) 386–400.
[54] M.Y. Jaffrin, L.H. Ding, O. Akoum, A. Brou, A hydrodynamic comparison between rotating disk and vibratory dynamic filtration systems, J. Memb. Sci. 242 (2004) 155–167.
[55] J. Postlethwaite, S.R. Lamping, G.C. Leach, M.F. Hurwitz, G.J. Lye, Flux and transmission characteristics of a vibrating microfiltration system operated at high biomass loading, J. Memb. Sci. 228 (2004) 89–101.
[56] S.P. Beier, M. Guerra, A. Garde, G. Jonsson, Dynamic microfiltration with a vibrating hollow fiber membrane module: filtration of yeast suspensions, J. Memb. Sci. 281 (2006) 281–287.

[57] K.H. Kroner, H. Schfitte, H. Hustedt, M.R. Kula, Crossflow filtration in the downstream processing of enzymes, Process Biochem. 19 (1984) 67–74.
[58] F. Meacle, A. Aunins, R. Thornton, A. Lee, Optimization of the membrane purification of a polysaccharide-protein conjugate vaccine using backpulsing, J. Memb. Sci. 161 (1999) 171–184.
[59] K. Matsumoto, S. Katsuyama, H. Ohya, Separation of yeast by crossflow filtration by backwashing, J. Ferment. Technol. 65 (1987) 77–83.
[60] K. Matsumoto, S. Katsuyama, H. Ohya, Crossflow filtration of yeast by microporous ceramic membrane with backwashing, J. Ferment. Technol. 66 (1988) 199–205.
[61] A. Nipkow, J.G. Zeikus, P. Gephardt, Microfiltration cell recycle pilot system for the continuous thermoaneorobic production of exo-β-amylase, Biotechnol. Bioeng. 34 (1989) 1075–1084.
[62] S.G. Redkar, R.H. Davis, Enhancement of crossflow microfiltration performance using high-frequency reverse filtration, AIChE J. 41 (1995) 501–508.
[63] H. Mallubhotla, G. Belfort, Semiempirical modeling of crossflow microfiltration with periodic reverse filtration, Ind. Eng. Chem. Res. 35 (1996) 2920–2928.
[64] V. Kuberkar, P. Czekaj, R.H. Davis, Flux enhancement for membrane filtration of bacterial suspensions using high frequency backpulsing, Biotechnol. Bioeng. 60 (1998) 77–87.
[65] W.D. Mores, C.N. Bowman, R.H. Davis, Theoretical and experimental flux maximization by optimization of backpulsing, J. Memb. Sci. 165 (2000) 229–240.
[66] C.S. Parnham, R.H. Davis, Protein recovery from bacterial cell debris using crossflow microfiltration with backpulsing, J. Memb. Sci. 118 (1996) 259–268.
[67] W.D. Mores, R.H. Davis, Yeast foulant removal by backpulses in crossflow microfiltration, J. Memb. Sci. 208 (2002) 389–404.
[68] J.N. Kuruzovich, P.R. Piergiovanni, Yeast cell microfiltration: optimization of backwashing for delicate membranes, J. Memb. Sci. 112 (1996) 241–247.
[69] V.T. Kuberkar, R.H. Davis, Microfiltration of protein-cell mixtures with crossflushing or backflushing, J. Memb. Sci. 183 (2001) 1–14.
[70] H. Mallubhotla, E. Nunes, G. Belfort, Microfiltration of yeast suspensions with self-cleaning spiral vortices: Possibilities for a new membrane module design, Biotechnol. Bioeng. 48 (1995) 375–385.
[71] P. Moulin, C. Serra, J.C. Rouch, M.J. Clifton, P. Aptel, Dean vortices in coiled tubular membranes, J. Memb. Sci. 114 (1996) 235–244.
[72] O. Al-akoum, M. Mercier-Bonin, L. Ding, C. Fonade, P. Aptel, M. Jaffrin, Comparison of three different systems used for flux enhancement: application to crossflow filtration of yeast suspensions, Desalination 147 (2002) 31–36.
[73] J.N. Ghogomu, C. Guigui, J.C. Rouch, M.J. Clifton, P. Aptel, Hollow-fibre membrane module design: comparison of different curved geometries with Dean vortices, J. Memb. Sci. 181 (2001) 71–80.
[74] T. Kluge, C. Rezende, D. Wood, G. Belfort, Protein transmission during Dean vortex microfiltration of yeast suspensions, Biotechnol. Bioeng. 65 (1999) 649–658.
[75] S. Luque, H. Mallubhotla, G. Gehlert, R. Kuriyel, S. Dzengeleski, S. Pearl, G. Belfort, A new coiled hollow-fiber module for enhanced microfiltration performance in biotechnology, Biotechnol. Bioeng. 63 (1999) 247–257.
[76] M. Schutyser, R. Rupp, J. Wideman, G. Belfort, Dean vortex membrane microfiltration and diafiltration of rBDNF *E. coli* inclusion bodies, Biotechnol. Prog. 18 (2002) 322–329.
[77] C. Visvanathan, R. Ben Aim, Application of an electric field for the reduction of particle and colloidal fouling in crossflow microfiltration, Sep. Sci. Technol. 24 (1989) 383–398.

[78] R.J. Wakeman, E.S. Tarleton, An experimental study of electroacoustic cross-flow microfiltration, Chem. Eng. Res. Des. 69 (1991) 386–397.
[79] R.J. Wakeman, Electrically enhanced microfiltration of albumin suspensions, Trans Inst. Chem. Eng. Part C 76 (1998) 53–59.
[80] R.W. Field, D. Wu, J.A. Howell, B.B. Gupta, Critical flux concept for microfiltration fouling, J. Memb. Sci. 100 (1995) 259–272.
[81] Y.P. Zhang, A.W.K. Law, A.G. Fane, Determination of critical flux by mass balance technique combined with direct observation image analysis, J. Memb. Sci. 365 (2010) 106–113.
[82] V. Chen, A.G. Fane, S. Madaeni, I.G. Wenten, Particle deposition during membrane filtration of colloids: transition between concentration polarization and cake formation, J. Memb. Sci. 125 (1997) 109–122.
[83] P.H. Hermans, H.L. Bredee, Principles of the mathematical treatment of constant-pressure filtration, J. Soc. Chem. Ind. 55T (1936) 1.
[84] M. Hlavacek, F. Bouchet, Constant flow-rate blocking laws and an example of their application to dead-end microfiltration of protein solutions, J. Memb. Sci. 82 (1993) 285–295.
[85] C.C. Ho, A.L. Zydney, Transmembrane pressure profiles during constant flux microfiltration of bovine serum albumin, J. Memb. Sci. 209 (2002) 363–377.
[86] C.C. Ho, A.L. Zydney, A combined pore blockage and cake filtration model for protein fouling during microfiltration, J. Colloid Interface Sci. 232 (2000) 389–399.
[87] M.T. Aspelund, G. Rozeboom, M. Heng, C.E. Glatz, Improving permeate flux and product transmission in the microfiltration of a bacterial cell suspension by flocculation with cationic polyelectrolytes, J. Memb. Sci. 324 (2008) 198–208.
[88] S.R. Wickramasinghe, Y.L. Wu, B.B. Han, Enhanced microfiltration of yeast by flocculation, Desalination 147 (2002) 25–30.
[89] S.R. Wickramasinghe, B.B. Han, S. Akeprathumchai, V. Chen, P. Neal, X. Qian, Improved permeate flux by flocculation of biological feeds: comparison between theory and experiment, J. Memb. Sci. 242 (2004) 57–71.
[90] J.S. Kim, S. Akeprathumchai, S.R. Wickramasinghe, Flocculation to enhance microfiltration, J. Memb. Sci. 182 (2001) 161–172.
[91] P. Narong, A.E. James, Effect of the ξ-potential on the micro/ultra-filtration of yeast suspensions using ceramic membranes, Sep. Purif. Technol. 49 (2006) 149–156.
[92] M.N. Karim, H. Graham, B. Han, A. Cibulskas, Flocculation enhanced microfiltration of *Escherichia coli* lysate, Biochem. Eng. J. 40 (2008) 512–519.
[93] Z.F. Cui, K.I.T. Wright, Air sparging to enhance microfiltration in waste water treatment, in Engineering in Membrane Processes II – Environmental Applications, Il Ciocco, Italy, 1994. 26–28 April.
[94] M. Mercier, C. Maranges, C. Fonade, Yeast suspension filtration, flux enhancement using an upward gas/liquid slug flow-application to continuous alcoholic fermentation with cell recycle, Biotechnol. Bioeng. 58 (1998) 47–57.
[95] H.W. Sur, Z.F. Cui, Experimental study on the enhancement of yeast microfiltration with gas sparging, J. Chem. Technol. Biotechnol. 76 (2001) 477–484.
[96] S.B. Gupta, M.A. Hashim, K.B. Ramachandran, I.S. Gupta, Z.F. Cui, The effect of gas sparging in cross-flow microfiltration of 2,3-butanediol fermentation broth, Eng. Life Sci. 5 (2005) 54–57.
[97] V.T. Kuberkar, R.H. Davis, Modeling of fouling reduction by secondary membranes, J. Memb. Sci. 168 (2000) 243–258.

[98] K.J. Hwang, H.C. Hwang, The purification of protein in cross-flow microfiltration of microbe/protein mixtures, Sep. Purif. Technol. 51 (2006) 416–423.
[99] N. Arora, R.H. Davis, Yeast cake layers as secondary membranes in dead-end microfiltration of bovine serum albumin, J. Memb. Sci. 92 (1994) 247–256.
[100] C. Guell, P. Czekaj, R.H. Davis, Microfiltration of protein mixtures using yeast to reduce membrane fouling, J. Memb. Sci. 155 (1999) 113–122.
[101] V.T. Kuberkar, R.H. Davis, Effects of added yeast on protein transmission and flux in crossflow membrane microfiltration, Biotechnol. Prog. 15 (1999) 472–479.
[102] A. Jokić, Z. Zavargo, Z. Šereš, M. Tekić, The effect of turbulence promoter on cross-flow microfiltration of yeast suspensions: a response surface methodology approach, J. Memb. Sci. 350 (2010) 269–278.
[103] B.B. Gupta, J.A. Howell, D. Wu, R.W. Field, A helical baffle for cross-flow microfiltration, J. Memb. Sci. 99 (1995) 31–42.
[104] S. Sundaram, M. Auriemma, G. Howard, H. Brandwein, F. Leo, Application of membrane filtration for removal of diminutive bioburden organisms in pharmaceutical products and processes, PDA.J, Pharm. Sci. Technol. 53 (1999) 186–201.
[105] P. Rajniak, S.C. Tsinontides, D. Pham, W.A. Hunke, S.D. Reynolds, R.T. Chern, Sterilizing filtration—Principles and practice for successful scale-up to manufacturing, J. Memb. Sci. 325 (2008) 223–237.
[106] A. Allegrezza, T. Ireland, W. Kools, M. Phillips, B. Raghunath, R. Wilkins, A. Xenopoulos, Membranes in the biopharmaceutical industry, in: K.V. Peinemann, S. Pereira-Nunes (Eds.), Membranes for Life Science, Wiley-VCH, Germany, 2008, pp. 91–154.
[107] G. Russotti, A.E. Osawa, R.D. Sitrin, B.C. Buckland, W.R. Adams, S.S. Lee, Pilot-scale harvest of recombinant yeast employing microfiltration: a case study, J. Biotechnol. 42 (1995) 235–246.
[108] T.S. Lee, A methodological approach to scaling up fermentation and primary recovery processes to the manufacturing scale for vaccine production, Vaccine 27 (2009) 6439–6443.
[109] M.K.H. Liew, A.G. Fane, P.L. Rogers, Fouling of microfiltration membranes by broth–free antifoam agents, Biotechnol. Bioeng. 56 (1997) 89–98.
[110] P.B. Conrad, S.S. Lee, Two-phase bioconversion product recovery by microfiltration 1. Steady state studies, Biotechnol. Bioeng. 57 (1998) 631–641.
[111] R. Datar, Studies on the separation of intracellular soluble enzymes from bacterial cell debris by tangential flow membrane filtration, Biotechnol. Lett. 7 (1985) 471–476.
[112] S.M. Forman, E.R. DeBernardez, R.S. Feldberg, R.W. Swartz, Crossflow filtration for the separation of inclusion bodies from soluble proteins in recombinant *Escherichia coli* cell lysate, J. Memb. Sci. 48 (1990) 263–279.
[113] M.S. Le, L.B. Spark, P.S. Ward, The separation of aryl acylamindase by cross flow microfiltration and the significance of enzyme/cell debris interaction, J. Memb. Sci. 21 (1984) 219–232.
[114] A.V. Quirk, J.R. Woodrow, Tangential flow filtration – A new method for the separation of bacterial enzymes from cell debris, Biotechnol. Lett. 5 (1983) 277–282.
[115] A.V. Quirk, J.R. Woodrow, Investigation of the parameters affecting the separation of bacterial enzyme from cell debris by tangential flow filtration, Enzyme. Microb. Technol. 6 (1984) 201–206.
[116] G.L. Baruah, G. Belfort, Optimized recovery of monoclonal antibodies from transgenic goat milk by microfiltration, Biotechnol. Bioeng. 87 (2004) 274–285.

[117] H.V. Adikane, R.K. Singh, S.N. Nene, Recovery of penicillin G from fermentation broth by microfiltration, J. Memb. Sci. 162 (1999) 119–123.
[118] J.L. Davies, F. Baganz, A.P. Ison, G.J. Lye, Studies on the interaction of fermentation and microfiltration operations: Erythromycin recovery from Saccharopolyspora erythraea fermentation broths, Biotechnol. Bioeng. 69 (2000) 429–439.
[119] H. Carrère, F. Blaszkowa, H. Roux de Balmann, Modelling the clarification of lactic acid fermentation broths by cross-flow microfiltration, J. Membr. Sci. 186 (2001) 219–230.
[120] H. Carrère, F. Blaszkowa, H. Roux de Balmann, Modelling the microfiltration of lactic acid fermentation broths and comparison of operating modes, Desalination 145 (2002) 201–206.
[121] A. Nishikawa, I.L. Dunn, Performance of a two-stage fermentor with cell recycle for continuous production of lactic acid, Bioprocess Eng. 21 (1999) 299–305.
[122] J. Sikder, C. Pereira, S. Palchoudhury, K. Vohra, D. Basumatary, P. Pal, Synthesis and characterization of cellulose acetate-polysulfone blend microfiltration membrane for separation of microbial cells from lactic acid fermentation broth, Desalination 249 (2009) 802–808.
[123] C. Harscoat, M.Y. Jaffrin, P. Paullier, B. Courtois, J. Courtois, Recovery of microbial polysaccharides from fermentation broths by microfiltration on ceramic membranes, J. Chem. Technol. Biotechnol. 74 (1999) 571–579.
[124] J.A. Levesley, M. Hoare, The effect of high frequency backflushing on the microfiltration of yeast homogenate suspensions for the recovery of soluble proteins, J. Memb. Sci. 158 (1999) 29–39.
[125] H. Zhou, J. Ni, W. Huanga, J. Zhang, Separation of hyaluronic acid from fermentation broth by tangential flow microfiltration and ultrafiltration, Sep. Purif. Technol. 52 (2006) 29–38.
[126] M.L. Hans, A.M. Lowman, Biodegradable nanoparticles for drug delivery and targeting, Curr. Opin. Solid State Mater. Sci. 6 (2002) 319–327.
[127] A. Rolland, Clinical pharmacokinetics of doxorubicin in hepatoma patients after a single intravenous injection of free or nanoparticle-bound anthracycline, Int. J. Pharm. 54 (1989) 113–121.
[128] P. Beck, D. Scherer, J. Kreuter, Separation of drug-loaded nanoparticles from free drug by gel filtration, J. Microencapsul. 7 (1990) 491–496.
[129] N. Al Khouri-Fallouh, L. Roblot-Treupel, H. Fessi, J.P. Devissaguet, F. Puisieux, Development of a new process for the manufacture of polyisobutylcyanoacrylate nanocapsules, Int. J. Pharm. 28 (1986) 125–132.
[130] P. Calvo, C. Remunan-Lopez, J.L. Vila-Jato, M.J. Alonso, Chitosan and chitosan/ethylene oxide-propylene oxide block copolymer nanoparticles as novel carriers for proteins and vaccines, Pharm. Res. 14 (1997) 1431–1436.
[131] T. Govender, S. Stolnik, M.C. Garnett, L. Illum, S.S. Davis, PLGA nanoparticles prepared by nanoprecipitation: drug loading and release studies of a water soluble drug, J. Control. Release 57 (1999) 171–185.
[132] S.K. Sahoo, J. Panyam, S. Prabha, V. Labhasetwar, Residual polyvinyl alcohol associated with poly (D, L-lactide-coglycolide) nanoparticles affects their physical properties and cellular uptake, J. Control. Release 82 (2002) 105–114.
[133] C.A. Nguyen, E. Allemann, G. Schwach, E. Doelker, R. Gurny, Synthesis of a novel fluorescent poly (D, L-lactide) endcapped with 1-pyrenebutanol used for the preparation of nanoparticles, Eur. J. Pharm. Sci. 20 (2003) 217–222.
[134] G. Dalwadi, H.A.E. Benson, Y. Chen, Comparison of diafiltration and tangential flow filtration for purification of nanoparticle suspensions, Pharm. Res. 22 (2005) 2152–2162.

[135] E. Allémann, E. Doelker, R. Gurny, Drug loaded poly(lactic acid) nanoparticles produced by a reversible salting-out process: purification of an injectable dosage form, Eur. J. Pharm. Biopharm. 39 (1993) 13–18.
[136] D. Quintanar-Guerrero, A. Ganem-Quintanar, E. Allémann, H. Fessi, E. Doelker, Influence of the stabilizer coating layer on the purification and freeze-drying of poly (D, L-lactic acid) nanoparticles prepared by an emulsion–diffusion technique, J. Microencapsul. 15 (1998) 107–119.
[137] I. Limayem, C. Charcosset, H. Fessi, Purification of nanoparticle suspensions by a concentration/diafiltration process, Sep. Purif. Technol. 38 (2004) 1–9.
[138] S.R. Wickramasinghe, B. Kalbfuß, A. Zimmermann, V. Thom, U. Reichl, Tangential flow microfiltration and ultrafiltration for human influenza A virus concentration and purification, Biotechnol. Bioeng. 92 (2005) 199–208.
[139] M.T. Aspelund, C.E. Glatz, Clarification of aqueous corn extracts by tangential flow microfiltration, J. Memb. Sci. 365 (2010) 123–129.
[140] W. Li, W. Xing, W. Jin, N. Xu, Effect of pH on microfiltration of Chinese herb aqueous extract by zirconia membrane, Sep. Purif. Technol. 50 (2006) 92–96.
[141] G. Green, G. Belfort, Fouling of ultrafiltration membranes: lateral migration and the particle trajectory model, Desalination 35 (1980) 129–147.
[142] F.W. Altena, G. Belfort, Literal migration of spherical particles in porous flow channels: application to membrane filtration, Chem. Eng. Sci. 39 (1984) 343–355.
[143] G. Belfort, Fluid mechanics in membrane filtration: recent developments, J. Memb. Sci. 40 (1989) 123–147.
[144] J.R. Otis, F.W. Altena, J.T. Mahar, G. Belfort, Measurement of single spherical particle trajectories with lateral migration in a slit with a porous wall under laminar flow conditions, Exp. Fluids 4 (1986) 1–10.
[145] A.L. Zydney, C.K. Colton, A concentration polarization model for the filtrate flux in crossflow microfiltration of particulate suspensions, Chem. Eng. Commun. 47 (1986) 1–21.
[146] E.G. Leonard, C.S. Vassilieff, The deposition of rejected matter in membrane separation processes, Chem. Eng. Commun. 30 (1984) 209–217.
[147] R.H. Davis, S.A. Birdsell, Hydrodynamic model and experiments for crossflow microfiltration, Chem. Eng. Commun. 49 (1987) 217–234.
[148] C.A. Romero, R.H. Davis, Transient model of crossflow microfiltration, Chem. Eng. Sci. 45 (1990) 13–25.
[149] S. Sethi, M.R. Wiesner, Modeling of transient permeate flux in crossflow membrane filtration incorporating multiple particle transport mechanisms, J. Memb. Sci. 136 (1997) 191–205.
[150] M.M. Kim, A.L. Zydney, Theoretical analysis of particle trajectories and sieving in a two-dimensional cross-flow filtration system, J. Memb. Sci. 281 (2006) 666–675.
[151] J.S. Knutsen, R.H. Davis, Deposition of foulant particles during tangential flow filtration, J. Memb. Sci. 271 (2006) 101–113.
[152] R.F. Probstein, Physicochemical Hydrodynamics, second ed., Wiley, New York, 1994.
[153] S.L. Goren, The hydrodynamic force resisting the approach of a sphere to a plane permeable wall, J. Colloid Interface Sci. 69 (1979) 78–85.
[154] J.N. Ryan, M. Elimelech, Colloid mobilization and transport in groundwater, Colloid Surf. A 107 (1996) 1–56.
[155] R.H. Davis, Modeling of fouling of cross-flow microfiltration membranes, Sep. Purif. Meth. 21 (1992) 75–126.
[156] M. Elimelech, J. Gregory, X. Jia, R.A. Williams, Particle Deposition and Aggregation: Measurement, Modeling and Simulation, Butterworths–Heinemann, Woburn, MA, 1995.

[157] R.I. Hogg, T.W. Healy, D.W. Fuerstenau, Mutual coagulation of colloidal dispersions, Trans. Faraday Soc. 62 (1966) 1638–1651.
[158] W.M. Lu, S.C. Ju, Selective particle deposition in cross-flow filtration, Sep. Sci. Technol. 24 (1989) 517–540.
[159] N.J. Blake, I.W. Cumming, M. Streat, Prediction of steady-state cross-flow filtration using a force balance model, J. Memb. Sci. 68 (1992) 205–216.
[160] K. Stamatakis, C. Tien, A simple-model of cross-flow filtration based on particle adhesion, AIChE J. 39 (1993) 1292–1302.
[161] D.J. Chang, F.C. Hsu, S.J. Hwang, Steady-state permeate flux of cross-flow microfiltration, J. Memb. Sci. 98 (1995) 97–106.
[162] D.J. Chang, S.J. Hwang, Unsteady-state permeate flux of cross-flow microfiltration effect of particle-size distribution, Sep. Sci. Technol. 30 (1995) 2917–2931.
[163] G. Foley, D.M. Malone, F. Macloughlin, Modeling the effects of particle polydispersity in cross-flow filtration, J. Memb. Sci. 99 (1995) 77–88.
[164] W. Wang, X.D. Jia, G.A. Davies, A theoretical study of transient cross-flow filtration using force balance analysis, Chem. Eng. J. Biochem. Eng. J. 60 (1995) 55–62.
[165] D. Elzo, P. Schmitz, D. Houi, S. Joscelyne, Measurement of particle/membrane interactions by a hydrodynamic method, J. Memb. Sci. 109 (1996) 43–53.
[166] S.J. Hwang, D.J. Chang, C.H. Chen, Steady state permeate flux for particle cross-flow filtration, Chem. Eng. J. Biochem. Eng. J. 61 (1996) 171–178.
[167] J. Altmann, S. Ripperger, Particle deposition and layer formation at the crossflow microfiltration, J. Memb. Sci. 124 (1997) 119–128.
[168] I.H. Huisman, E. Vellenga, G. Tragardh, C. Tragardh, The influence of the membrane zeta potential on the critical flux for crossflow microfiltration of particle suspensions, J. Memb. Sci. 156 (1999) 153–158.
[169] H.K. Vyas, R.J. Bennett, A.D. Marshall, Cake resistance and force balance mechanism in the crossflow microfiltration of lactalbumin particles, J. Memb. Sci. 192 (2001) 165–176.
[170] S. Chellam, M.R. Wiesner, Particle-transport in clean membrane filters in laminar-flow, Environ. Sci. Technol. 26 (1992) 1611–1621.
[171] S. Chellam, Artificial neural network model for transient crossflow microfiltration of polydispersed suspensions, J. Memb. Sci. 258 (2005) 35–42.
[172] A.S. Kim, Y. Liu, Critical flux of hard sphere suspensions in crossflow filtration: Hydrodynamic force bias Monte Carlo simulations, J. Memb. Sci. 323 (2008) 67–76.

4 Virus filtration

Chapter Outline

Virus filtration is an essential component of the overall viral clearance strategy in the production of biopharmaceutical products derived from human or animal origin. Obtaining virus clearance data is essential in the manufacture of protein-based biopharmaceuticals, as well as in the plasma-derived blood products industries. Manufacturers are required to validate virus clearance before regulatory authorities will approve a product for release to market. Membrane processes play an important role in virus removal and are now found in almost all biopharmaceutical manufacturing processes.

Viruses are quite large when compared with biomolecules such as proteins, peptides, glycoproteins, sugars and nucleic acids. While proteins typically range from 5 to 100 kDa, viruses generally exceed 5000 kDa. Virus membranes must remove

Membrane Processes in Biotechnology and Pharmaceutics. DOI: 10.1016/B978-0-444-56334-7.00004-6

99.9% of virus particles while passing almost all proteins. As smaller viruses (e.g. parvoviruses) have become important, the difference in size between the protein and the virus has decreased, making size-based separations more difficult. This chapter focuses on membranes that remove viruses by size exclusion. Adsorptive membranes, which remove viruses through electrostatic interaction, are presented in Chapter 5.

4.1 Principles

4.1.1 Introduction

In the manufacture of biological pharmaceuticals, the elimination of potentially infectious agents, such as viruses, is a critical component in the purification process. Human or animal cell lines are frequently used in the production of recombinant proteins. However, cell lines may be contaminated with virus or virus-like particles. In addition to endogenous contaminants, viruses may also be introduced by addition of supplements and other constituents introduced into the fermentation process, or through handling and other manipulations during processing. Over the last decade, US and European regulatory agencies have started to require that the risk of virus contamination in biotechnology products derived from cell lines be minimized [1].

Virus clearance can be accomplished by a variety of means. The effectiveness of a specific technique varies depending on the type of virus targeted. Heat and irradiation are generally effective [2], although for practical reasons they are not commonly employed in bioprocesses to produce therapeutic proteins. Chemical treatment with a low-pH solution is very effective for lipid-enveloped viruses [3], and is commonly employed in the production of monoclonal antibodies (MAbs), where it can be integrated at the end of a Protein A chromatography step. Solvent–detergent solutions have also been shown to inactivate enveloped viruses, such as HIV 1+2, HBV and HCV [4,5]. Virus filtration provides an effective, size-based removal mechanism that complements other physical removal techniques (chromatography) and inactivation methods (irradiation, temperature, pH and solvent–detergent mixtures). Viral removal studies have been performed using plasma-borne and model viruses, including those listed in Table 4.1.

Virus filtration is also involved in the purification of virus particles and viral vectors for gene therapy applications and viral vaccines [6]. Today, viral vectors and viral vaccines play an important role in current medical approaches. Gene vectors like adenoviruses, adeno-associated viruses or retroviruses are developed for delivering genetic material to the target cell in gene therapy. Viral vaccines, like attenuated or inactivated rabies virus, influenza virus or hepatitis virus vaccines, are used to limit serious viral infections and pandemics. Higher safety demands by regulatory authorities like Food and Drug Administration (FDA) and European Agency for the Evaluation of Medicinal Products (EMEA) imply improvement in the production and purification processes. Therefore, manufacturers begin to switch from egg cultivation to mammalian cell culture systems. Tangential-flow filtration is increasingly reported for these applications [7].

Table 4.1 Size of plasma-borne and model viruses commonly used for validation of virus filtration processes [8]

	Size (nm)
Plasma-borne viruses	
Human immunodeficiency virus (HIV)	100–120
Hepatitis B virus (HBV)	50
Hepatitis C virus (HCV)	30–45
Hepatitis A virus (HAV)	25–30
Parvovirus B19	18–24
Model viruses used for validation studies	
Pseudorabies virus (PRV)	120–200
Herpes simplex virus (HSV)	150
Murine leukemia virus (MuLV)	80–130
Reovirus	60–80
Bovine viral diarrhoea virus (BVDV)	40–60
Sindbis virus	60–70
SV40	45–55
Feline calicivirus	35–39
Poliovirus	25–30
Animal parvoviruses (bovine, canine or porcine)	18–24
Encephalomyocarditis virus (EMC)	25–30

4.1.2 Previous reviews

Burnouf and Radosevich [8] have reviewed the use of virus filtration in the preparation of plasma-derived products, including coagulation factors, immunoglobulins, transferrins and protease inhibitors. Reviews of the use of membranes for virus filtration in the biopharmaceutical industry have also been provided by Levy et al. [9] and Carter and Lutz [10]. A specific section on virus filtration can be found in van Reis and Zydney's paper [11] on bioprocess membrane technologies. Principles of virus retention and fouling phenomena are presented as well as membranes, modules, equipment and processes including industrial-scale processes. Allegrezza et al. [12] provided a very detailed description of practical aspects of virus filtration process design and implementation. The authors focused on membrane selection, process development and optimization, capacity, small-scale and pilot scales, validation studies and implementation. Wickramasinghe [7] considered tangential-flow filtration for virus capture. He recalled that virus capture is needed in the biotechnology industry for two main goals: viral clearance and production of viral vectors and vaccines for gene therapy applications. He distinguished filtration of enveloped viruses and non-enveloped viruses. Enveloped viruses contain an outer lipoprotein bilayer membrane derived from the membranes of the host cell, and non-enveloped viruses lack this outer lipoprotein membrane. These properties influence the filtration performance.

4.1.3 Basic parameters

Filtrate flux and filter capacity

Like in ultrafiltration (UF) and microfiltration (MF), the filtrate flux, J_v, is a crucial parameter which is measured during constant pressure or constant flux filtration. The capacity of the virus filter can be defined as the cumulative filtrate volume when the flux drops to 20% of the clean membrane value. Other definitions of the capacity can be used, for example, the cumulative filtrate volume obtained when the flux declines to 10% or 30% of its initial value [13]. Typical experimental data for the filtrate flux, J_v, during the constant pressure filtration of a 1 g l^{-1} solution of cysteinylated-bovine serum albumin (BSA) through the Viresolve 180 membrane with the skin-side up (facing the feed) are shown in Figure 4.1 [13]. Data are shown for three experiments performed at transmembrane pressures (TMP) of 69, 103 and 172 kPa, with the results plotted as a function of the cumulative filtrate volume. The initial filtrate flux was greatest at the highest pressure of 172 kPa. At longer times, the highest filtrate flux was obtained for the run with the lowest applied pressure of 69 kPa. The capacity, defined as the cumulative filtrate volume when the flux drops to 20% of the clean membrane value, was 9 l m^{-2} at 172 kPa, which was less than the 17 l m^{-2} obtained at a pressure of 103 kPa and only 15% of the 57 l m^{-2} capacity at 69 kPa.

Viral reduction ability

Scaled-down process models are used for viral clearance validations usually referred to as 'spiking studies'. Regulatory agencies require spiking studies to be

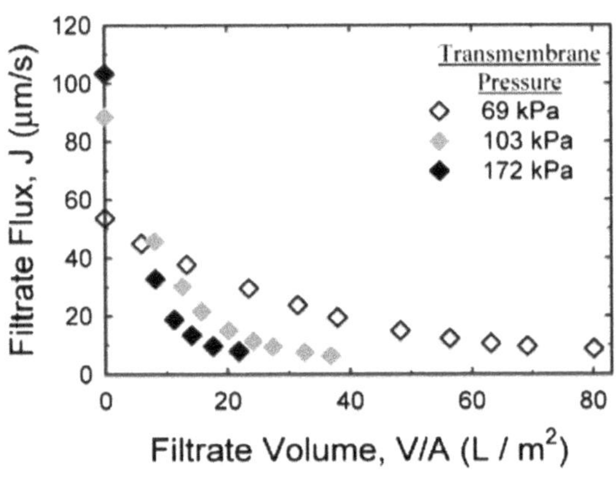

Figure 4.1 Filtrate flux as a function of the cumulative filtrate volume per unit membrane area for the Viresolve 180 membrane in the skin-side up orientation for operation at constant transmembrane pressure [13].
This article was published in Biotechnol. Prog., Vol. 22, Z.H. Syedain, D.M. Bohonak, A.L. Zydney, Protein fouling of virus filtration membranes: Effects of membrane orientation and operating conditions, pp. 1163–1169, Copyright John Wiley and Sons (2006).

representative of the process being evaluated, including process intermediate, buffer formulation, protein concentration, operational pressure, volumetric throughput and filter area [14]. The virus clearance is typically, measured from a small quantity of a concentrated virus, less than 5% of the volume of the process fluid, spiked into a representative process feedstock [15]. The concentration of the virus in the inlet and outlet streams is measured and the viral reduction ability of the process is expressed as a $\log_{10}$ reduction value:

$$LRV = \log\left(\frac{C_F}{C_P}\right) \tag{4.1}$$

where C_F and C_P are the virus concentrations in the feed and permeate streams, respectively. For small viruses (such as parvovirus), log reduction of virus (LRV) data are higher than 3–4. For larger virus (such as retrovirus), LRV data are usually higher than 6.

Virus filtration studies can be performed with a select panel of model viruses as those listed in Table 4.1. The regulation authorities recognize model viruses that represent endogenous viruses and non-specific model viruses to validate 'general viral clearance' for adventitious contamination. Murine Leukemia Virus (MuLV) is the generally accepted model for endogenous virus tests. The non-enveloped parvoviruses (20 nm) are often accepted as a worst case for filtration. The test usually comprises a four or five virus panel that represents viruses of different genomes (DNA and RNA), sizes and surface properties (enveloped and non-enveloped).

Phages, viruses that replicate in bacteria, can also be used as surrogates for mammalian viruses [16,17]. Phages are about the same size as mammalian parvoviruses and can be safely and easily handled, grown to higher titres and quantified quickly and easily. Phages tested for filtration studies include the coliphage ΦX-174 (mean diameter 27 ± 1 nm comparable with the size of mammalian parvoviruses) and phage pp7, which has nearly the same size [16]. In a recent study, the MS2 bacteriophage was selected [18]. A series of experiments were carried out to quantify the adsorption of viruses on various containers and hollow fibre membranes used in drinking water production. The MS2 bacteriophage is a non-enveloped RNA virus with an icosahedral capsid. Its diameter is between 23–30 nm and its isoelectric point is 3.9. The authors underlined the advantages of the MS2 bacteriophage: a very small virus, close in size, shape (icosahedron) and nucleic acid (single-stranded RNA) to the enteric virus hepatitis A and poliovirus, and especially non-pathogenic.

The virus spiking method has been discussed by several authors and a number of improvements have been proposed. An inline spiking method was developed recently to allow the use of inline prefiltration with direct measurement of virus filter removal capabilities [17]. The method was tested with three different protein feeds, two different parvovirus filters, two virus injection rates, a salt spike, a bacteriophage spike and two mammalian virus spikes. This inline method is recommended for applications where prefiltration is needed to improve throughput, prefiltration significantly reduces virus titre, and virus filter throughput is significantly reduced using batch versus inline prefiltration.

Khan et al. [14] underlined that the addition of impurities from virus spike preparations used in validation studies can significantly reduce filter capacity, resulting in an oversized virus filtration step. As shown in Figure 4.2A, the BSA flux versus throughput performance of the virus filter was a function of virus spike purity. Spiking protein with a relatively crude murine minute virus preparation accelerated flow decay and limited volumetric throughput. In contrast, spiking with a high titre, high-purity murine minute virus preparation did not noticeably change filter hydraulic performance as compared with protein alone. Similarly, spiking with a high-purity bacteriophage ΦX174 had negligible impact on filter hydraulic performance as compared with protein alone (Figure 4.2B), whereas addition of a mock preparation of crude murine minute virus to the ΦX174 and protein accelerated filter flow decay and limited volumetric capacity. To bypass this limitation, the authors developed a different approach to virus filter validation using filter flow decay as a relevant processing endpoint. This method entailed processing non-spiked feed to a volumetric throughput target, followed by processing virus-spiked feed to a final flow decay endpoint to determine viral clearance.

'V_{max}' model

Virus clearance is usually performed in a dead-end (normal flow) configuration. In a dead-end the 'V_{max}' model may be used to determine the maximum volume of fluid that can be filtered before the membrane is completely plugged (V_{max}). Details on the method can be found in various references [12,19,20]. The experiment consists in measuring the flow rate decrease over a short filtration time (typically, 10–15 min), with the data extrapolated to longer filtration times using the linearized form of the pore constriction model [21]:

$$\frac{t}{V} = \frac{1}{Q_0} + \left(\frac{1}{V_{max}}\right)t \qquad (4.2)$$

where V is the total filtrate volume collected over time t and Q_0 is the initial volumetric filtrate flow rate. V_{max} is evaluated directly from the flux decay data as the inverse of the slope on a plot of t/V as a function of t (equation (4.2)). The system capacity is then calculated using the expression for the flow rate as a function of time for the pore constriction model:

$$\frac{Q}{Q_0} = \left(1 + \frac{Q_0 t}{V_{max}}\right)^{-2} \qquad (4.3)$$

to give:

$$V_{capacity} = V_{max}\left(1 - \sqrt{\frac{Q_{min}}{Q_0}}\right) \qquad (4.4)$$

where Q_{min} is the minimum specified flow rate (typically 10 or 20% of Q_0).

The model is based on the assumption of gradual cylindrical pore blockage by monodisperse particles suspended homogenously in a liquid. Each particle fully or partially blocks the membrane pores, so that there is a progressively smaller area for

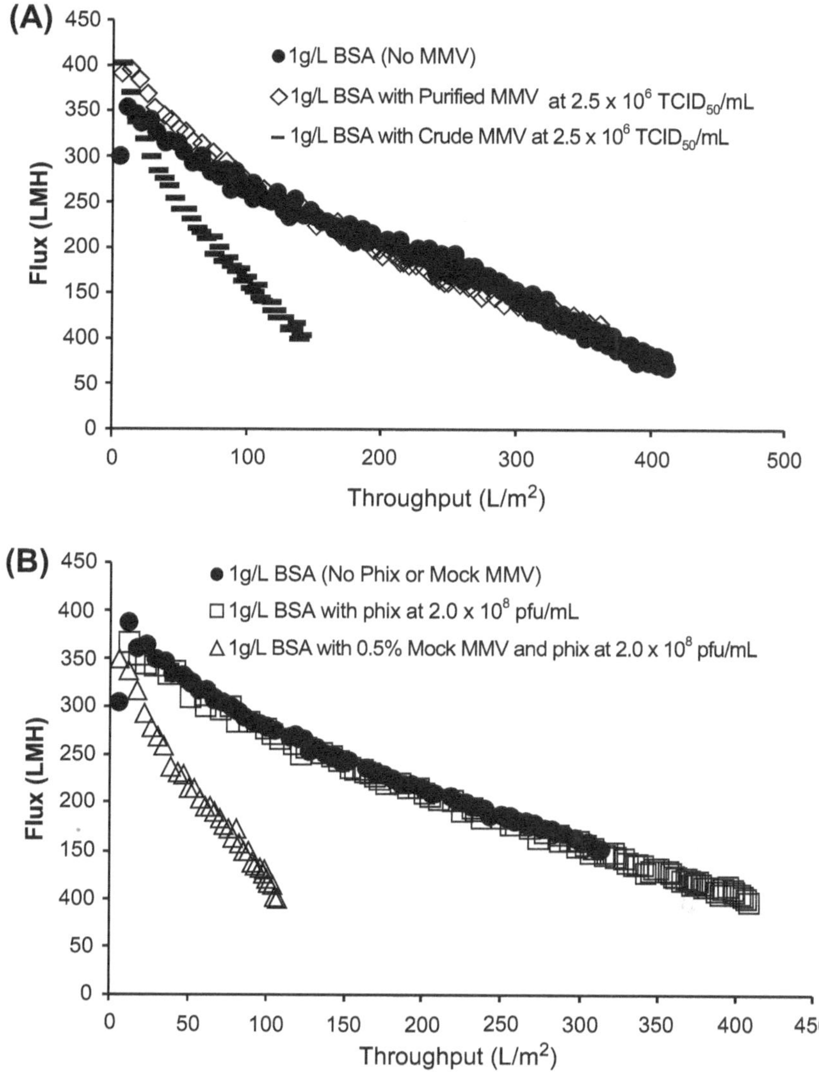

Figure 4.2 Virus spike impurities accelerate membrane fouling. (A) 1.0 g l^{-1} BSA was spiked with either purified or crude murine minute virus (MMV) stocks to a final concentration of 2.5×10^6 plaque-forming unit $TCID_{50}$/ml and filtered. (B) 1.0 g l^{-1} bovine serum albumin (BSA) was spiked to a final concentration of 2.0×10^8 pfu (plaque-forming units) ml^{-1} with either ΦX174 or ΦX174 plus 0.5% mock MMV stock, and filtered [14].
Reproduced with permission, from N.Z. Khan, J.J. Parrella, P.W. Genest, M.S. Colman, (2009), Biotechnol. Appl. Biochem., Vol. 52, Filter preconditioning enables representative scaled-down modelling of filter capacity and viral clearance by mitigating the impact of virus spike impurities, pp. 293–301. © The Authors.

the passage of the liquid phase. As a result, the flow rate decreases. Zydney and Ho [19] have shown that the V_{max} model can either underestimate or overestimate the area requirements depending on the feed protein solution. The analysis significantly overestimated the system capacity for proteins like BSA that fouled primarily by pore blockage, but it underestimated the capacity for lysozyme which fouled primarily by cake formation. The authors proposed a different fouling model that described the flux decline due to simultaneous pore blockage and cake formation, leading to more accurate sizing and scale-up of normal flow filtration devices. Davies and Smith [20] underlined the need to perform a direct scale-up from laboratory-scale experiments through large-scale pilot instead of using the V_{max} analysis.

4.2 Membranes and configurations

4.2.1 Membranes

In the early stage development, UF membranes were used for virus removal [9]. However, UF membranes contain mainly small pores, but also some large pores which do not retain viruses. Currently, Millipore, Pall Corp., Asahi Kasei and Sartorius manufacture commercial membranes specifically designed for virus removal applications (Table 4.2). These membranes typically have a complex multilayer structure to ensure that there are no defect paths, which is essential to obtain the high levels of virus removal. Virus filters are generally available in two classes of pore size: more open filters designed to remove retroviruses ($\approx$80–120 nm diameter) and tighter filters designed to remove smaller parvovirus (>18 nm diameter). Retrovirus filter devices are typically designed to remove greater than 6 logs of retrovirus, and parvovirus filter devices to remove greater than 4 logs of parvovirus [10].

Table 4.2 Properties of some commercialized virus filters

Filter	Manufacturer	Type, material and pore size	Effective filtration area
Planova BioEX	Asahi Kasei	Hollow fibre, PVDF, NA	10 cm^2 and 1 m^2
Planova 15 N 20 N, 35 N, 75 N	Asahi Kasei	Hollow fibre, regenerated cellulose different pore size available: 15 nm to 72 nm	10 cm^2 to 4 m^2
Viresolve NFP and Pro	Millipore	Flat-sheet, modified PVDF, NA	3.5 cm^2 to 1.26 m^2
Ultipor DV20 and DV50	Pall	Flat-sheet, modified PVDF, NA	0.074 m^2 to 1.63 m^2
Virosart	Sartorius	Flat-sheet, PES, 20 nm	5.3 cm^2 to 2.1 m^2

For example, the Planova membrane, commercialized by Asahi Kasei, consists of cellulose hollow fibres of Bemberg microporous membrane (BMM) manufactured by means of a microphase separation method using a regenerated cellulose cuprammonium solution [22]. The high reduction in the level of virus particles from suspensions is usually attributed to the structure of the membrane. The BMM consists of 100–200 layers of lamellar membranes. The fine structure of the wall is a channel network system consisting of voids and capillaries which produce a large space to retain the virus particles. The diameter of the capillary void is essential in determining the mean pore size of the membrane resulting in its high removal capacity.

Virus filters have much more stringent manufacturing requirements than typical MF and UF membranes. The virus filtration membranes contain a very narrow pore-size distribution to ensure maximum rejection of virus particles and passage of product species. Membranes should contain a slight gradation of pore size, the larger pore size surface being in contact with the feed stream [23]. This more open surface acts as a prefilter removing any aggregates. For symmetric membranes, the upstream surface and the membrane structure may remove aggregates present, though there is no specific retentive surface to protect. For both asymmetric and symmetric membranes, pore interconnectivity is essential to ensure a maximum permeate flux by allowing fluid to flow around blocked pores. Absence of macro-defects is usually measured and controlled by appropriate integrity tests. The retention of viruses can also be observed using electron microscopy (Figure 4.3) [22,24].

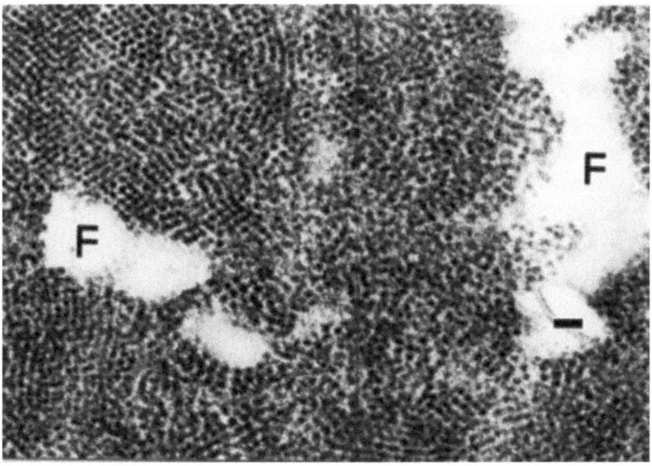

Figure 4.3 Electron micrograph of human parvovirus B19 particles packed around Bemberg microporous membrane fibres of Planova filters (Asahi Kasei, Japan) near the inner surface. The bar indicates 100 nm [22].
This article was published in J. Membr. Sci., Vol. 298, K. Yamaguchi, E. Miyagawa, H. Takahashi, T. Miyazaki, H. Ikeda, Electron microscopic estimation of removal of parvovirus B19 (HPVB19) by nanofiltration with a novel filter membrane, pp. 99–109, Copyright Elsevier (2007).

Virus filtration membranes can be used with either the skin-side up or with the skin-side facing away from the feed. Syedain et al. [13] have shown that this orientation had a strong effect on the filtration performance. Experiments were performed using Viresolve 180 membrane (Millipore) under both constant pressure and constant flux operation with sulfhydryl-modified BSA used as a model protein. The Viresolve 180 membranes have a composite asymmetric structure with a 1 μm thick skin layer on top of a fairly dense intermediate layer that is supported on a more open MF membrane [25]. Syedain et al. [13] have shown that the capacity with the skin-side up was greater during operation with constant flux and at low TMP, with the flux decline or pressure rise due primarily to osmotic pressure effects. Data obtained with the skin-side down showed a slower, steady increase in total resistance with the cumulative filtrate volume, with minimal contribution from osmotic pressure. The capacity with the skin-side down was significantly greater than that with the skin-side up. The authors concluded that different fouling mechanisms occurred for the two membrane orientations.

Virus filters are available in a variety of areas for scalability. In virus filtration, scalability affects not just process economics but also validation of virus spiking studies [12]. Small area filters are available for process development and virus validation studies, and larger area filters up to 4 m^2 for pilot and production-scale operations. In addition, multiple filter module assemblies can be used to achieve even larger filtration areas.

4.2.2 Configurations

Dead-end and cross-flow

Virus filters are used in dead-end flow filtration, which is also referred to in the literature as normal flow filtration or direct flow filtration, and in cross-flow or tangential-flow filtration. Although higher filtrate flux is typically achieved in cross-flow mode, due to a lower degree of concentration polarization, dead-end virus filtration is more popular in industry due to easier operation and lower capital requirements [1].

Cross-flow virus filters are available, for example, the Viresolve™ 70 and 180 modules commercialized by Millipore. These devices can be scaled to accommodate larger process volumes. They are of particular interest for validation of parvovirus clearance [7]. These non-enveloped viruses are small (18–24 nm). Therefore, when the product of interest is large, use of normal flow filtration can lead to significant product rejection. The Viresolve 70 filter is rated to pass 90% of proteins with molecular weight lower than 70 kDa and reject more than 3.5 log of poliovirus. The Viresolve 180 is rated to pass more than 90% of proteins with molecular weight higher than 180 kDa and reject more than 2 log of poliovirus.

Prefilters

The use of prefilters in series has been shown to improve the robustness and performance of virus filters [26,27]. The prefilters enhance the virus filter performance by removing fouling components in feed streams, thus reducing the impact on the virus filter capacity.

Prefilters are commercialized, for example, by Millipore under the trade name of Viresolve™ Prefilter and are available as disposable capsule filters and Pod disposable devices (0.1 m^2, 0.5 m^2 or 1 m^2 filtration areas). The Viresolve™ Prefilters utilize entrapped diatomaceous earth to hydrophobically bind fouling species. Botlon et al. [26] investigated their ability to remove the species that foul Viresolve™ normal-flow parvovirus (NFP) filters. A 1 mg ml^{-1} human polyclonal immunoglobulin G (IgG) solution was filtered through various membranes. Permeate samples were then collected and filtered through NFP Optiscale-25 devices (Millipore) to determine capacity and protein yield. The adsorptive Viresolve™ Prefilter medium significantly improved NFP capacity, while the other membranes provided only limited improvement (Figure 4.4). The Viresolve™ Prefilter also provided an increase in virus filter capacity for solutions containing a variety of MAbs. The authors performed SDS/PAGE, isoelectric-focussing, size-exclusion chromatography, CD and ANS (1-anilinonaphthalene-8-sulfonate) titration to show that the component of the human IgG stream that bound to the prefilter consisted of monomeric IgG variants containing more exposed hydrophobic surfaces.

Prefilters specifically designed for virus filtration are also commercialized by Asahi Kasei. Planova™ 35 N filters can be used as effective prefilters prior to final virus filtration. Planova 75 N filters are prefilters designed to remove impurities or aggregated proteins prior to final virus filtration.

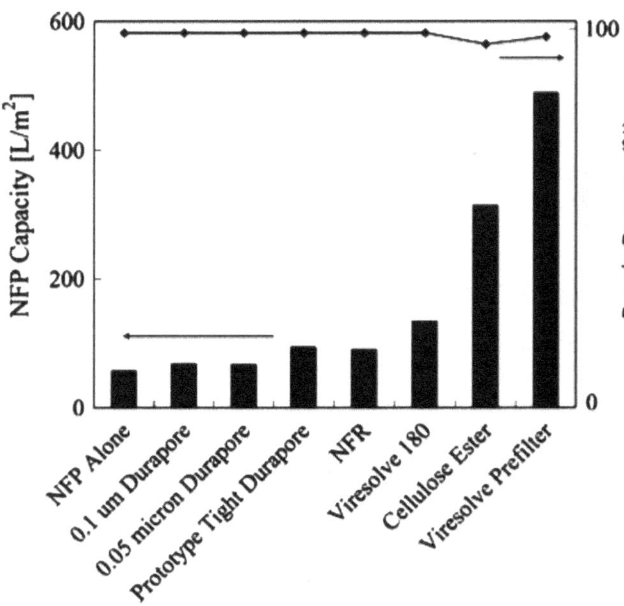

Figure 4.4 Effect of prefilter type on NFP capacity and percentage protein passage [26]. Reproduced with permission, from G.R. Bolton, S. Spector, D. Lacasse, (2009), Biotechnol. Appl. Biochem., Vol. 43, Increasing the capacity of parvovirus retentive membranes: performance of the Viresolve™ prefilter, pp. 55–63. © the Biochemical Society.

Alternative to prefilters have been proposed such as ion exchange membranes, used as inline prefilters, to improve parvovirus filter performance [28]. This technique has been applied successfully on a wide range of antibodies. The method was claimed to provide a clean, well defined, chemically stable option for enhancing throughput. Elution samples revealed the presence of high-molecular-weight species ranging from 8 to 13 nm in hydrodynamic radius, which are similar in size to parvoviruses and thus would be expected to plug the pores of a parvovirus filter. This technique can be an interesting application of membrane chromatography, leading to a cost-effective and robust approach to parvovirus filtration for the production of MAbs.

Process parametres

Various parametres were tested to achieve high throughputs of an antibody or therapeutic protein solution, such as the effect of virus spiking conditions [29,30] and the antibody solution conditions (ionic strength, pH, IgG concentration, buffer composition and aggregate level) [31]. Bolton et al. [27] have investigated a combination of methods including prefilter, the addition of additives, and the optimization of protein concentration, temperature, buffer composition and solution pH. The effect of temperature on the mass throughput of MAb1was evaluated, with the mass throughput determined at the end of a 4 h experiment (Figure 4.5). Three virus filters were tested: the Virosart CPV (Sartorius), Viresolve Pro (Millipore) and Planova 20 N (Asahi Kasei Medical) filters. The mass throughput increased with increasing temperature using the Planova 20 N filter. The highest Planova 20 N mass throughput was 7.3 kg m^{-2} at the optimum temperature of 45 °C. The increased throughput was attributed to the lower viscosity at higher temperature. The mass throughput through the Virosart CPV and Viresolve Pro filters did not increase uniformly with increasing temperature. The highest mass throughputs using the Virosart CPV and Viresolve Pro filters were achieved at the optimum temperature of 35 °C.

4.2.3 Method validation

For method validation, the determination of membrane integrity is critical. Although conventional bubble-point and air-diffusion tests are widely accepted for integrity testing of bacteria retentive membranes, these tests have severe limitations to assess the integrity of virus retentive membranes. Because of the combination of an extremely large surface tension at the air–liquid (usually water, isopropyl alcohol or a combination) interface and the small pore sizes of conventional virus filtration membranes, the pressures required to observe the bubble-point can exceed 500 psi [32]. Conducting tests at such high pressures is not possible and potentially damaging to the membrane structure.

A non-destructive liquid–liquid intrusion integrity test has been developed to be employed both before and after membrane use to validate virus removal. The CorrTest™ (Millipore) is a modified liquid–liquid porosimetric technique which is employed to correlate and predict the bacteriophage ϕ X174 log reduction value on

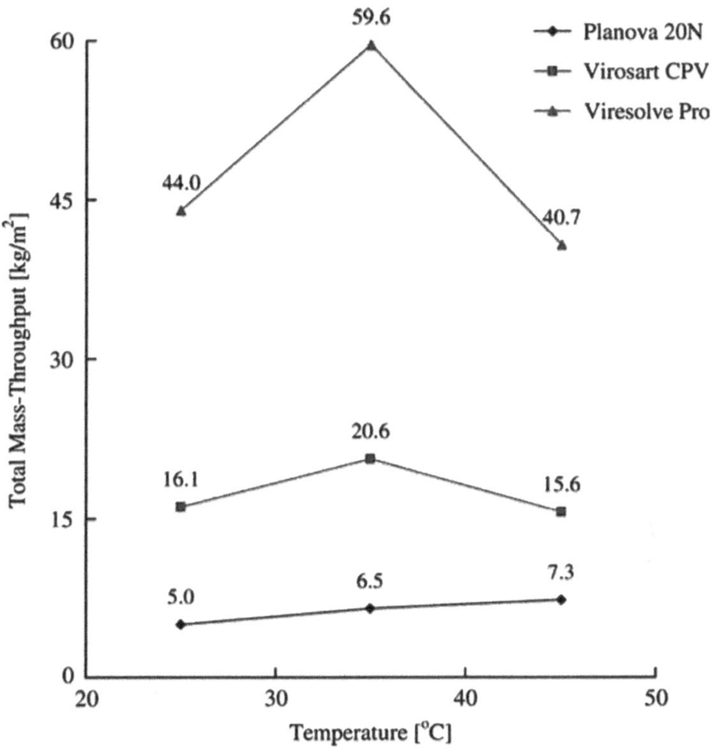

Figure 4.5 Effect of temperature on the MAb1 total mass throughput through the Viresolve Pro, Virosart CPV and Planova 20 N filters in 4 h passage [27]. Original in colour.
This article was published in Biotechnol. Prog., Vol. 26, G.R. Bolton, J. Basha, D.P. LaCasse, Achieving high mass-throughput of therapeutic proteins through parvovirus retentive filters, pp. 1671–1677, Copyright John Wiley and Sons (2010).

Viresolve filters [32]. Two fluids are formed upon mixing polyethylene glycol 8000 (PEG-8000), ammonium sulphate and water in specific proportions. The resulting PEG-rich phase and ammonium sulphate-rich phase correspond to the wetting and intrusion fluids, respectively. The components used to form the two immiscible fluids are removed from the membrane structure with a simple flushing protocol and are acceptable reagents for pharmaceutical applications. Methods for cleaning the membrane prior to a post-use integrity test are also available.

4.2.4 Other filtration methods

Other filtration techniques for virus clearance have been proposed; however, they are not typical in industry. MF membranes with nominal membrane pore sizes of 0.1 and 0.65 μm have been used for MuLV removal in flocculated Chinese hamster ovary (CHO) cell suspensions after addition of a cationic polyelectrolyte [33]. The results suggested that virus clearance involved interactions between the virus particles, the cationic polyelectrolyte, and the CHO cells present [34]. The addition of the cationic polyelectrolyte resulted in significant clearance of viruses and also led to an increase in permeate flux.

Yang et al. [35] proposed a nanoporous membrane suitable for virus filtration with good dimensional stability under high pressures maintaining high selectivity. The membrane consisted of a double layer: the upper layer was a nanoporous film with pore size of ≈ 17 nm and a thickness of ≈ 160 nm, which was prepared by polystyrene-block-poly(methyl methacrylate) copolymer (PS-b-PMMA). The nanoporous block copolymer film was combined with a conventional MF membrane to enhance mechanical strength.

4.3 Fouling

One of the critical issues in the operation of normal flow virus filtration systems is membrane fouling [13]. Fouling can significantly limit the filter capacity. Specific limits on the maximum pressure are determined by the membrane manufacturer or by the end-user.

Since virus filtration is usually performed near the end of the downstream purification, the process stream contains primarily the protein of interest with low levels of contaminants and very low levels of virus. Thus, the dominant foulant of most virus filters is the protein itself. Studies on protein fouling during MF and UF are difficult to apply to virus filtration. The complex multilayer structure of the virus filtration membranes feature selective pores that are only slightly larger than the size of the proteins, allowing proteins to pass through the membrane and into the filtrate (protein loss is typically much less than 10%). Virus filters thus lie between conventional UF membranes, which are fully retentive to most proteins, and conventional MF membranes, which have pores that are an order of magnitude or more greater than the size of the proteins of interest.

Some authors have evaluated the effect of protein fouling during virus filtration. Brough et al. [36] tested the Viresolve NFR membrane (Millipore) for retrovirus removal in a normal flow configuration. Protein passage was shown to be high (95–100%) for a variety of therapeutic protein products, including MAbs. Higuchi et al. [37] obtained data for the filtrate flux during constant pressure filtration of 5 g l^{-1} γ-globulin solutions through Planova 35 N membranes (Asahi Kasei). In the absence of salt, the flux was only 10 L m^{-2} as a result of the rapid fouling under these conditions. The flux was significantly larger in the presence of salt [37] or upon treatment with DNase [38], this effect being attributed to the dissociation of large complexes that formed between the γ-globulin and trace amounts of DNA present in the feed solution. These complexes were assumed to block the pores in the multilayer structure of the Planova membranes. The importance of large protein aggregates on the flux decline with both albumin and γ-globulin solutions was further demonstrated [39].

Syedain et al. [13] examined the effect of operating conditions and membrane orientation on BSA fouling during virus filtration. The capacity of the virus filter was defined as the total cumulative filtrate volume collected up until the point where the flux had declined to 10% of the clean membrane flux. The capacity for the Viresolve 180 membrane (Millipore) with the skin-side up increased with decreasing pressure and when operated at low flux (for constant flux experiments). The capacity was also much greater when the membrane was used with the skin-side down, which the authors attributed to the removal of protein aggregates within the macroporous substructure, thereby reducing the rate of fouling of the membrane skin. Similar results were obtained by Ireland et al. [40] for NFP filters during filtration of a 10 g l^{-1} human IgG solution, with changes in the filter capacity attributed to variations in the properties and quantity of large protein aggregates at different solution pH and ionic strength.

Recently, Bakhshayeshi and Zydney [41] showed that membrane capacity and protein yield were minimal at the protein isoelectric point, which was due to the greater degree of concentration polarization associated with the smaller protein diffusion coefficient at this pH. In contrast, the actual protein sieving coefficient was maximum at the protein isoelectric point due to the absence of any strong electrostatic exclusion under these conditions. Typical experimental data for the filtrate flux during the constant pressure filtration of a 1 g l^{-1} BSA solution through a Viresolve 180 membrane with the skin-side up are shown in Figure 4.6 for several values of the solution pH. The initial flux was lowest at pH 4.7, which is the isoelectric point of BSA, and increased at pH 7. The flux declined quite rapidly at the start of the filtration,

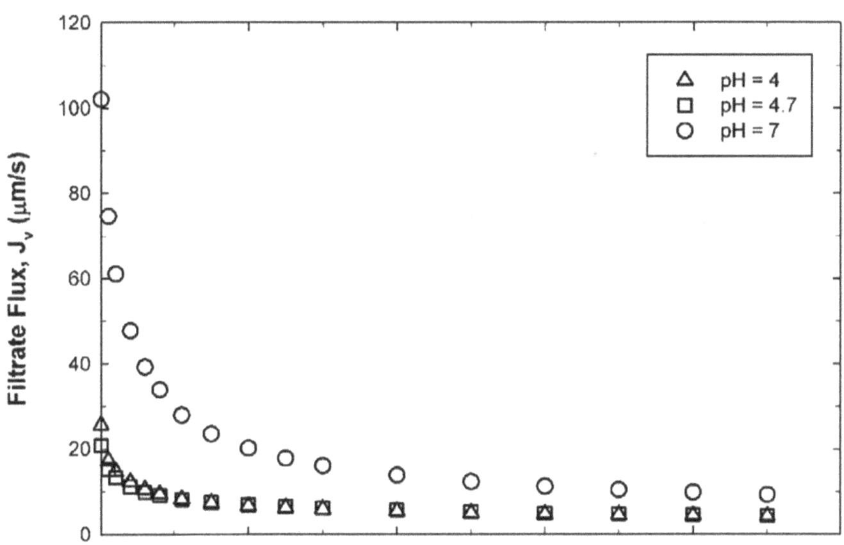

Figure 4.6 Filtrate flux as a function of time for the unstirred filtration of a 1 g l^{-1} solution of BSA at different pH through the Viresolve 180 membrane with the skin-side up at a constant pressure of 103 kPa [41].

This article was published in Biotechnol. Bioeng., Vol. 100, M. Bakhshayeshi, A.L. Zydney, Effect of solution pH on protein transmission and membrane capacity during virus filtration, pp. 108–117, Copyright John Wiley and Sons (2008).

decreasing to 20% of its initial value after 20 min at pH 7 and 90 min at pH 4.7. In contrast, the permeability measured after the protein filtration at pH 7 was only 67% of the initial value and this dropped to 60% for the experiment performed at pH 4.7.

The freezing of a material can also have a significant impact on virus filtration performance [12]. For example, it has been observed that the required filter area was 5- to 6-fold higher when measured using material that has been previously frozen compared to fresh feed [40]. While the actual purification process may not have a freezing step, feed samples required for virus validation testing are often provided in a frozen form due to material stability and availability. If freeze–thaw is observed to produce an adverse effect on filtration performance, a prefilter can be used to restore performance similar to the unfrozen material.

4.4 Applications

4.4.1 Blood and plasma products

Virus filtration of plasma products has been implemented at a production scale in the early 1990s, as a complement to the viral reduction treatments, such as solvent–detergent and heat, already applied for the inactivation of human immunodeficiency virus, hepatitis B and C virus [8]. Plasma products treated by virus filtration are clinically used or in the process of licensing, such as coagulation factors, factor IX, prothrombin complex, factor VIII and factor XI. Other plasma products include immunoglobulins, protease inhibitors and transferrin.

Factor IX has been among the first product successfully being virus filtered [42]. The factor IX product was prepared by three chromatographic steps and viral inactivation was performed by solvent–detergent treatment. The final factor IX eluate was used as the material for virus filtration experiments. Virus validations indicated that virus filtration membranes removed more than 5.7–7.8 $\log_{10}$ of HIV, bovine viral diarrhoea virus, porcine pseudorabies virus, reovirus type 3 and simian virus 40, as well as the small non-enveloped viruses, poliovirus Sabin type 1 and bovine parvovirus.

Virus filtration of prothrombin complex has been more difficult to achieve because of the complexity of this type of preparation that has a high protein content (25–40 $g\,l^{-1}$), and contains several coagulation factors (MW between 55 and 72 kDa) and proteins of higher MW, such as kininogen (110 kDa) [8]. A prothrombin complex manufactured in Austria is virally inactivated by solvent–detergent and virus filtered [43]. No thrombogenicity was evidenced by in vitro or animal models and during clinical use. A prothrombin complex concentrates (PCC) produced in the Netherlands is nanofiltered with two serially connected Planova 15 nm filters (Asahi Kasei) to increase the virus removal capacity. The virus validation study indicated removal of over 5.9 logs of canine parvovirus [8].

Ogle and Azari [44] and Azari et al. [45] gave a detailed example of viral clearance of human haemoglobin from red blood cell lysate for use in the production of the acellular oxygen transport solution diaspirin cross-linked haemoglobin (DCLHb). The production is an established process, for which virus inactivation/removal steps are included to eliminate the potential for undetected contaminants from red blood cells. This step also serves to obtain the virus removal requirements of regularly agencies. The process includes cell wash, cell lysis and haemoglobin purification (ultrafiltration/virus removal, concentration, diafiltration). The purified haemoglobin is then treated for deoxygenation, chemical modification, heat treatment/virus inactivation). Finally, the DCLHb is ultrafiltred, concentrated, and diafiltered. Two steps were implemented and validated with respect to virus removal or inactivation: (1) stroma removal from haemolysate and (2) cross-linked haemoglobin heat-treatment. A hollow fibre module (A/G Technology Corp., 500,000 MWCO) was chosen for use in the cell membrane fragments (stroma) and viruses step, because of acceptable filtrate flux (60 to 100 mL min^{-1} ft^{-2}). Two non-destructive methods were evaluated as integrity tests: air diffusion rate testing, and poly(*N*-vinyl-2-pyrrolidone) (PVP) retention. In addition, the validation of the process was described in detail. The authors concluded that the 500 kD hollow filter device was a robust and reliable method for viral clearance of human haemoglobin from red blood cell lysate.

4.4.2 Monoclonal antibody products

In antibody production, unprocessed bulk from a stirred tank fermentation process typically contains no infectious viruses. However, retrovirus-like particles may be detected by electronic microscopy. The quantification of these particles is used to estimate the amount of retrovirus-like particles in a single-dose-equivalent of unprocessed bulk [46]. Depending on the antibody titre and the amount of antibody in a dose, the viruses in the single-dose-equivalent are estimated in the ranges from 10^{10} to 10^{15} particles/ml. Therefore, a purification process must be capable of eliminating more than 15 to 20 logs of viruses. To achieve such a high clearance factor, several methods can be used in the purification process, including virus filtration, chromatography, extreme pH condition, detergent and heat treatment.

Figure 4.7 shows a typical virus removal filtration step in downstream antibody production. A wide range of feed solution conditions can be found including antibody concentration, ionic strength, pH, buffer composition and aggregate level. These variations in feed solution influence the filter performance. Hongo-Hirasaki et al. [31] investigated recently the effect of antibody solution conditions (ionic strength, pH, IgG concentration, buffer composition and aggregate level (dimer content)) on virus removal filtration using the Planova 20 N filter (Asahi Kasei). A high flux was maintained at an ionic strength greater than 10 mM and at pH 4–8 under a typical buffer composition (sodium chloride, citrate, acetate and phosphate). Optimum IgG concentration was found to be 10–20 mg ml^{-1} allowing for high throughput (kg m^{-2} of IgG). Dimer content negligibly affected the flux level.

Aranha-Creado et al. [47] evaluated virus removal for monoclonal IgG products with the hydrophilic polyvinylidene fluoride (PVDF) microporous membrane (Ultipor VF grade DV50, Pall, USA). Tests conducted with murine leukaemia virus, a model virus used in retroviruses removal studies, demonstrated a greater than 7 log clearance. The filterability test data (flux rates and product transmission) established

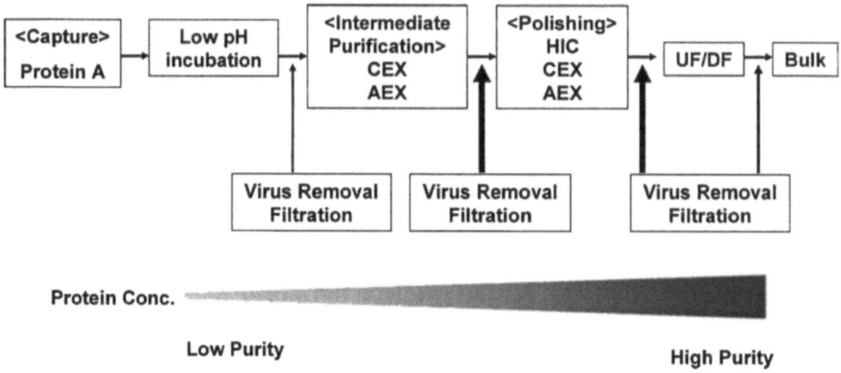

Figure 4.7 The typical virus filtration step in downstream antibody production [31]. This article was published in Biotechnol. Prog., Vol. 26, T. Hongo-Hirasaki, M. Komuro, S. Ide, Effect of antibody solution conditions on filter performance for virus removal filter Planova™ 20N, pp. 1080–1087, Copyright John Wiley and Sons (2010).

the feasibility of positioning the filter at either the eluate or the formulated bulk stage.

Davies and Smith [20] provided a detailed study of virus clearance in MAb production. The authors investigated the location of the virus filtration step within the purification process and the influence of process parameters on the operation and economics of virus filtration at 'ultralarge' scale. Xenotropic murine leukemia retrovirus (X-MuLV) and the parvovirus minute virus of mice (MVM) were selected. The X-MuLV (80–110 nm) is a model for type A and C retroviruses, which are expressed in mammalian cell lines. MVM (18–24 nm) is a potential contaminant, which may be introduced into cell culture processes via raw materials such as amino acids and glucose. Because of its small size and physicochemical stability, this virus is a good model for novel and unpredictable viral contaminants. The IgG feed streams were spiked with either 1% X-MuLV and filtered with 50 nm pore size filters or spiked with 0.2% MVM and filtered with 20 nm filters. Virus clearance and flux-decay profiles were recorded for various process parameters. The authors underlined that the location of the virus step was one of the most important parameter to minimize filtration area and costs and to maximize efficiency of virus clearance.

Recently, Marques et al. [1] presented the screening of various virus filters using MAb solutions with a concentration range of 4–20 g l^{-1}. For the MAbs of interest (two different humanized IgG1s) flux decay was not observed up to a filter loading of 200 l m^{-2} with a Planova 20 N virus removal filter. In contrast, the Viresolve NFP and Virosart filters were plugged by solutions of the same MAbs with concentrations > 4 g l^{-1} well before 50 l m^{-2}. Differences in filtration flux performance between the two MAbs under similar operating conditions indicate the sensitivity of the system to small differences in protein structure, presumably due to the impact of these differences on non-specific interactions between the protein and the membrane. Virus clearance data with two model viruses, xenotropic MuLV or murine minute virus,

Table 4.3 Log reduction of virus (LRV) with regenerated cellulose hollow fibre membranes at different MAb A concentrations [1]

MAb A Concentration (g l^{-1})	XMuLV (LRV)	MMV (LRV)
4	$\geq$3.18 $\pm$ 0.32*	$\geq$4.19 $\pm$ 0.32*
8	$\geq$4.08 $\pm$ 0.25*	3.10 $\pm$ 0.10
14	$\geq$3.44 $\pm$ 0.37*	3.97 $\pm$ 0.63

Virus spike = 0.1% (v/v) (single virus per experiment). Filter loading = 200 l m^{-2}. P_{inlet} = 14 psi.
* Virus titre was below limit of detection of the $TCID_{50}$ assay for this combination of virus spike level and sampling volume. The theoretical titre was determined based on the Poisson distribution.

confirmed the ability of hollow fibre membranes with 19 $\pm$ 2 nm pore size to achieve at least 3–4 LRV, independent of MAb concentration, over the range examined (Table 4.3).

4.5 Purification of viral vectors and vaccines

Purification methods of viral vectors and vaccines are aimed at eliminating contaminants originating from host cells or culture media and producing large volumes of concentrated, biologically active viruses. As impurities and contaminants can induce immunological and biological responses, they must be removed to comply with regulatory guidelines. For viruses, there are additional requirements to separate infective vectors from non-infective and/or empty capsids. Morenweiser [6] discussed general purification methods for viral vectors and vaccines and techniques used for downstream processing of recombinant retroviruses, adenoviruses and adeno-associated viruses. He underlined that process development for production of viral gene therapy vectors and vaccines consists of three integrated stages: upstream processing (selection of appropriate producer cell culture lines), defining optimal growth conditions and downstream processing (isolation and purification steps), which have to be performed in parallel.

Filtration of viral vectors and viral vaccines is a technique close to UF and MF. UF and MF membranes are usually preferred to membranes specifically designed for virus clearance and cross-flow filtration is often used. Grzenia et al. [48] and Wickramasinghe [7] summarized some of the purification studies of viruses for gene therapy applications and viral vaccines using cross-flow filtration.

4.5.1 Non-enveloped viruses

In addition to their use in gene therapy applications, non-enveloped adenoviruses show great promise as gene delivery vehicles for vaccination against infectious diseases such as HIV-1 (AIDS). Large-scale production and recovery methods for virus bulk are critical for commercialization of such viral vaccines. Subramaniam et al. [49] have described a hollow fibre-based recovery and concentration process for adenoviral vectors. Hollow fibre membranes were used to subject the feed to high levels of shear for cell lysis and virus release, while retaining and concentrating the virus. Lysis and concentration were thus integrated in one unit operation, reducing process time and increasing virus yield. Using a 5.6 m^2 cartridge to process approximately 240 l of harvest, the cell disruption and virus release were as efficient as observed in the lab-scale experiments. Virus release was observed to reach a plateau after 28 recirculations, or an exposure time of approximately 45 s, similar to that observed in lab-scale experiments (Figure 4.8). This interesting process can be used for other viral vectors.

4.5.2 *Enveloped virus*

Many studies have reported the use of cross-flow filtration to recover, purify and concentrate viruses. In a short communication, Saha et al. [50] reported the concentration of MuLV using 100 kDa UF membranes. The authors were able to concentrate viruses about 30-fold without losing any infectivity. In comparison, traditional sucrose density gradient purified viruses lost a significant portion of their infectivity. Braas et al. [51] described a two-stage membrane filtration process to recover and purify retrovirus particles. The membrane pore size in the first cross-flow

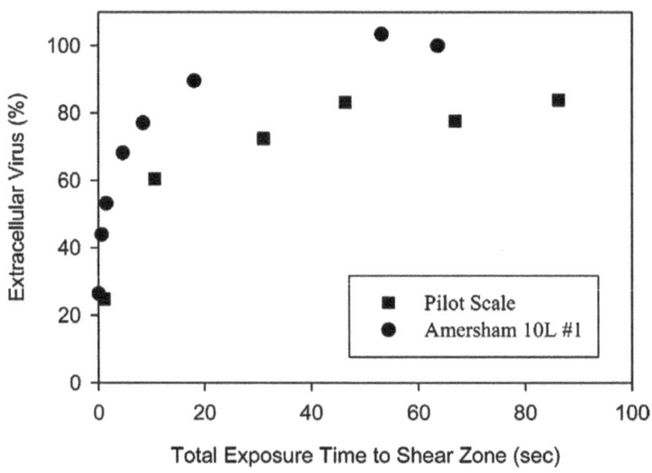

Figure 4.8 Plot of virus release with respect to the total exposure time to shear zone in a 240-L scale recovery process. Virus release is compared between representative lab-scale and pilot-scale experiments and is plotted against exposure time to shear zone. Kinetics of virus release are similar at both scales [49].
This article was published in Biotechnol. Prog., Vol. 21, S. Subramanian, G.M. Altaras, J. Chen, B.S. Hughes, W. Zhou, N.E. Altaras, Pilot scale adenovirus seed production through concurrent virus release and concentration using hollow fiber membranes, pp. 851–859, Copyright John Wiley and Sons (2005).

filtration step was larger than the virus particles. Consequently, the virus particles were recovered in the permeate. The membrane pore size in the second filtration step was smaller than the virus particles, thus concentrating the virus in the retentate.

Wickramasinghe et al. [52] have investigated the use of cross-flow filtration to recover and concentrate human influenza A virus particles grown using a cell culture technique. Influenza A virus is a lipid-enveloped RNA virus that belongs to the *Orthomyxoviridae* family. Inactivated vaccines are mainly produced in the allantoic cavity of embryonated hen eggs [53]. Development of a cell culture–based production system would overcome many of the disadvantages of the current process such as the difficulty in increasing production rapidly to meet peak vaccine demands, production of a vaccine based on viruses that are often less representative of the natural virus strains and the presence of egg proteins that can cause allergic reactions. Since human influenza virus particles are 80–120 nm in size, fractionation of the virus particles was possible using membranes with a nominal pore size of 0.1 μm. Fractionation of virus particle could be of practical benefit, allowing for separation of viral fragments and aggregated virus particles.

HIV-derived lentiviral vectors are efficient to deliver genes into the brain and hold great promise for future gene therapy of neurodegenerative disorders [54]. Application of these vectors requires standardization and up-scaling of lentiviral vector production methods. Serum-free HIV-1 vector production was efficiently up-scaled by use of cell factories and of cross-flow UF prior to centrifugation [55]. Cross-flow UF allowed a 66-fold concentration of the vectors. Further concentration of the vector (30-fold) was achieved either by low speed centrifugation or by ultracentrifugation. Combination of cross-flow UF and ultracentrifugation resulted in a vector recovery of 90–100%.

Grzenia et al. [48] investigated tangential flow UF of parvoviruses (*Aedes aegypti* densonucleosis virus with 18–26 nm mean size). Membranes with 30, 50 and 100 kDa rejected the virus particles, whereas the 300 kDa membranes allowed some virus particles to pass into the permeate. For the 300 kDa UF membrane, the decrease in permeate flux was much higher than for the 30, 50 and 100 kDa membranes, indicating possible entrapment of virus particle in the membrane pores. The permeate flux and protein rejection were strongly dependent on the cell culture growth medium used to grow the virus particles. By carefully selecting the molecular weight cut off of the membrane and maintaining a constant TMP (the permeate was pumped co-current to the feed), virus particles were purified and concentrated by UF [56].

4.6 Theory of virus clearance

Very few models are available to predict virus clearance. The model proposed by Bolton et al. [15] was based on the relationship between flow and sieving through individual pores [57]. The pore-size distribution of the membrane was divided into two classes, small pores retentive to viruses and large pores which allowed virus passage. It was assumed that a fraction δ of the total initial flow rate Q_0 passes through the population of large pores and the remaining fraction $(1 - \delta)$ passes through the population of small pores. The initial virus retention, LRV_0, was calculated as follows:

$$LRV_0 = -\log\left((S_s(1-\delta)Q_0 + S_l\delta Q_0)/Q_0\right) = -\log(S_l\delta) \tag{4.5}$$

where S_s is the sieving coefficient for the combined flow through the small pores ([virus filtrate from small pores]/[virus charge]) and S_l is the sieving coefficient for the combined flow through the large pores ([virus filtrate from large pores]/[virus filtrate]). Equation (4.5) was simplified by assuming that the passage of virus through the small-pore population was zero (i.e. $S_s = 0$). If the small pores become selectively plugged, the total flow through the membrane, Q, decreases and instantaneous LRV may be obtained in terms of LRV_0, Q and Q_0. Assuming that fluid flow through the large pores remains unchanged, the rearrangement of equation (4.5) gives:

$$LRV = -\log(S_l\delta Q_0/Q) = LRV_0 - \log(Q_0/Q) \tag{4.6}$$

Equation (4.6) indicates that a 90% decline in flow from its initial value gives a 10-fold reduction in virus retention. It should be noted that this model is based on the assumption that the larger-pore population remains unaffected. This is reasonable for selective plugging by virus-sized particles, but would not be expected in the case of adsorptive fouling, because these pores would also be constricted by protein build-up. In this case, these 'large' pores may behave as 'small' pores. Thus, a more sophisticated modelling is needed to predict the behaviour of adsorption-fouled filters.

References

[1] B.F. Marques, D.J. Roush, K.E. Göklen, Virus filtration of high-concentration monoclonal antibody solutions, Biotechnol. Prog. 25 (2009) 483–491.
[2] P. Roberts, A. Hope, Virus inactivation by high intensity broad spectrum pulsed light, J. Virol. Methods 110 (2003) 61–65.
[3] K. Brorson, S. Krejci, K. Lee, E. Hamilton, K. Stein, Y. Xu, Bracketed generic inactivation of rodent retroviruses by low pH treatment for monoclonal antibodies and recombinant proteins, Biotechnol. Bioeng. 8 (2003) 321–329.
[4] L. Biesert, H. Suhartono, Solvent/detergent treatment of human plasma. A very robust method for virus inactivation, Validated virus safety of OCTAPLAS, Vox Sang. 74 (1998) 207–212.
[5] B. Horowitz, A. Lazo, H. Grossberg, G. Page, A. Lippin, G. Swan, Virus inactivation by solvent/detergent treatment and the manufacture of SD-plasma, Vox Sang. 74 (1998) 203–206.
[6] R. Morenweiser, Downstream processing of viral vectors and vaccines, Gene Ther. 12 (2005) S103–S110.
[7] S.R. Wickramasinghe, Tangential flow filtration for virus capture, in: N.N. Li, A.G. Fane, W.S.W. Ho, T. Matsuura (Eds.), Advanced Membrane Technology and Applications, Wiley, Hoboken, NJ, 2008, pp. 541–556.
[8] T. Burnouf, M. Radosevich, Nanofiltration of plasma-derived biopharmaceutical products, Haemophilia 9 (2003) 24–37.

[9] R.V. Levy, M.W. Phililps, H. Lutz, Filtration and the removal of viruses from biopharmaceuticals, in: T.H. Meltzer, M.W. Jornitz (Eds.), Filtration in the Biopharmaceutical Industry, Marcel Dekker, New York, 1998, pp. 619–646.
[10] J. Carter, H. Lutz, An overview of viral filtration in biopharmaceutical manufacturing, Eur. J. Parenter. Sci. 7 (2002) 72–78.
[11] R. van Reis, A.L. Zydney, Bioprocess membrane technology, J. Membr. Sci. 297 (2007) 16–50.
[12] A. Allegrezza, T. Ireland, W. Kools, M. Phillips, B. Raghunath, R. Wilkins, A. Xenopoulos, Membranes in the biopharmaceutical industry, in: K.V. Peinemann, S. Pereira-Nunes (Eds.), Membranes for Life Science, Wiley-VCH, Germany, 2008, pp. 91–154.
[13] Z.H. Syedain, D.M. Bohonak, A.L. Zydney, Protein fouling of virus filtration membranes: effects of membrane orientation and operating conditions, Biotechnol. Prog. 22 (2006) 1163–1169.
[14] N.Z. Khan, J.J. Parrella, P.W. Genest, M.S. Colman, Filter preconditioning enables representative scaled-down modelling of filter capacity and viral clearance by mitigating the impact of virus spike impurities, Biotechnol. Appl. Biochem. 52 (2009) 293–301.
[15] G. Bolton, M. Cabatingan, M. Rubino, S. Lute, K. Brorson, M. Bailey, Normal-flow virus filtration: detection and assessment of the endpoint in bioprocessing, Biotechnol. Appl. Biochem. 42 (2005) 133–142.
[16] S. Lute, M. Bailey, J. Combs, M. Sukumar, K. Brorson, Phage passage after extended processing in small-virus retentive filters, Biotechnol. Appl. Biochem. 47 (2007) 141–151.
[17] H. Lutz, W. Chang, T. Blandl, G. Ramsey, J. Parella, J. Fisher, E. Gefroh, Qualification of a novel inline spiking method for virus filter validation, Biotechnol. Prog. 27 (2011) 121–128.
[18] G. Pierre, C. Causserand, C. Roques, P. Aimar, Adsorption of MS2 bacteriophage on ultrafiltration membrane laboratory equipments, Desalination 250 (2010) 762–766.
[19] A.L. Zydney, C.C. Ho, Scale-up of microfiltration systems: fouling phenomena and V_{max} analysis, Desalination 146 (2002) 75–81.
[20] J.L. Davies, M.P. Smith, Membrane applications in monoclonal antibody production, in: Z.F. Cui, H.S. Muralidhara (Eds.), Membrane Technology, Oxford, 2010, pp. 79–120.
[21] J. Hermia, Constant pressure blocking filtration laws: application to power-law non-Newtonian fluids, Trans. Inst. Chem. Eng. 60 (1982) 183–187.
[22] K. Yamaguchi, E. Miyagawa, H. Takahashi, T. Miyazaki, H. Ikeda, Electron microscopic estimation of removal of parvovirus B19 (HPVB19) by nanofiltration with a novel filter membrane, J. Membr. Sci. 298 (2007) 99–109.
[23] S.R. Wickramasinghe, E.D. Stump, D.L. Grzenia, S.M. Husson, J. Pellegrino, Understanding virus filtration membrane performance, J. Membr. Sci. 365 (2010) 160–169.
[24] K. Yamaguchi, Y. Hamamoto, S. Manabe, N. Yamamoto, Microparticle removability of the regenerated cellulose hollow fiber (BMM). An electron microscopic evaluation, J. Electron. Microsc. 40 (1991) 337–345.
[25] A. DiLeo, A. Allegrezza, S. Builder, High-resolution removal of virus from protein solutions using a membrane of unique structure, Bio/Technology 10 (1992) 182–188.
[26] G.R. Bolton, S. Spector, D. Lacasse, Increasing the capacity of parvovirus retentive membranes: performance of the Viresolve™ prefilter, Biotechnol. Appl. Biochem. 43 (2006) 55–63.

[27] G.R. Bolton, J. Basha, D.P. LaCasse, Achieving high mass-throughput of therapeutic proteins through parvovirus retentive filters, Biotechnol. Prog. 26 (2010) 1671–1677.
[28] A. Brown, C. Bechtel, J. Bill, H. Liu, J. Liu, D. McDonald, S. Pai, A. Radhamohan, R. Renslow, B. Thayer, S. Yohe, C. Dowd, Increasing parvovirus filter throughput of monoclonal antibodies using ion exchange membrane adsorptive pre-filtration, Biotechnol. Bioeng. 106 (2010) 627–637.
[29] T. Hongo-Hirasaki, K. Yamaguchi, K. Yanagida, K. Okuyama, Removal of small viruses (parvovirus) from IgG solution by virus removal filter Planova® 20N, J. Membr. Sci. 278 (2006) 3–9.
[30] T. Hongo-Hirasaki, K. Yamaguchi, K. Yanagida, H. Hayashida, S. Ide, Effects of varying virus-spiking conditions on a virus-removal filter Planova™ 20N in a virus validation study of antibody solutions, Biotechnol. Prog. 27 (2011) 162–169.
[31] T. Hongo-Hirasaki, M. Komuro, S. Ide, Effect of antibody solution conditions on filter performance for virus removal filter Planova™ 20N, Biotechnol. Progress 26 (2010) 1080–1087.
[32] M.W. Phillips, A.J. DiLeo, A validatible porosimetric technique for verifying the integrity of virus-retentive membranes, Biologicals 24 (1996) 243–253.
[33] S. Akeprathumchai, B. Han, S.R. Wickramasinghe, J.O. Carlson, P. Czermak, K. Preiss, Murine leukemia virus clearance by flocculation and microfiltration, Biotechnol. Bioeng. 88 (2004) 880–889.
[34] S.R. Wickramasinghe, B. Han, J.O. Carlson, S.M. Powers, Clearance of minute virus of mice by flocculation and microfiltration, Biotechnol. Bioeng. 86 (2004) 612–621.
[35] S.Y. Yang, J. Park, J. Yoon, M. Ree, S.K. Jang, J.K. Kim, Virus filtration membranes prepared from nanoporous block copolymers with good dimensional stability under high pressures and excellent solvent resistance, Adv. Funct. Mater. 18 (2008) 1371–1377.
[36] H. Brough, C. Antoniou, J. Carter, J. Jakubik, Y. Xu, H. Lutz, Performance of a novel Viresolve NFR virus filter, Biotechnol. Prog. 18 (2002) 782–795.
[37] A. Higuchi, A. Komuro, K. Hirano, B.-O. Yoon, M. Hara, T. Hirasaki, M. Nishimoto, M. Yokogi, S. Manabe, Permeation of γ-globulin through microporous membranes in the presence of trace DNA, J. Membr. Sci. 186 (2001) 9–18.
[38] A. Higuchi, M. Nemoto, H. Koyama, K. Hirano, B.-O. Yoon, M. Hara, M. Yokogi, S. Manabe, Enhanced microfiltration of γ-globulin solution upon treatment of NaCl addition and/or DNase digestion, J. Membr. Sci. 210 (2002) 369–378.
[39] A. Higuchi, M. Kyokon, S. Murayama, M. Yokogi, T. Hirasaki, S. Manabe, Effect of aggregated protein sizes on the flux of protein solution through microporous membranes, J. Membr. Sci. 236 (2004) 137–144.
[40] T. Ireland, H. Lutz, M. Siwak, G. Bolton, Viral filtration of plasma derived human IgG: a case study using Viresolve NFP, Biopharm. Int. 17 (2004) 38–44.
[41] M. Bakhshayeshi, A.L. Zydney, Effect of solution pH on protein transmission and membrane capacity during virus filtration, Biotechnol. Bioeng. 100 (2008) 108–117.
[42] M. Burnouf-Radosevich, P. Appourchaux, J.J. Huart, T. Burnouf, Nanofiltration, a new specific virus elimination method applied to high-purity factor IX and factor XI concentrates, Vox Sang. 67 (1994) 132–138.
[43] D. Josic, L. Hoffer, A. Buchacher, H. Schwinn, W. Frenzel, L. Biesert, H.P. Klocking, R. Hellstern, R. Rokicka-Milewska, A. Klukowska, Manufacturing of a prothrombin complex concentrate aiming at low thrombogenicity, Thromb. Res. 100 (2000) 433–441.
[44] K.F. Ogle, M.R. Azari, Virus removal by ultrafiltration: A case study with diaspirin crosslinked hemoglobin (DCLHb), in: W.K. Wang (Ed.), Membrane Separations in Biotechnology, Marcel Dekker, New York, 2001, pp. 299–326.

[45] M. Azari, J.A. Boose, K.E. Burhop, T. Camacho, J. Catarello, A. Darling, A.A. Ebeling, T.N. Estep, L. Pearson, S. Guzder, J. Herren, K. Ogle, J. Paine, K. Rohn, R. Sarajari, C.S. Sun, L. Zhang, Evaluation and validation of virus removal by ultrafiltration during the production of diaspirin crosslinked haemoglobin (DCLHb), Biologicals 28 (2000) 81–94.
[46] P.Y. Huang, J. Peterson, Scaleup and virus clearance studies on viral filtration in monoclonal antibody manufacture, in: W.K. Wang (Ed.), Membrane Separations in Biotechnology, Marcel Dekker Inc., New York, 2001, pp. 327–350.
[47] H. Aranha-Creado, J. Peterson, P.Y. Huang, Clearance of murine leukaemia virus from monoclonal antibody solutions by a hydrophilic PVDF microporous membrane filter, Biologicals 26 (1998) 167–172.
[48] D.L. Grzenia, J.O. Carlson, P. Czermak, B. Han, R.K. Specht, S.R. Wickramasinghe, Purification of densonucleosis virus by tangential flow ultrafiltration, Biotechnol. Prog. 22 (2006) 1346–1353.
[49] S. Subramanian, G.M. Altaras, J. Chen, B.S. Hughes, W. Zhou, N.E. Altaras, Pilot scale adenovirus seed production through concurrent virus release and concentration using hollow fiber membranes, Biotechnol. Prog. 21 (2005) 851–859.
[50] K. Saha, Y.-C. Lin, P.K.Y. Wong, Short communication: a simple method for obtaining highly viable virus form culture supernatant, J. Virol. Methods 46 (1994) 349–352.
[51] G. Braas, P.F. Searle, N.K.H. Slater, A. Lyddiatt, Strategies for the isolation and purification of retroviral vectors for gene therapy, Bioseparation 6 (1996) 211–228.
[52] S.R. Wickramasinghe, B. Kalbfuss, A. Zimmerman, V. Thom, U. Reichl, Tangential flow microfiltration and ultrafiltration for human influenza A virus concentration and purification, Biotechnol. Bioeng. 92 (2005) 199–208.
[53] C. Gerdil, The annual production cycle for influenza vaccine, Vaccine 21 (2003) 1776–1779.
[54] V. Baekelandt, K. Eggermont, M. Michiels, B. Nuttin, Z. Debyser, Optimized lentiviral vector production and purification procedure prevents immune response after transduction of mouse brain, Gene Ther. 10 (2003) 1933–1940.
[55] M. Geraerts, M. Michiels, V. Baekelandt, Z. Debyser, R. Gijsbers, Upscaling of lentiviral vector production by tangential flow filtration, J. Gene Med. 7 (2005) 1299–1310.
[56] D.L. Grzenia, J.O. Carlson, S.R. Wickramasinghe, Tangential flow filtration for virus purification, J. Membr. Sci. 321 (2008) 373–380.
[57] S. Mochizuki, A.L. Zydney, Theoretical analysis of pore-size distribution effects on membrane-transport, J. Membr. Sci. 82 (1993) 211–227.

5 Membrane chromatography

Chapter Outline

Membrane Processes in Biotechnology and Pharmaceutics. DOI: 10.1016/B978-0-444-56334-7.00005-8

5.1 Principles

5.1.1 Introduction

Traditional bead chromatography is a widely utilized and a well-studied technology for the purification of biological products. Adsorptive membranes, also termed chromatographic membranes, became available almost two decades ago, and competed in the past decade with bead chromatography. The interaction modes investigated with microporous membranes include affinity interaction, ion exchange, hydrophobic interaction, reversed-phase and multistage chromatography. Adsorptive membranes have been designed and evaluated in the form of flat sheet systems and stacks of membranes, hollow fibres, radial flow catridges and tubes.

The benefit of adsorptive membranes is the shorter diffusion times than those obtained in bead chromatography, as the interactions between molecules and active sites on the membrane occur in convective through-pores rather than in stagnant fluid inside the pores of an adsorbent particle (Figure 5.1). Therefore, the adsorptive membranes have the potential to purify large biomolecules at high flow rates, reducing biomolecule degradation and denaturation. In addition, adsorptive membranes can withstand high flow rates without significant loss in capacity, with clear advantages for scale-up.

5.1.2 Previous reviews

Many review articles have focused on the development of adsorptive membranes, with specific interest towards membrane materials, ligand immobilization, coupling chemistries and applications [1–6]. Recent articles have reviewed specific aspects of membrane chromatography. Ghosh [7] presented the separation of proteins using adsorptive membranes. The transport phenomenon of an adsorptive membrane was discussed, along with various separation chemistries and different applications. Suen et al. [9] have discussed the features of immobilized metal affinity membranes by

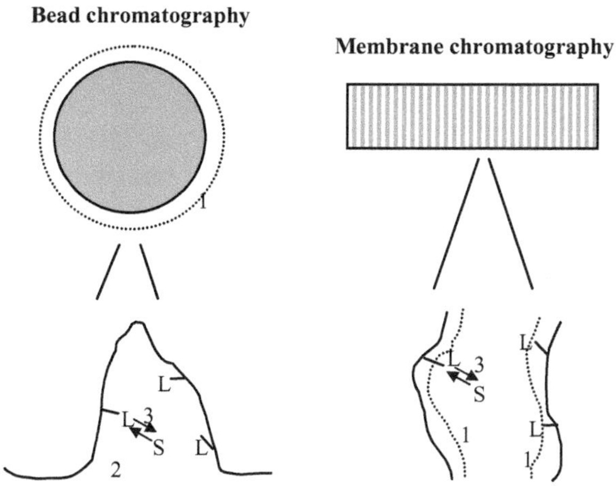

Limitations:
1. Film diffusion
2. Pore diffusion
3. Binding kinetics

Limitations:
1. Film diffusion
3. Binding kinetics

Figure 5.1 Comparison between bead chromatography and membrane chromatography. L: ligand, S: solute [8].
This article was published in Comprehensive Biotechnology, Second Edition, Vol. 2, C. Charcosset, Downstream Processing and Product Recovery/Membrane Systems and Technology, pp. 603–618, Copyright Elsevier (2011).

focussing on membrane matrices, immobilized metal affinity method and module designs. Several examples of immobilized metal affinity membranes for the purification of potential therapeutics were presented. Zhou and Tressel [10] have reviewed the Q (quaternary ammonium) anion-exchange membrane for process-scale antibody production. The advantages and disadvantages of using Q adsorptive membranes as a purification unit in large-scale production were discussed. The review included a cost analysis, which illustrated that Q membrane chromatography was a viable alternative to Q column chromatography as a polishing step for process-scale antibody production.

Boi [11] has recently summarized the use of adsorptive membrane as an alternative technology for capture and polishing steps for the purification of monoclonal antibodies. Both affinity membranes and ion-exchange membranes were discussed. In a recent book chapter, Avramescu et al. [12] discussed recent advances in membrane chromatography with special attention to adsorptive membrane preparation, including membrane preparation of affinity membranes, monolithic stationary phases and particle-loaded membranes. Monolithic stationary phases are another alternative to traditional packed-bed columns with advantages similar to membrane chromatography. They differ from membranes in terms of material, preparation and morphology [13,14]. The monoliths are prepared mainly by in situ polymerization of organic species or silicon alkoxides and consolidation inside the column. Monolithic stationary phases are discussed at the end of this chapter.

5.1.3 Basic definitions

Loading, washing, elution and regeneration steps

An adsorptive membrane process is typically carried out in four steps: loading, washing, elution and regeneration (Figure 5.2). During the loading step, the species to be purified is pumped into the device to produce the breakthrough curve. Unbound species are then rinsed from the device using the loading buffer. Bound species are eluted using the elution buffer. After elution is complete, the device is washed with the loading buffer to restore it to its initial conditions before carrying out the next experiment. The concentration of the species in the feed and filtrate streams can be monitored by continuous measurement of the optical adsorption using a UV flow cell detector. Readings from the feed UV detector are checked to ensure that it returns to baseline at the end of the washing step, indicating that all unbound species are removed, and at the end of the elution step, indicating that all adsorbed species are removed from the membrane binding sites.

Breakthrough curve

The breakthrough curve is the concentration profile during the loading step, as a function of the mass of species loaded or the effluent volume (Figure 5.2). Breakthrough is defined as the point at which the solute being purified appears in the effluent solution. Most breakthrough curves present an S shape. Initially, the species concentration in the filtrate is zero, reflecting complete adsorption of the molecules by

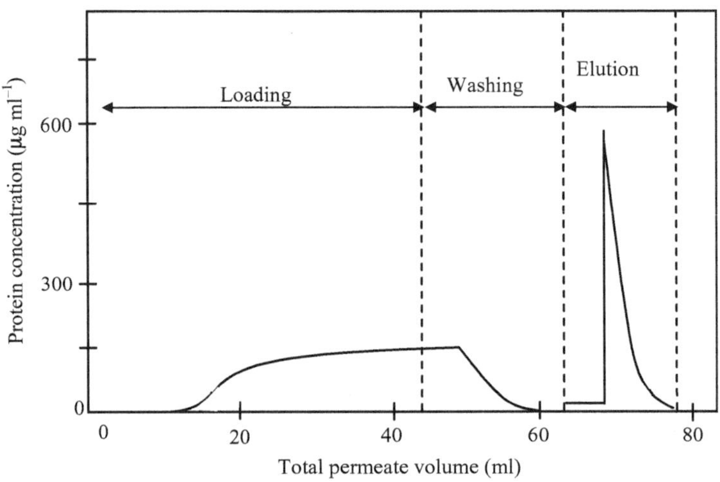

Figure 5.2 Typical loading, washing and elution steps in the operation of a membrane adsorber illustrated with an experimental chromatogram [8].
This article was published in Comprehensive Biotechnology, Second Edition, Vol. 2, C. Charcosset, Downstream Processing and Product Recovery/Membrane Systems and Technology, pp. 603–618, Copyright Elsevier (2011).

the adsorptive membrane. As the loading step proceeds and binding sites become occupied, the species appear in the filtrate. Following breakthrough, the species concentration, C, in the filtrate increases rapidly and then more slowly as C asymptotically approaches the feed concentration, C_0. At this point, steady state is achieved and no further adsorption occurs.

Ideal breakthrough occurs only after the species being purified and ligand reach saturation, at which point the effluent concentration would instantaneously increase from zero to that of the feed solution. Actual breakthrough curves are broadened by the nonidealities of real flow systems, e.g. dead volume mixing and flow maldistribution, and by slow intrinsic sorption kinetics (Figure 5.2). In industrial purification of high-value proteins, it is desirable to minimize the loss of product into the filtrate during the loading step. Consequently, adsorption is allowed to proceed until breakthrough occurs, at which point the loading step is terminated. In contrast, in medical therapeutic applications the objective is to remove a pathogenic agent from blood or plasma, and adsorption is allowed to continue until the matrix is nearly saturated.

Adsorption isotherm

The adsorption isotherm for binding of the species to the adsorptive membrane can be measured by static incubation experiments. Typically, the chromatographic membrane is cut into small pieces and added to flasks containing various concentrations of species in the same final volume. The suspensions are gently agitated at room temperature for the system to come to equilibrium. An aliquot is then removed and its concentration, C_{eq}, is determined.

A typical adsorption isotherm is shown in Figure 5.3. The adsorbed species concentration increases rapidly with increasing free species concentration below

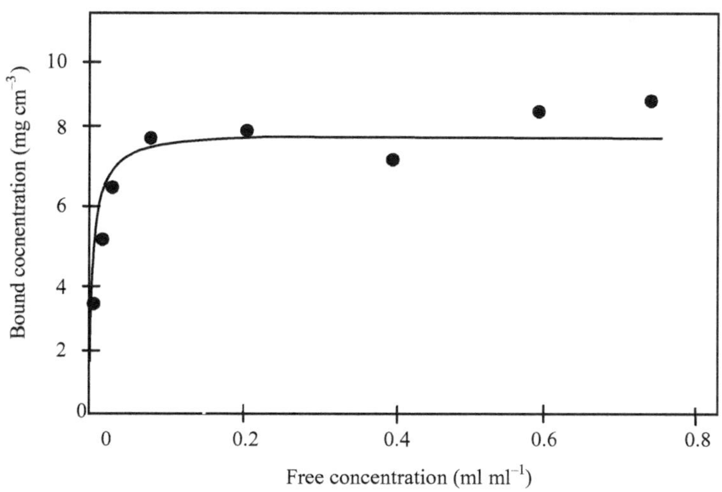

Figure 5.3 Typical adsorption isotherm obtained from static incubation experiments or dynamic elution breakthrough curves.

a certain concentration. At higher free concentration, the bound species concentration is less sensitive to free species concentration. The binding data may be fitted to classical isotherms, such as the Langmuir isotherm, for which the bound species concentration, Q_{eq}, at equilibrium varies with the free ligate concentration, C_{eq}, according to:

$$Q_{eq} = Q_m \frac{C_{eq}}{C_{eq} + K_d} \tag{5.1}$$

where Q_m is the maximum data for the static binding capacity, and K_d is the equilibrium dissociation binding constant. The static binding capacity, Q_m, is expressed per membrane volume.

Dynamic binding capacity

The adsorption isotherm for the species may also be obtained from measurement of the amount eluted during breakthrough experiments performed at different feed concentrations. The isotherm data obtained from dynamic experiments generally agree with the data determined by static incubation, as did the estimate of Q_m. The binding capacities given by manufacturers are generally the maximum data for the dynamic binding capacity, Q_m.

5.2 Membrane support

In this section, the required characteristics for an adsorptive membrane are presented, along with the membrane materials and devices. Some commercialized adsorptive membranes are detailed in Table 5.1.

Table 5.1 Properties of some commercialized adsorptive membranes

Adsorbers	Manufacturer	Type, material and pore size interaction mode	Effective filtration area or membrane bed volume
Sartobind membrane adsorbers	Sartorius (Germany)	Syringe disposable unit, small centrifugal devices, large to very large devices, Regenerated cellulose, 3–5 μm, Ion-exchange, metal chelate, Affinity protein A and hydrophobic interaction	0.6 cm^2–6 m^2
Mustang membranes	Pall (USA)	Polyethersulfone, 0.2 or 0.8 μm, Syringe disposable units Ion-exchange, and E surface-modified interaction	0.18 mL

5.2.1 Required characteristics

An ideal membrane support for successful application in chromatography has the following features: (1) microporous to accommodate the free interaction of high-molecular-weight biomolecules with ligands, which could themselves be macromolecules; (2) hydrophilic and neutral to prevent the biomolecules from interacting non-specifically with the membrane itself; (3) contain functional groups to allow activation by a wide variety of reactions; (4) be chemically stable to withstand harsh conditions during adsorption, elution and regeneration; (5) be physically stable to withstand transmembrane pressure and, when applicable, sterilization by autoclaving; and (6) readily available at low cost to facilitate industrial applications. The limitation for membrane support applications in chromatography lies generally in the necessity for chemically active sites in the membrane matrix for bonding the ligand.

5.2.2 Membrane material

Membrane materials tested for chromatographic applications include cellulose and regenerated cellulose, polysulfone and its derivatives, polyamide, hydrazide and composite membranes.

Cellulose and regenerated cellulose membranes

The first basic material is cellulose, which has long been used in membrane preparation and is a good matrix for affinity ligand coupling. However, the commercially available cellulose derivative membranes have relatively small pores and are not suitable for chromatography owing to the high pressure drop through the stacked membranes. To overcome this limit, a new type cellulose microfiltration membrane was prepared by Guo et al. [15] and used as an affinity matrix. The membrane, consisting of coarse fibres, had a higher porosity and larger pore size (1–2 μm), but was less uniform in structure. To enhance the mechanical and chemical stabilities, cross-linking by epoxypropane chloride was carried out prior to immobilization of triazine dyes.

Recently, cellulose membranes were prepared by the phase inversion method using the ionic liquid [BMIM][Cl] as the solvent and water as the antisolvent [16]. Ionic liquids were chosen instead of toxic solvents in order to achieve a greener preparation process. After functionalization with a synthetic ligand 2-(3-aminophenol)-6-(4-amino-1-naphthol)-4-chloro-*s*-triazine, these adsorptive membranes were evaluated for human immunoglobulin G (IgG) adsorption. Bovine serum albumin (BSA), chosen as a model impurity, did not adsorb onto the membrane while up to 6 mg IgG g^{-1} was bound and 2 mg IgG g^{-1} recovered. Another recent example is a cellulose nanofibrous membrane (fibre with 200-μm average diameters) prepared by electrospinning of cellulose acetate followed by alkaline hydrolysis [17]. Cibacron Blue was immobilized through the nucleophilic reaction of cellulose hydroxyl with the triazinyl chloride of the ligand. A lipase loading capacity of 16.21 mg g^{-1} was obtained and could be optimized to reach at least 150 mg g^{-1}. The authors concluded that these nanofibrous membranes were highly efficient for lipase binding and that they were potentially attractive for immobilizing other enzymes and ligands.

A large range of adsorptive membranes made from regenerated cellulose are actually commercialized by Sartorius (Göttigen, Germany) grafted or loaded with conventional ligands. These membranes are available under the trademark Sartobind including: strong ion exchanger with quaternary ammonium groups (Q) or sulfonic acid groups (S), weak ion exchange with carboxylic acid groups (C) or diethylamine groups (D), metal chelate (iminodiacetic acid), coupling (epoxy or aldehyde), affinity (protein A) and hydrophobic interaction (phenyl). The membranes are available in different configurations, from syringe disposable unit, to small centrifugal devices and large to very large devices. Syringe disposable units for one-time use have the major advantage to avoid the need for validation of membrane cleaning after use.

Commercial Sartobind cation-exchanger membranes were investigated in detail using conventional scanning electron microscopy (SEM), environmental scanning electron microscopy (ESEM) and confocal laser scanning microscopy (CLSM) [18]. SEM, ESEM and CLSM show similar morphologies at different depths of membrane (Figure 5.4a–c). The Sartobind membranes have an anisotropic macroporous structure. Their mechanical stability is mainly due to thick fibres with a diameter of about 10 μm located in the middle section of the membrane. From SEM pictures, the typical morphology for the bottom side of Sartobind S and C membrane in dry and wet state is a very irregular and complex structure. The morphology from CLSM image for the dye-stained wet membrane is similar to that in wet state measured using ESEM.

In the same study, visualization by CLSM of pore morphology and protein binding pattern were obtained after binding of labelled protein to fluorescence-stained membranes. For the Sartobind S membranes, the protein binding pattern was similar to the pore morphology. The protein filled in a relatively thick and homogeneous layer significant fraction of the macropores, leaving the interior of these pores empty. In addition, the bound protein density was much lower in the region of the thick cellulose fibres. For the Sartobind C membranes, a more clustered protein pattern in the macropore space was observed, again with very little detectable protein on the thick cellulose fibres.

Polysulfone membranes and their derivatives

Polysulfone is a suitable membrane polymer for chromatography because of its good film-forming properties and its thermal and biological resistance. Polysulfone is chemically fairly stable and, on the other hand, can be converted by some special reactions that do not alter the polymer properties. One reaction is the metalation of the aromatic ring with *n*-butyllithium [19]. The membrane may be then substituted with chelating groups, which can bind metal ions. Other reactions with polysulfone membranes include acrylation–amination and chloromethylation–amination, prior to immobilization of triazine dyes [20].

Polyethersulfone (PES)-adsorptive membranes are commercialized by Pall (USA) in the form of syringe disposable units. The available membranes include Q strong

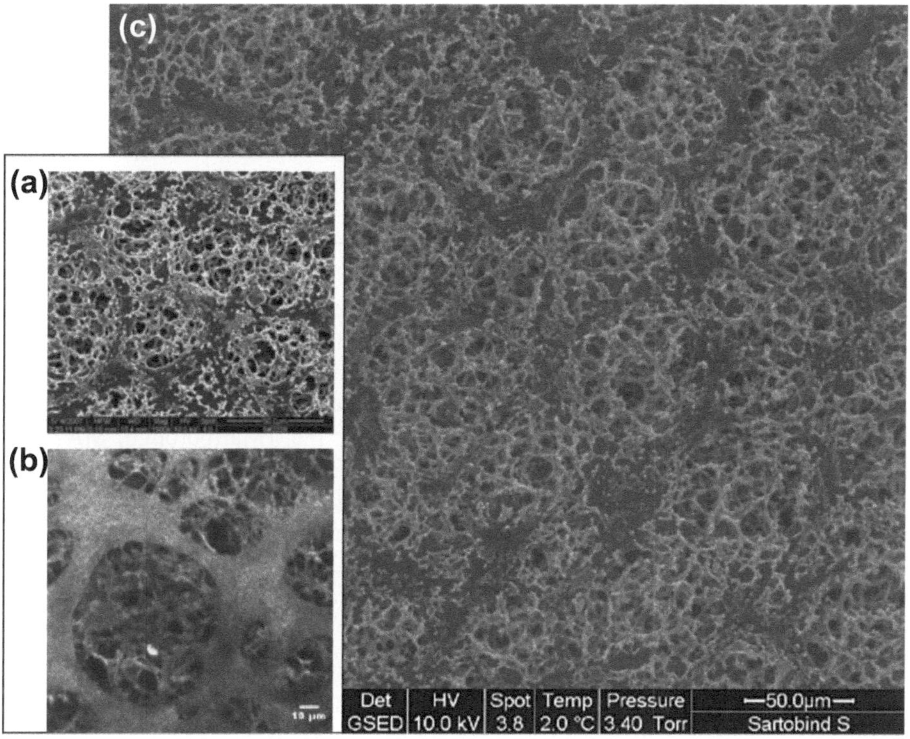

Figure 5.4 Morphology of membrane Sartobind S, bottom side: (a) SEM image of outer surface in dry state; (b) CLSM image in a depth of 10 μm 50 mM phosphate buffer; (c) ESEM image of outer surface at 2 °C and 3.4 Torr after equilibration in 10 mM phosphate buffer for 24 h [18]. This article was published in J. Membr. Sci., Vol. 320, J. Wang, F. Dismerb, J. Hubbuch, M. Ulbricht, Detailed analysis of membrane adsorber pore structure and protein binding by advanced microscopy, pp. 456–467, Copyright Elsevier (2008).

anion exchanger, S strong cation exchanger, and E surface-modified membrane. Non-woven polyethersulfone membranes were also prepared by electrospinning [21].

Polyamide membranes

Microporous polyamide membranes offer narrow pore-size distribution and mechanical rigidity. However, polyamide membranes have only a low concentration of terminal amino groups; therefore, the direct activation of the polyamide matrix leads to low ligand densities using metal chelators or triazine dyes [22]. Another problem of polyamide membranes is the non-specific adsorption of proteins. Immobilization of hydroxyethyl cellulose (HEC) onto polyamide membranes was demonstrated to exhibit low non-specific binding of proteins and proper binding capacities for affinity ligands [23].

Composite membranes

Other adsorptive membranes were prepared in a two-step procedure by grafting a copolymer onto an inert microporous membrane, followed by the introduction of the functional groups. The inert microporous membrane served as the mechanical support of a specific pore size and the copolymer served as a chemical functional group carrier. Some composite membranes for chromatography are described below.

A composite membrane of cellulose with grafted acrylic polymers was produced by developing a glycidyl methacrylate (GMA) polymer in the presence of dispersed cellulosic fibres, followed by an in situ covalent binding of the acrylic polymer to the cellulose [24–26]. The composite membrane thus formed consisted of a cellulosic core as the mechanical support and the acrylic sheath as a chemical functional group carrier. The composite membrane can be further activated to the required functional groups, such as sulfopropyl or dicarboxyl groups and metal chelators.

The composite membrane produced by Sepracor Inc. (Marlborough, MA) consisted of a blend of PES and polyethylene oxide (PEO) coated on all surfaces with a covalently bound layer of HEC [27,28]. The PES backbone provided physical strength and the HEC coating provided activable surface hydroxyl groups to which ligands might be covalently attached. Both the PEO and HEC conveyed low non-specific protein binding characteristics. The hydroxyl groups contained in the membrane-bound HEC might be further activated for immobilization of protein A by covalent linkage.

Adsorptive membranes were also prepared from hydrolysed polyglicidyl methacrylate (PGMA)–ethylene dimethacrylate (EDMA) copolymer grafted with PGMA oligomers [29–31]. PGMA–EDMA membranes are fragile and may cause non-specific adsorption of proteins. These defects were reduced for PGMA–EDMA membranes on which hydrophobic PGMA oligomers were grafted. The functional groups were further introduced into the flexible oligomer chain.

Radiation-induced graft polymerization can introduce a high density of grafted chains homogeneously onto polymeric materials by using an e-beam or γ-ray irradiation [32–35]. By applying the radiation-induced graft polymerization method, the polymer chains containing a sulfonic acid group or diethylamino group were grafted onto a porous hollow fibre membrane. These grafted polymer chains were found to function as tentacle polymer chains capable of holding the proteins in multilayers. Recently, Kumar et al. [36] used a mutual radiation-induced grafting technique with a Co-60 γ-radiation source to graft (vinylbenzyl)trimethylammonium chloride onto cotton cellulose. These authors found that the equilibrium binding capacity and elution percentage of BSA on the grafted anion-exchange matrix were 40 mg g^{-1} and 94%, respectively, from analysis of the breakthrough curve.

Photopolymerization is another graft polymerization method to produce ion-exchange membranes. Ulbricht and Yang [37] obtained a lysozyme binding capacity of 20 mg ml^{-1} using polypropylene membranes grafted with poly(acrylic acid) (PAA). Modification was done by photoinitiated, surface selective graft copolymerization. Capacities were increased to 60 mg ml^{-1} by changing the effective thickness and internal structure of the grafted PAA-based layer [38].

Atom transfer radical polymerization (ATRP) is also reported for the functionalization of commercially available regenerated cellulose membranes with weak cation-exchange groups [39–42]. Singh et al. [39] have used ATRP to modify polyvinylidene fluoride (PVDF) membranes with weak-base ion-exchange groups, and found that pore-size polydispersity decreased and ion-exchange capacity increased with increasing polymerization time. Surface-initiated ATRP was used to grow the cation-exchanger PAA on the surfaces of regenerated cellulose membranes (average pore diameter 1 μm) [40]. By varying the polymerization time, the thickness of the nanolayer was controlled. Static and dynamic binding capacities were measured for lysozyme on the PAA-modified regenerated cellulose membranes. Capacities initially increase with increasing modification times and reach maximum values of 98.5 mg ml^{-1} (static) and 71.2 mg ml^{-1} (dynamic). Using ATRP, Bhut and Husson [41] reported weak anion-exchange membranes with very high protein binding capacities (dynamic capacities up to 130 mg mL^{-1} for BSA) at high linear flow velocities (>350 cm h^{-1}) and low transmembrane pressure drop (<3 bar). Separation performance was evaluated by fractionating a mixture of BSA and haemoglobin. The purities of recovered proteins were high (>97%), as well as the recoveries (>98%).

Inorganic–organic membranes

Hybrid polymeric/ceramic membranes were recently proposed to perform the purification of proteins by metal affinity chromatography [43]. The ceramic support conferred to the membrane its excellent mechanical properties, especially attractive for industrial applications. Chitosan was chosen for coating the ceramic support because of its biocompatibility and non-toxicity properties.

Other inorganic–organic anion-exchange membranes were prepared by coating a precursor, the product of *N*-[3-(trimethoxysilyl)propyl] ethylene diamine reacted with 3-(triethoxysilyl)propyl isocyanate, on macroporous glass fibre and alumina membranes, followed by bromoethane treatment [44]. A similar method was used with alumina and Teflon plates as substrates [45,46]. Another example is the preparation of macroporous glass membranes from glass fibre filters via chemical cross linking and modification [47].

5.2.3 Membrane devices

The efficiency of membrane chromatography is critically dependent on membrane module design. Inlet fluid distribution and outlet fluid collection inadequacies result in broad breakthrough curves and poor utilization of membrane binding capacity. To solve these problems, new designs have been developed. For example, Ghosh and Wong [48] proposed the incorporation of fluid distributors/collectors on both the inlet and the outlet sides of the module. Membrane binding capacity utilization was shown to be significantly improved.

Commercial adsorptive membranes have usually a flat sheet format. In small membrane chromatographic units, membrane layers are stacked inside a circular holder. Scale-up of these cartridges is difficult because increasing the number of membrane layers results in an increase in the pressure drop [11]. At the same time, increasing the membrane diameter makes the fluid distribution a critical issue. For chromatographic membranes sold by Sartorius (Goettingen, Germany), scale-up is realized by wrapping membrane layers around a porous core. Further scale-up can be accomplished by combination of these units in series or parallel.

Chromatographic membranes were first developed in a hollow fibre configuration by Sepracor (USA). Hollow fibre membranes allow the design of devices with high surface area and high binding capacity. Scale-up is accomplished by increasing the number and length of the fibres. Hollow fibre chromatographic membranes are no more commercially available but they are still developed by different research groups [49,50].

5.3 Membrane activation

For specific applications, it may be of interest to prepare homemade adsorptive membranes. Some general rules are given below. The preparation of an adsorptive membrane is a two-step reaction, consisting of activation of the chemically inert membrane and coupling of the ligand to this modified support. The chemistry employed is dictated mostly by the nature and stability of the membrane, and by the ligand being immobilized. One can couple the ligand directly onto the membrane support, or to a spacer group, which immobilizes the ligand away from the matrix. Methods for activation of the membrane and coupling of the ligand are identical to methods used for packed columns of gel beads, which have been extensively reviewed [51,52]. Preparation of affinity membranes has been extensively reviewed by Klein [6].

5.3.1 Immobilization methods

Activation of membrane supports has been adapted from protocols used for chromatographic beads. Common activating reagents are carbonyl diimidazole (CDI), 2-fluoro-1-methyl pyridinium toluene-4 sulfonate (FMP), cyanogen bromide (CNBr), glutaraldehyde, cyanuric chloride (trichlorostriazine), *N*-hydroxy succinimide esters (NHS), 1-ethyl-3-(3-dimethylaminopropyl) carbodiimide (EDC) and epoxy. CDI, bisoxiranes, cyanuric acid, NHS and FMP are used to activate hydroxyl groups. CDI and EDC are used to activate carboxyl groups. Epichlorohydrine, glutaraldeide and epoxide are used to activate amino groups. Some data on matrix activation and coupling chemistry are given below. Further details will be found in the literature [3,4,6].

Carbonyl diimidazole activation

The CDI reagent can be used to activate hydroxyl-containing matrices. The imidazoyl carbonate groups introduced into the matrix react with amino-containing ligands with formation of carbamates. The activation has to be performed in organic solvent. The ligand coupling can be run in aqueous systems at high pH. Examples of membrane activation using CDI include immobilization of protamine on modifed nylon fibres [22] and immobilization of protein A and human IgG on poly(ether-urethane-urea) membranes [53].

2-Fluoro-1-methyl pyridinium toluene-4-sulfonate activation

The FMP reagent is used to activate hydroxyl-containing matrices in the presence of a slight excess of a tertiary amine [54]. The activated group reacted with nucleophilic groups (e.g. primary amines) on the ligand in a slightly alkaline aqueous solution (pH = 8–9), displacing the pyridinium moiety as 1-methyl-2-pyridone to form a stable, non-ionic secondary amine. The FMP method was used by Unarska et al. [55] for immobilization of protein A on nylon membranes.

Cyanogen bromide activation

Cyanogen bromide (CNBr) activation is suitable for all hydroxyl-containing support materials [56]. A matrix is reacted with CNBr at high pH (pH = 11–12). This leads to the introduction of cyanate-ester and imidocarbonate groups into the matrix. Amino-containing ligands are covalently linked to the activated matrix in aqueous medium at pH 7–8. The bonds formed between the ligand and the matrix are mainly of the isourea type. Examples of microporous membrane activation using CNBr include poly(L-lysine) immobilization on poly(ethylene vinyl alcohol)-coated polyethylene, cellulose diacetate and Hemophan hollow fibres [57].

Glutaraldehyde activation

The last example is glutaraldehyde activation, which is used with amine- or amide-containing support materials. Glutaraldehyde may react as a monomer, as a polymeric product of aldol condensation or as a trimeric product [58]. Therefore, different binding mechanisms are also possible: a primary amide may react with an activated double bond or by Schiff base formation. Glutaraldehyde reagent was used successfully to activate cellulose and acrylic composite membranes before coupling trypsin inhibitor and *p*-aminobenzamidine and recombinant protein A and protein G [25,59].

5.3.2 Spacer arms

A ligand might show poor function due to low steric availability. This rarely happens with high-molecular-weight ligands, but may occur with low-molecular-weight ligands. The use of a 'spacer arm' may solve this problem. The role of the spacer arm in protein A ligand coupling onto a composite membrane was reported by Hou et al. [25]. A spacer longer than 12 atoms appeared to offer favourable results at high ligand concentration, but the possibility of arm folding with such a long spacer led to a sharp reduction in capacity and should be avoided during the ligate purification. Thus, a spacer arm length of 11 atoms was chosen for ligand extension from the solid surface. Bueno et al. [60] also indicated that a longer spacer arm might render the ligand more accessible and facilitate adsorption for the separation of IgG from human serum using L-histidine affinity membranes. Today, grafted copolymer chains are more common in use than spacer arms to provide ligand accessibility as well as hydrophilicity and capacity.

5.4 Interaction mode

Adsorptive membranes are used for a variety of mechanisms, such as affinity, ion-exchange, hydrophobic and reversed-phase. Ion-exchange membranes have been investigated with strongly acidic (sulfonic acid), strongly basic (quaternary ammonium), weakly acid (carboxylic acid) and weakly basic (diethylamine) groups. Affinity membranes have been tested for a large range of ligands, such as immunoaffinity ligands, Proteins A and G, low-molecular-mass ligands (Cibacron Blue, histidine, tryptophan) and other ligands (peptide, Cu^{2+}).

5.4.1 Affinity chromatography

Affinity chromatography is one of the most selective methods of purification. This method relies on the attractive force between different molecules or functional groups, such as an antibody and an antigen, enzymes and substrates and receptors and ligands. The mobile phase contains the molecule to be purified and the stationary support contains a species (ligand) with preference to retain only the desired molecule (Figure 5.5).

Antibodies

Techniques of immunoaffinity bead chromatography were transferred to hollow fibre membrane devices. For example, immobilized antibodies were used in the industrial-scale production of three recombinant proteins: human interferon-a2a, interleukin-2 and interleukin-2 receptor [60,61]. These membranes had the ability to process proteins from dilute feed streams at residence times only in the order of seconds, with antigen breakthrough occurring near the saturation of the immunoaffinity membrane.

Protein A and protein G

Protein A and protein G are common ligands used for IgG purification [62]. Protein A is a cell wall protein component produced by several strains of *Staphylococcus aureus*. It contains four high-affinity binding sites capable of interacting with the Fc region from IgG. Protein G is also a bacterial cell wall protein isolated from group G *Streptococci*. Protein G contains two IgG-binding domains with higher affinity than protein A. Protein A/G combines the four Fc-binding domains from protein A and the two from protein G with the albumin-binding site of protein G eliminated, making it a very interesting tool for IgG purification. Protein A/G is a genetically engineered gene fusion product.

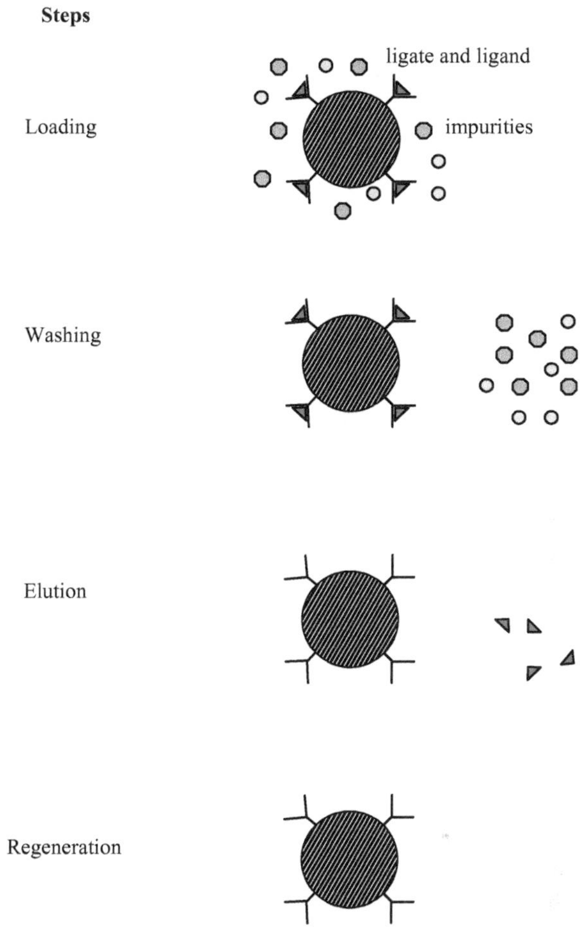

Figure 5.5 Schematic representation of the various steps during an affinity purification.

The first paper in the field of membrane chromatography was published by Brandt et al. [63] and is particularly interesting. Protein A hollow fibre devices were proposed for purification of fibronectin from blood plasma and purification of IgG. The high throughput rate and the efficient ligand use of this device permitted fast loading/elution cycles. Because the volume of a typical agarose affinity system was 100–1000 times that of the affinity membrane device, the membrane device required only about 0.1% as much ligand to handle the same throughput at the same mass transfer efficiency.

Protein A and protein G affinity membranes worked successfully at high speed as exemplified by an experiment in which 1 dm^3 of dilute hybridoma culture supernatant was processed [64]. In a single pass, at a high flow rate (150 cm^3 min^{-1}) and a low pressure drop (8.3×10^4 Pa), the IgG was purified in only 10 min. The highest binding capacities obtained with microporous membranes containing immobilized protein A were comparable to that attainable with bead matrices [27]. Immunoaffinity membranes also have potential use in medical therapeutic applications, that is, for the selective removal of human IgG from plasma or serum in vitro for clinical apheresis application [60]. Castilho et al. [65] prepared adsorptive membranes using various ligands, including protein A, peptide, histidine, thiophilic ligand and iminodiacetic acid (IDA) complexed with Zn(II). Protein A-based polysulfone and regenerated cellulose membranes showed the best selectivity for IgG purification from cell culture supernatant and stability throughout repeated adsorption–elution cycles. Recently, Boi et al. [66] characterized new protein A affinity membranes, developed by Sartorius, in batch and in dynamic conditions using pure polyclonal IgG solutions as well as a cell culture supernatant containing monoclonal IgG1.

Dyes and histidine

Pseudoaffinity ligands, such as dyes and histidine, are well-established in conventional packed column chromatography. Most dye ligands bind proteins non-specifically by a combination of electrostatic, hydrophobic, hydrogen bonding and charge-transfer interactions [67]. Some dyes can interact more specifically with the active sites of proteins by mimicking the structure of the substrates and cofactors or the binding agents of the proteins. About 30 dye-ligands have been reported for the separation and purification of various types of proteins. Among them, Cibacron Blue F3GA (CB) is a common ligand used for purification of many enzymes.

Dyes and histidine adsorptive membranes have been proposed in several studies. Cibacron Blue F3GA dye membranes allowed the purification of glucose-6-phosphate dehydrogenase from *Saccharomyces cerevisiae* within 10 min at a high filtration flow rate [68]. The binding capacities obtained were comparable with that attainable with packed beads [69,70]. A scale-up of two enzyme purifications (formate dehydrogenase from *Candida boidinii* and pyruvate decarboxylase from *Zymomonas mobilis*) with Cibacron Blue F3GA membranes and their incorporation in the recovery procedure were also presented [71]. Membrane modules with 135 cm^2 surface allowed the purification of 4.8-kU *formate dehydrogenase* and 480-kU *pyruvate decarboxylase* per hour. The flow rate for all chromatographic steps was 5 cm min^{-1} resulting in a cycle times of less than 5 min instead of several hours if packed columns were used.

Several studies have reported the variation of physicochemical parameters, such as pH, ionic strength, temperature, extraction time, initial protein concentrate and protein yield [72–74]. Su et al. [75] focused on the effects of operating parameters, such as initial pH, temperature and initial concentration for the adsorption of papain on the dye affinity membrane. The data obtained were analysed by the response surface methodology, a mathematical technique useful for designing experiments and analysing the effects of the several independent variables. Other recent studies include purification of protease by Yellow HE-4R, Red HE-3B and Cibacron Blue F3G adsorptive membranes [76], purification of lactoferrin from bovine whey and colostrum on Red HE-3B hollow fibres affinity membranes [77], and molecular studies of rheumatoid factor using pseudobioaffinity ligand L-histidine membrane chromatography [78].

Metal affinity chromatography

The immobilized metal affinity method uses chelating agents coupled on the supporting matrix to immobilize metal ions, which specifically interact with electron donating amino acid residues (such as histidine, cysteine, tryptophan, tyrosine, aspartic acid or glutamic acid) of the biomolecule [9]. IDA is the most popular chelating agent due to its lower price and availability. In addition to chelating agents, suitable metal ions are selected to ensure a stable immobilization and a high adsorption of the target biomolecule. When applying IDA as the chelating agent, the affinities to the retained biomolecules are usually in the following order: $Cu^{2+} > Ni^{2+} > Zn^{2+} \geq Co^{2+}$. Cu^{2+} and Ni^{2+} are the most commonly used metal ions in immobilized metal affinity chromatography.

Immobilized metal affinity membranes were reported to be applied in isolating or purifying enzymes, albumins, immunoglobulins, haemoglobin, ribonuclease and growth factors. Because of the affinity of histidine and tryptophan for the complexed Cu^{2+} ions, these two amino acids could easily be separated from solutions containing other amino acids [19]. Metal affinity chromatography allows proteins to be separated provided they bear histidine or tryptophan at the outer sphere. As an application study of metal affinity chromatography, a large volume of S-oxynitrilase from *Sorghum bicolour* was purified on immobilized IDA-Cu^{2+} ion membrane [79]. The IDA-Cu^{2+} membrane exhibited a capacity of 0.15 mg *S-oxynitrilase* per cm^2, which was slightly weaker than the capacity of a gel. The immobilized metal membrane allowed a 200-fold increase in flow rate compared with a packed column. In addition, polyhistidine tags (such as His6) are usually used for biospecies without directly accessible surface-exposed special residues. A successful example was the isolation of His6-tagged *Eco*RV, a recombinant fusion protein and an *Escherichia coli* (*E. coli*) restriction endonuclease, using an IDA-chelating immobilized metal affinity membrane adsorber from Sartorius with Ni^{2+} immobilized [80].

Recent applications include a novel purification process for the purification of recombinant adenovirus combining an anion-exchange chromatography using Q Sepharose XL resin and a Sartobind IDA membrane unit charged with Zn^{2+} ions as affinity [81]. The metal affinity membrane efficiently removed residual host cell impurities that co-eluted with adenovirus during the previous anion-exchange chromatography step and separated defective adenovirus particles from the infectious adenovirus fraction. Other recent examples include the purification of an IgG1 monoclonal antibody [82] and the purification of cell culture-derived influenza virus A/Puerto Rico/8/34 [83].

5.4.2 Ion exchange

Ion-exchange chromatography is a widely used chromatographic method for separation and purification of biomolecules [84]. This process involves the separation of biomolecules based on their ionic charge. After the desired anions are retained on the support, they are eluted out of the stationary phase by passing a buffer with a stronger anionic strength, commonly NaCl or NaOH. The rate of migration along the stationary phase is dependent on the charge density of the biomolecules, thus enabling their separation. The high salt concentration solution containing the desired product is further treated by diafiltration with water to remove the salt.

Anion- and cation-exchange membranes allowed the purification of monoclonal antibodies and human recombinant antithrombin III (rAT III) from cell culture supernatant. Purities of 80% for monoclonal antibodies and 75% for rAT III were achieved [85]. The capacity for monoclonal antibodies was found to be 1 mg cm^{-2} with recoveries up to 96% and that for rAT III was 0.15 mg cm^{-2} with recoveries up to 91%. Ion-exchange membranes were also used for the isolation at pilot scale of recombinant immunofusion proteins produced extracellularly by recombinant *E. coli* [86]. A complete cycle for processing 200 dm^3 at a flow rate of 1 dm^3 min^{-1} of clarified diluted fermentation broth (including equilibration, loading, elution and regeneration) was completed in less than 6 h with a membrane surface area of 1.12 m^2. Ion-exchange membranes have also potential use in efficient gradient elution of protein mixtures under elevated flow rate and mass loading conditions. High purity plasma fractions were obtained from 1 dm^3 of human plasma in a few hours, with a flow rate of 400 cm^3 min^{-1}, demonstrating the feasibility of ion-exchange membranes for large-scale protein fractionation [87].

The work of Belanich et al. [88] is an example of a successful application of ion-exchange adsorptive membranes in the pharmaceutical manufacturing. The scale-up of strong anion-exchange adsorptive membrane that removed endotoxin from bacterial extracts while preserving enzyme activity in the protein mixture was demonstrated. The endotoxin removal procedure was directly adapted from the small-scale Q-100 MA cartridge of Sartorius (100-cm^2 working surface area) to the large-scale Q-550 MA sheets (5500-cm^2 working surface area). The characteristics of endotoxin removal, protein adsorption and photolyase purification were similar in the two systems.

Other applications of ion-exchange membranes include the isolation of antibacterial peptides from lactoferrin [89], the purification of alphaviruses with a cation-exchange membrane [90], the adsorption of DNA with an anion-exchange membrane [91], high-speed protein purification by cation-exchange hollow-fibre cartridges [92], and the production and concentration of HIV-1-based lentiviral vectors using HYPERFlask vessels anion-exchange membrane chromatography [93].

5.4.3 Hydrophobic interaction and reversed phase

Hydrophobic interaction chromatography is a less widely used method in the purification of biomolecules compared to ion-exchange chromatography [84]. This technique involves passing a mobile phase across a stationary phase immobilized with hydrophobic interaction ligands, causing hydrophobic species in the mobile phase to interact with that on the stationary phase. This process is usually used for the purification of proteins. It can also be applied to the separation of nucleic acids, viruses, cells and carbohydrates. The retained molecules are eluted in increasing order of hydrophobicity. The biomolecule separation depends on the hydrophobicity of their surface and density, as well as the type of hydrophobic ligands attached to the surface of the support.

Hydrophobic interaction membrane chromatography was used for the separations of model mixtures of proteins on membranes modified with C4 and C8 groups [29]. The separations were similar to those obtained using a packed column. The advantage of adsorptive membranes was that the pressure used was lower by as much as two orders of magnitude than that required to reach the same flow rate in a packed column, although a high binding capacity was still achieved. Separations obtained with a continuous linear gradient were used to design a stepwise gradient profile, which decreased both time and mobile phase in the separations of proteins [94].

In a recent application, Wang and Ghosh [95] described a hydrophobic interaction membrane chromatography-based method for rapid and efficient separation and analysis of monoclonal antibody aggregates. This method was shown to be able to resolve Campath-1H monomer, dimer, trimer, tetramer and pentamer. Other applications of hydrophobic interaction membrane chromatography include the separation of model proteins (CAMPATH-1G monoclonal antibody and BSA) [96], the purification of humanized monoclonal antibody [97] and the fractionation of human plasma proteins [98].

5.4.4 Multistage chromatography

Combinations of different chromatographic supports are often needed to achieve higher degrees of selectivity. Therefore, combinations of affinity, ion-exchange, hydrophobic interaction and reversed-phase chromatography have been proposed for various purification purposes. For example, a sequence of cation-exchange, dye-ligand and anion-exchange membranes enabled the purification of formate dehydrogenase from *Candida boidinii* [68]. The sequence of ligands and the buffer conditions were chosen in such a way that the fraction eluted from one membrane could be directly loaded on the next membrane. By using a sequence of ultrafiltration (UF), diafiltration, Cibacron Blue, anion exchange and heparin membrane adsorber, an electrophoretically pure recombinant human Antithrombin III from hamster cell culture supernatants was obtained [99]. Purities of 94% were reached, even in the presence of high serum concentrations.

Another example is the purification of humanized monoclonal antibody expressed in tobacco juice by ion exchange followed by hydrophobic interaction membrane chromatography [100,101]. Ion exchange was used for capture and purification, hydrophobic interaction membrane chromatography for high-resolution purification, followed by UF for polishing, desalting and buffer exchange. Using this scheme, both high monoclonal antibody purity and high recovery (77% of monoclonal antibody spiked into the tobacco extract) were achieved. Cross-flow filtration may be used as a pretreatment of the feed solution, as reported for the purification of plasmid DNA from *E. coli* ferments using anion-exchange membrane chromatography as a second step [102].

5.4.5 Hybrid processes

In some recent processes, membrane chromatography is closely linked to other techniques. The combination of sieving and hydrophobic interaction-based membrane chromatography using the same membrane filtration technique was used for purification of humanized monoclonal antibodies from mammalian cell culture supernatant [103]. A fraction of the monoclonal antibodies in the feed sample was precipitated using high ammonium sulfate concentration; the precipitates were then retained by sieving using a membrane. The membrane also served as a selective adsorption media and bound the monoclonal antibodies remaining in solution by hydrophobic interaction. The retained monoclonal antibodies were subsequently released by reducing the ammonium sulfate concentration, which resulted in the simultaneous dissolution of the precipitated monoclonal antibody fraction and elution of the membrane bound fraction.

The mixed matrix membrane (MMM) is another example of a hybrid membrane system, including membrane chromatography. The MMM incorporates adsorptive particles during membrane casting and is reported to have performances competitive with other adsorptive membranes. Wessling and co-workers prepared MMM using an ethylene vinyl alcohol-based membrane with several types of Lewatit-type cation- and anion-exchange particles. They successfully applied these adsorptive membranes to the separation of BSA, lysozyme and bilirubin [104–106] and to the fractionation of β-lactoglobulin from bovine whey [107]. Hollow fibre membranes were prepared according to the same dry-wet phase inversion process [108]. A solution containing dissolved polymers and dispersed Lewatit-type cation-exchange particles was spun into a microporous fibre via a phase inversion process. The prepared membranes contained ion-exchange particles tightly held together within the polymeric matrix (Figure 5.6). The hollow fibres were connected in series to increase the effective thickness and the protein residence time within the module. The separation of lysozyme was carried out, showing high static and dynamic adsorption capacities of more than 60 mg ml^{-1} membrane.

5.5 Applications

Membrane chromatography is currently being employed on both laboratory and industrial scales for the purification and polishing of different biomolecular species. In the following examples, some applications are detailed, including purification of monoclonal antibodies and DNA, and virus capture.

5.5.1 Monoclonal antibodies

Among the different methods available for monoclonal antibody purification, protein A affinity chromatography is one of the most employed method despite its cost and is generally included in the purification platforms developed at industrial scale [11]. Substantial efforts are currently being devoted to the research of suitable alternatives to protein A affinity chromatography and membrane adsorbers are among the

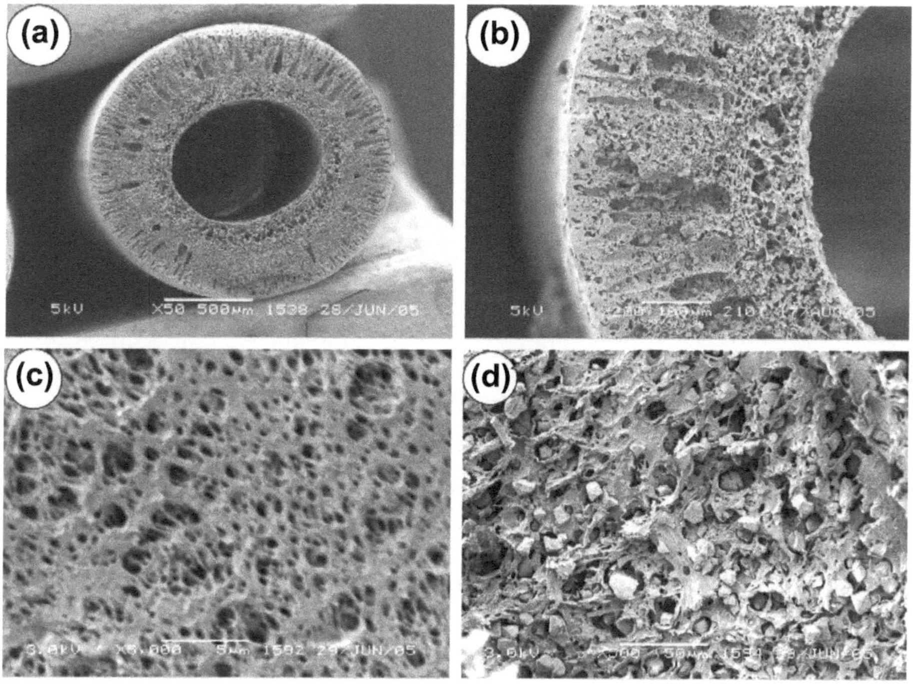

Figure 5.6 SEM micrographs of the mixed matrix hollow-fibre adsorbers prepared by incorporation of 50% Lewatit CNP80WS ion-exchange particles into a PES-based polymeric support: (a) cross-section, magnification ×50, the white horizontal bar indicates 500 μm; (b) cross-section, magnification ×200, the white horizontal bar indicates 100 μm; (c) outer surface, magnification ×5000, the white horizontal bar indicates 5 μm; (d) inner surface, magnification ×500, the white horizontal bar indicates 50 μm [108].
This article was published in J. Membr. Sci., Vol. 322, M.E. Avramescu, Z. Borneman, M. Wessling, Particle-loaded hollow-fiber membrane adsorbers for lysozyme separation, pp. 306–313, Copyright Elsevier (2008).

processes considered. Different affinity membranes have been studied for the separation of monoclonal antibodies and IgG, indicating several advantages of the method [27,64,65,109].

Recent studies include development of new affinity membranes with improved properties. For example, a new protein A affinity membrane, developed by Sartorius, was shown to have high binding capacity, significantly improved with respect to the commercially available supports [66]. The membranes were characterized in batch and in dynamic conditions using pure polyclonal IgG solutions, as well as a cell culture supernatant containing monoclonal IgG1. Experiments were performed with the cell culture supernatant, which was used after a prefiltration step through a 0.2-μm low protein adsorption membrane filter to remove aggregates and precipitates that could have been formed during storage. SDS-PAGE gel electrophoresis was performed on samples obtained from the experiment illustrated in Figure 5.7a. From the

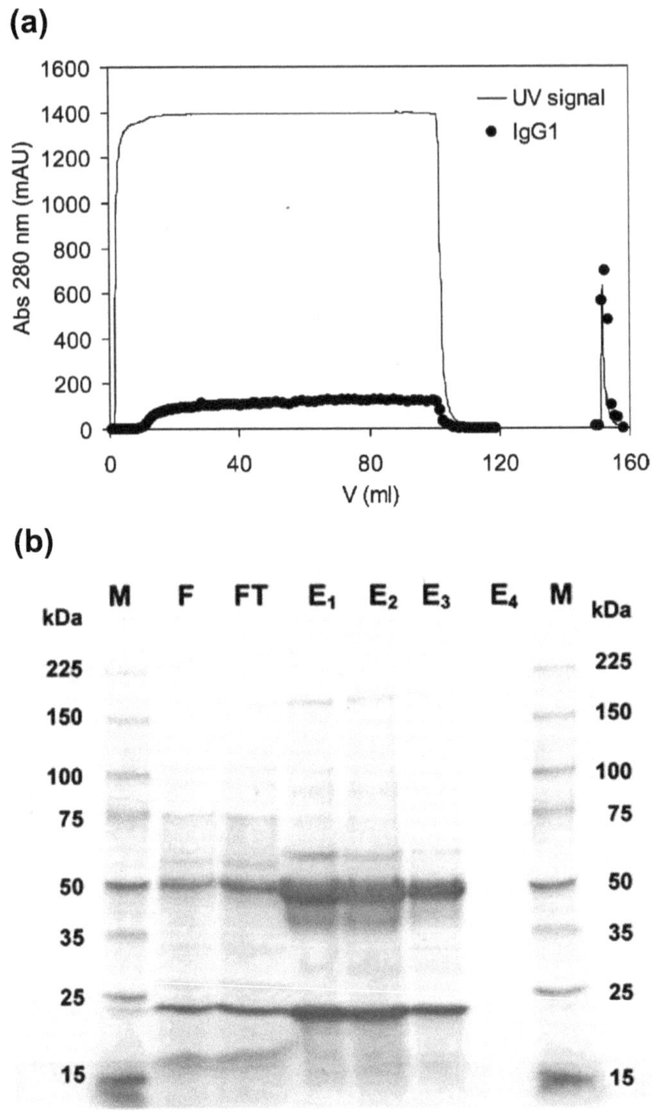

Figure 5.7 (a) UV signal of a typical cell culture supernatant during the capturing step and the corresponding IgG_1 profile obtained with HPLC protein A analysis of the collected fractions [66]. (b) Results of SDS-PAGE electrophoresis under reducing conditions of dynamic experiments with the cell culture supernatant.

M: markers, F: feed, FT: flow through at 100% BT, E_1: eluted fraction collected at 153 mL, E_2: eluted fraction collected at 154 mL, E_3: eluted fraction collected at 155 mL, E_4: eluted fraction collected at 158 mL [66]. Original in colour.

gel shown in Figure 5.7b, it can be observed that the membranes were selective for IgG. The concentration of the fractions collected at the centre of the peak (E1, E2 and E3) were higher than the concentration of the peak tail (E4), in which the bands of the heavy and light chains of IgG could be barely detected.

Combination of affinity membranes and ion-exchange may also be used, such a cation-exchange and protein A membranes for the purification of an anti-*Pseudomonas aerugenosa* human IgG1 monoclonal antibody from transgenic tobacco [100]. In that order, both high purity and recovery were achieved with high permeability. The protein A membrane alone showed severe fouling tendency and generated high back pressure. Using cation-exchange membrane chromatography as a first step, minimal membrane fouling and high permeability were obtained.

5.5.2 DNA

Large quantities of plasmid DNA are demanded for cloning, large-scale protein production and DNA vaccine/gene therapy applications. Purification of large quantities of plasmid DNA is a particularly difficult issue because of the relatively low capacity, flow rate and purity obtained using bead chromatography. It is generally admitted that membrane chromatography affords greater capacity for DNA than analogous bead chromatography. For example, Mustang-Q anion-exchange membranes (Pall, USA) showed a maximum binding capacity of 10 mg plasmid ml^{-1} membrane volume, an order of magnitude greater than typical values reported for porous beads [110]. Endres et al. [111] evaluated the performance of anion-exchange membranes for the purification of pDNA from *Escherichia coli* lysate solution. The dynamic binding capacity for pDNA was 20–25 times greater, and the flow rate was 55–550 times greater, than values observed for beads, demonstrating the advantages of ion-exchange membranes for the purification of DNA. Chang et al. [44] reported the separation of plasmid DNA and RNA from the feed of 10-μg plasmid DNA + 10-μg RNA mixture or cell lysate by a modified glass fibre membrane or a SB6407 adsorptive membrane (Pall, USA). The overall plasmid DNA recovery for the modified glass fibre membrane was found to be between 98% and 106%, higher than that of SB6407 membrane (91–96%). Zhang et al. [112] reported plasmid DNA purification from large volumes of clarified *E. coli* lysate using a 260-ml anion-exchange membrane (Pall, USA). The eluted plasmid DNA was concentrated 10-fold with a reduction in endotoxin of four orders of magnitude. The capture of 1.52 g of plasmid DNA was performed at a flow rate of 16 membrane vol. min^{-1}. In comparison, anion-exchange resins can bind about 1 mg of plasmid DNA ml^{-1} of resin at flow rates that are typically lower than 0.5 column vol. min^{-1} [112].

In addition, ion-exchange membranes were able to resolve various isoforms of DNA, surpassing the capabilities of analogous beads [113]. In contrast to purification of smaller biomolecules such as proteins and antibodies, the dynamic capacity for DNA was found to be highly dependent on flow rates and concentrations [113]. Increasing flow rates induced DNA elongation, which increased the surface coverage and, therefore, lowered the capacity (Figure 5.8).

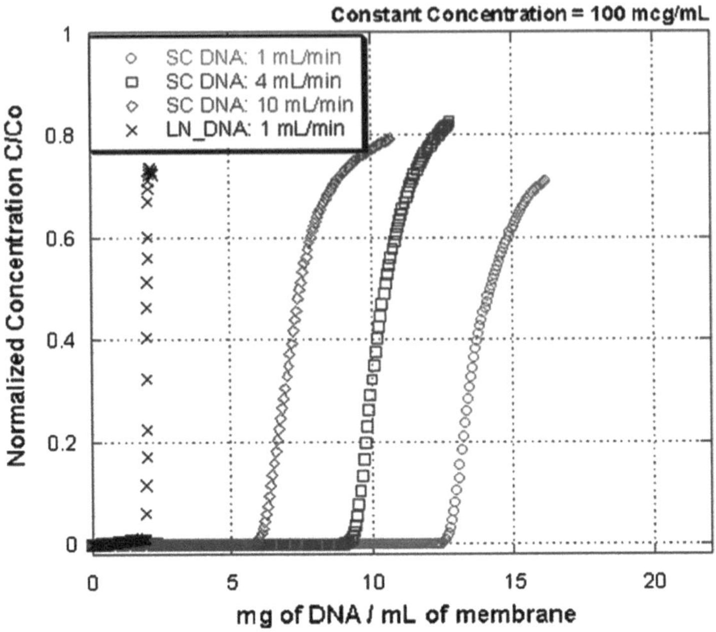

Figure 5.8 Dynamic loading capacity of membrane module at three different flow rates: 1, 4 and 7 mL/min at constant DNA concentration of 100 mcg/mL. Capacity is reduced due to the flow-induced extension of DNA. λ-Phage DNA, a long and linear DNA confirms this hypothesis (mcg/mL = μg/mL) [113]. Original in colour.
This article was published in Biotechnol. Bioeng., Vol. 88, C. Haber, J. Skupsky, A. Lee, R. Lander, Membrane chromatography of DNA: Conformation-induced capacity and selectivity, pp. 26–34, Copyright John Wiley and Sons (2004).

Advantages of adsorptive membranes for purification of large biomolecules such as DNA may be summarized as follows. Porous beads may see capacity improvements by decreasing the particle diameter, which comes at the expense of very large pressure drops, or by increasing the size, which leads to significant mass transfer limitations as the large biomolecules must diffuse through the stagnant fluid. Adsorptive membranes do not have these limitations, as the large convective through pores are accessible to large biomolecules, diffusional limitations are minimal, and the high porosity, large cross-sectional area and minimal thickness allow operation at low pressure drops. In order to increase production capacity, it is possible to use a larger membrane area and/or add devices in parallel or in series. Therefore, the method may allow an efficient capture step for plasmid DNA on the kilogram scale with high throughputs [112].

5.5.3 *Viruses*

Large-scale purification of virus vectors is needed in gene therapy and vaccine production. In addition, validation of virus clearance is critical in the manufacture of biopharmaceuticals. Viral particles are essentially nucleic acids inside a protein capsid, which is surrounded by a lipid bilayer and glycoproteins. Their structure gives the particle a charge allowing them to be separated by ion-exchange chromatography.

Ion-exchange membrane chromatography displaces potential viral clearance. Several studies have demonstrated the potentiality of Q membrane chromatography for viral clearance in monoclonal antibody production [114,115]. The charge-based Q membrane can efficiently remove different model viruses at high flow rates in a broad range of operational pressures [10]. The viral clearance capacity of these supports was as high as 3000 g m^{-2} membrane area. A recent study also demonstrated the purification of rotavirus-like particles by anion-exchange membrane chromatography as a step in the overall downstream processing [116] (Figure 5.9).

Peixoto et al. [117] proposed a process for purification of adenovirus vectors from cell-culture bulks based only on membrane devices. The process was divided into three parts: (1) clarification by MF and concentration by cross-flow UF, (2) purification using anion-exchange membrane adsorbers and (3) concentration and buffer exchange for final delivery formulation. UF membranes with molecular weight cutoffs of 300, 500 and 750 kDa were shown to concentrate the virus and to remove 90% of host cell DNA and proteins in the retentate. Two membrane adsorbers (Sartobind Q and Sartobind anion direct) were evaluated for adenovirus vectors capture and purification. Sartobind anion allowed higher recovery yields (up to 62%) of infectious adenoviruses than Sartobind Q. Identical ratios between total and infectious adenoviruses were achieved with both membrane adsorbers. The overall recovery yield of the process was 52%.

Other examples include the adsorption of *Aedes aegypti* densonucleosis virus (AeDNV) particles using Sartobind (Sartorius) anion- and cation-exchange membranes [118,119]. AeDNV is a mosquito-specific parvovirus (non-enveloped, single-stranded DNA virus). The solution pH was adjusted to maximize virus binding and minimize binding of other charged species present in the feed suspension. AeDNV particles were adsorbed by anion- and cation-exchange membranes providing the virus particles and membranes were oppositely charged. Strongly basic anion-exchangers Q membranes were found to be the most effective. Purification of two aplaherpes virus from cell culture supernatants was reported using a cation-exchange membrane (Sartorius) [90]. Over 85% of the virus was eluted within a single fraction and specific infectivity of the resulting virus preparation was over 10-fold higher than that of sucrose gradient-purified viruses. Human and equine influenza A virus in cell culture supernatant was adsorbed to Sartobind Q and D anion-exchange membranes [120]. Elution of adsorbed virus from Sartobind Q and Sartobind D resulted in average yields of 86% and 38%, respectively. A productivity of 67 L of culture broth m^{-2} h^{-1} was achieved and the average virus yield was 72%. The authors concluded that Sartobind Q anion-exchanger can be a valuable method for the large-scale purification of cell culture-derived influenza virus, due to its high productivity, ease of operation and acceptable yield.

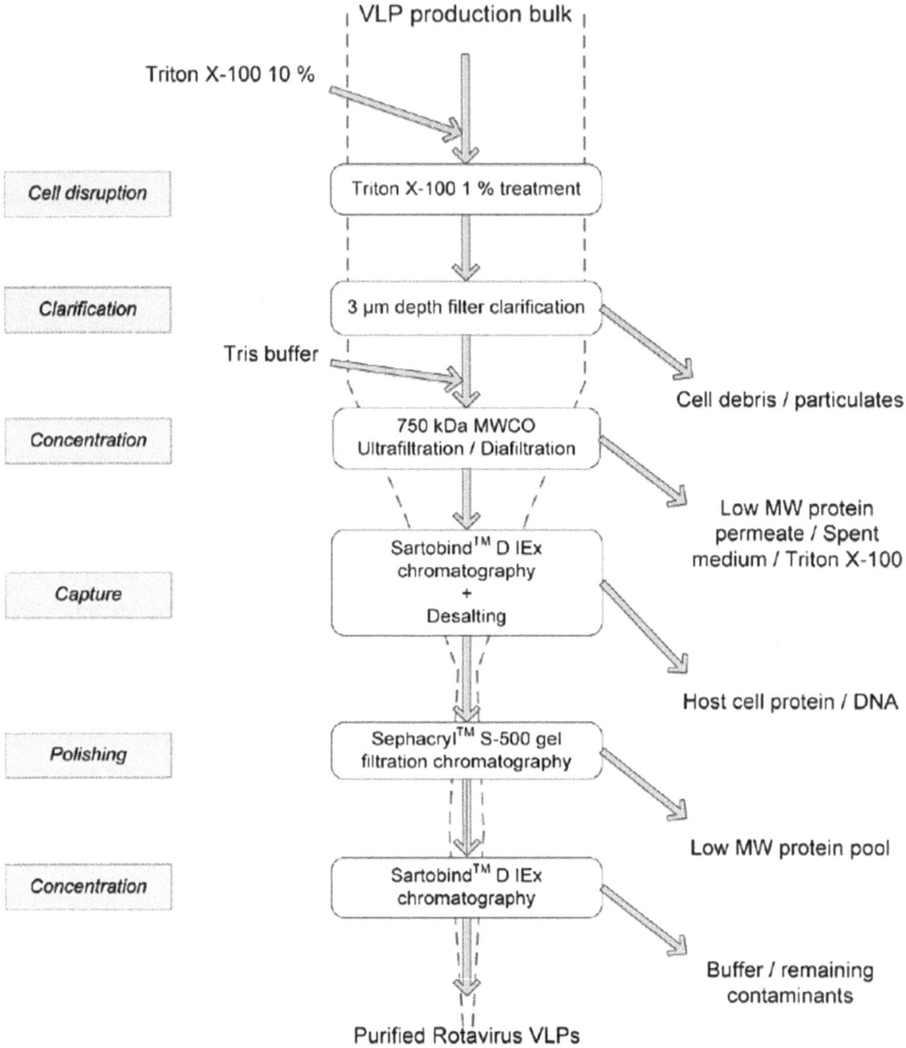

Figure 5.9 Flow diagram for the integrated use of anion-exchange membrane chromatography in the downstream processing of rotavirus-like particles.
VLP: virus-like particle, IEX: ion-exchange [116].
This article was published in J. Membr. Sci., Vol. 311, T. Vicente, M.F.Q. Sousa, C. Peixoto, J.P.B. Motab, P.M. Alves, M.J.T. Carrondo, Anion-exchange membrane chromatography for purification of rotavirus-like particles, pp. 270–283, Copyright Elsevier (2008).

5.6 Monolith chromatography

5.6.1 Introduction

Monolith chromatography has been presented as a variant of membrane chromatography, although it has now got its own specifications. Technical difficulties may arise when membranes with large surfaces are used, especially with regard to feed distribution in the device. Monoliths may solve this limitation, as they have been developed with varying thickness, thin disks as well as monolithic columns of various lengths [121]. In addition, a cylinder-shaped monolithic column may be used for scaling-up the system. The first polymethacrylate monoliths were described by Tennikova and Švec, who pointed out the advantages of convective mass transport in continuous bed support with a short length column [29,122]. From the middle of the 1990s, the interest for monolith chromatography has increased strongly.

A monolith is a single piece of highly porous material characterized by a highly interconnected network of channels with a diameter in the range of 10–4000 nm or even larger. Therefore, mass transport in monoliths is mainly based on convection. For the optimal separation performance of high-molecular-weight biomolecules, the chromatographic monolithic column is short. This enhances the speed of the separation process and reduces the backpressure, unspecific binding and product degradation, without reducing resolution. Their preparation is another advantage [123]. In contrast to particle preparation, where particle-size classification is required after polymerization is completed, monoliths are prepared with a bulk polymerization without further processing. Their structures are defined by the monomer composition and polymerization temperature.

Many reviews are available on specific aspects of monolithic chromatography. For example, Josić and Buchacher [121] focused on applications of monoliths as supports for affinity chromatography and fast enzymatic conversion. Josić et al. [13] discussed monoliths as stationary phases for separation of proteins and polynucleotides and enzymatic conversion. Mallik and Hage [124] focused on monoliths in affinity chromatography. Immobilization of ligands was considered including methods based on covalent immobilization, biospecific adsorption, entrapment, and the formation of coordination complexes. Guiochon [125] reviewed the origin of the concept, the differences between their characteristics and those of traditional packed columns, their advantages and drawbacks, the methods of preparation of monoliths of different forms, and the current status of the field. Chambers et al. [126] discussed the use and potential of both silica and polymeric monoliths in ion chromatography. Jungbauer and Hahn [127] focused on polymethacrylate monoliths for preparative and industrial chromatography. They discussed the pressure drop characteristics, mass transfer properties, scale-up and applications of monoliths for conventional chromatography. Vlakh and Tennikova [128] presented HPLC applications of polymethacrylate-based monoliths according to the mode of chromatographic separation (ion-exchange chromatography, reversed-phase chromatography, hydrophobic interaction chromatography and hydrophilic interaction chromatography).

5.6.2 Materials

Some commercially available monoliths are listed in Table 5.2 [13]. Figure 5.10 compares the structures of some of them [127]. Although their structures have different morphologies, their flow characteristics are similar. The channels are highly interconnected and all macropores are reached by convective flow. The difference is in the channel diameter, determining the range of the biomolecules, which can be separated. The silica monolith has also a skeleton with mesopores (Figure 5.10c). For small molecules, this monolith has a similar separation characteristic as a conventional chromatography column packed with a porous material [127]. However, the pressure drop is lower due to the higher bead porosity.

Monoliths have been prepared from synthetic polymer (polymethacrylate, polyacrylamide and polystyrene), natural polymer (agarose and cellulose) and silica. Polymer monoliths are produced by polymerization of organic monomers, including cross-linkers. The macroporosity of these materials is achieved by addition of porogenic solvents or pore-forming reagents, e.g. polymer solutions, to the polymerization mixture. Currently, a wide range of monomers are used for the synthesis of polymer monoliths. The most widely investigated are polyacrylamide [129], and polymethacrylate monoliths [123,128,130]. Polymethacrylate-based monoliths are polymerized from GMA and EDMA in the presence of porogens and an initiator. The polymer is chemically and mechanically very stable and contains epoxy groups that can be further modified to prepare the separation units suitable for ion exchange, hydrophobic interaction, reversed-phase or affinity separations. GMA-EDMA monoliths are available under the trademark Convective Interaction Media (CIM®,

Table 5.2 Commercially available monoliths [13].

Monolith	Manufacturer	Type, material and pore-size interaction mode	Effective filtration area or membrane bed volume
CIM	BIASeparations	Methacrylate, 1500 nm DEAE, QA, S, CM, EDA, C4, C2, epoxy, affinity	0.34-ml disk, 8-ml, 80-ml tube
UNO	Bio-Rad	Acrylate, estimated 1000 nm, Q, S	Columns 0.2, 1, 6, 12 ml
Chromolith	Merck	Silica, macropore 2000 nm, Mesopore 13 nm, C_{18}	Columns 50 × 4.6 mm^2
Seprasorb	Sepragen	Cellulose, 50–300 μm DEAE, QA, CM, S	Cartridges (10 ml, 2.5 × 2.0 cm^2)
Conchrom	Conchrom	Silica, mesopores 30 nm, Micropores 5 nm, RP 18, RP 8, ion-exchange	Different shapes

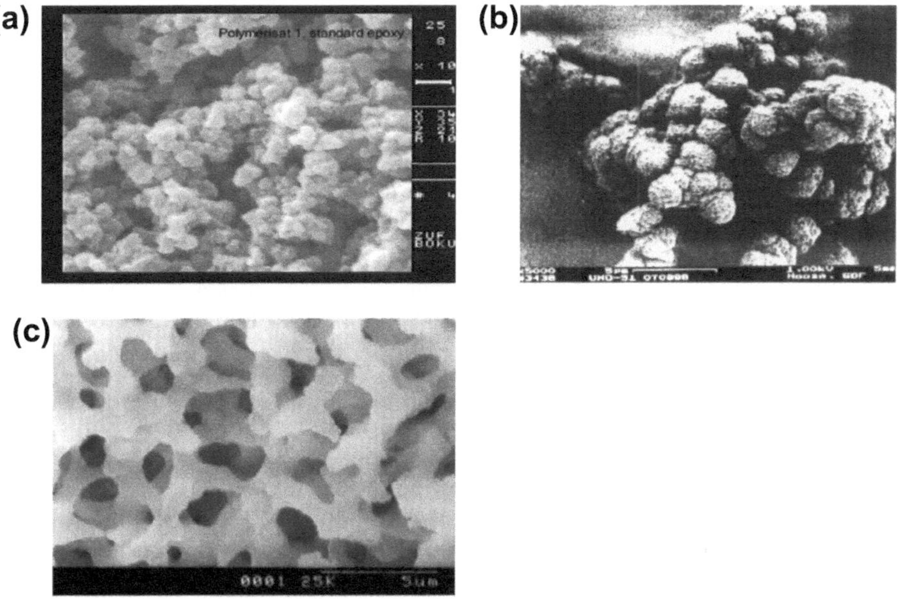

Figure 5.10 Electron micrographs of monoliths: (a) polymethacrylate (CIM disks and tubes, BIASeparations), (b) polyacryamide (UNO column, Bio-Rad), (c) silica-based monoliths (Chromolith, Merck) [127].
This article was published in J. Chromatogr. A, Vol. 1184, A. Jungbauer, R. Hahn, Polymethacrylate monoliths for preparative and industrial separation of biomolecular assemblies, pp. 62–79, Copyright Elsevier (2008).

BIA Separations, Ljubljana, Slovenia). Two geometries are available: disks that offer diameter higher than length and rod monoliths with length higher than diameter.

Silica monoliths have homogenously spaced and sized through-pores, providing a surface area larger than polymer monoliths [126]. Commercially available silica monolith columns (Merck) have macropores and mesopores of 2 μm and 13 nm, respectively. The total porosity is greater than 80% and the surface area is about 300 $m^2\ g^{-1}$. Compared to polymer, silica phases are less resistant to pH and temperature. At pH $>$ 8, silica dissolves, resulting in column packing instability, poor reproducibility, poor efficiencies, poor peak shapes and high back pressures. In addition, silica monoliths have the problem of cracking and shrinking after being prepared.

5.6.3 Applications

Monoliths with different modes, such as ion-exchange, hydrophobic interaction, hydrophilic interaction, reversed phase, have been used in the separation of biomolecules in analytical as well as in preparative scale [126,131]. Monoliths have also been used for affinity chromatography, immunoaffinity chromatography, immobilized metal ion affinity chromatography, dye ligand affinity chromatography and biomimetic chromatography [124,129,132] for the isolation of peptides, proteins, enzymes, viruses and plasmid DNA, as well as antibodies and antigens [133,134]. For example, Mönster et al. [134] demonstrated the separation of BGA proteins from cell culture supernatant on a CIM tube. The term 'blood group antigen' (BGA) refers to a set of proteins that are found on the surface of human red blood cells or platelets. The isolation of the three proteins – Knops, JMH and Scianna – by the CIM support is shown in Figure 5.11. The chromatogram verified that the binding and elution of the three proteins was specific. In the three cases, contaminating proteins were successfully removed and therefore no other peaks could be detected in the elution fractions.

Since large channels can easily accommodate large plasmid DNA, the CIM monoliths may have a very high binding capacity for these biomolecules [135–137]. CIM anion-exchange monolithic columns have been successfully used for the industrial-scale purification of pharmaceutical grade plasmid DNA of size up to 10 kb [136]. Other studies showed that monolithic supports were also suitable for the purification of very large plasmids with up to 93 kb. For example, Krajnc et al. [137] obtained up to 13 mg of plasmid per mL of the monolithic column capacity by varying salt concentration in loading buffer. The separation of linear and supercoiled isoforms

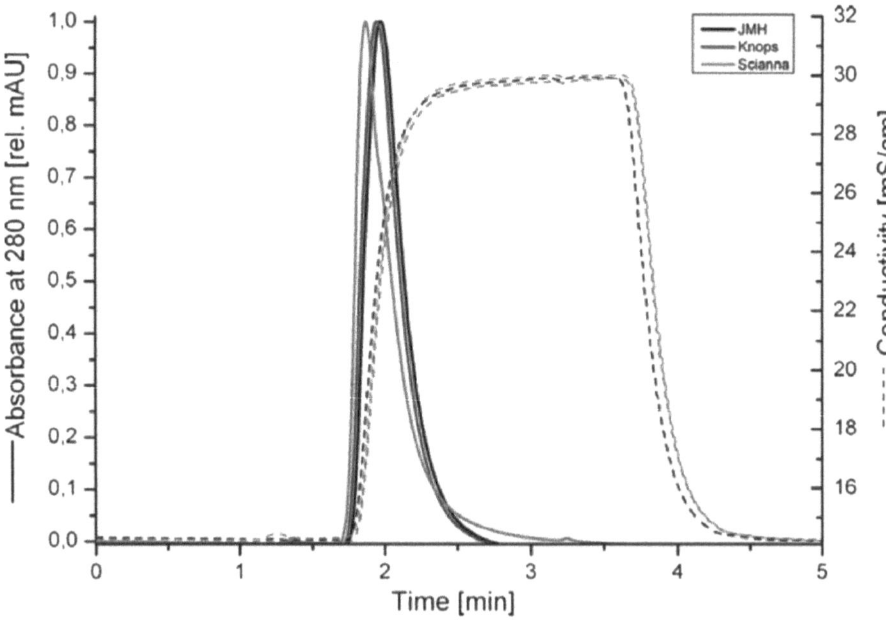

Figure 5.11 Separation of BGA proteins from cell culture supernatant on CIM Tube® at a flow rate of 16 mL min^{-1}. BGA proteins were eluted by 250 mM NaCl in 50 mM NaPP, pH 11.5 [134].
This article was published in J. Chromatogr. A, Vol. 1218, A. Mönster, O. Hiller, D. Grüger, R. Blasczyk, C. Kasper, Isolation and purification of blood group antigens using immuno-affinity chromatography on short monolithic columns, pp. 706–710, Copyright Elsevier (2011).

of large plasmids was obtained. In addition, the one-step purification on larger CIM DEAE 8-mL tube monolithic column was performed successfully.

Another successful application is the separation of viral particles by ion-exchange monolith chromatography. For example, Branovic et al. [138] used CIM DEAE disks for the concentration of measles and mumps viruses from virus vaccines containing attenuated viruses and some preservatives. The ability of monolithic columns to concentrate highly diluted plant viruses was also tested with a model plant virus, a rod-shaped tomato mosaic virus [139,140]. The virus was concentrated by several orders of magnitude in one-step procedure using a strong anion-exchanger CIM QA disk. The method could be applied to concentrate diluted virus samples before quantitative analysis.

5.7 Theory

The theoretical description of breakthrough curves is obtained by a mass balance over a section of the membrane. Kinetic equations, mixing and dead volume effects are then added to the model.

5.7.1 Mass balance

The model used to describe affinity membrane breakthrough curves is shown in Figure 5.12 [141]. It considers the isothermal adsorption of a single species during laminar flow through a membrane onto which a ligand is immobilized. The feed species concentration is noted c_0. The species solution has a concentration $c(z,t)$ with an axial diffusion coefficient D. It flows with a constant interstitial flow velocity v through a membrane of thickness L and void porosity ε. The accessible total ligand capacity in the solid phase is c_l, and the concentration of species–ligand complex in the solid phase is $c_s(z,t)$.

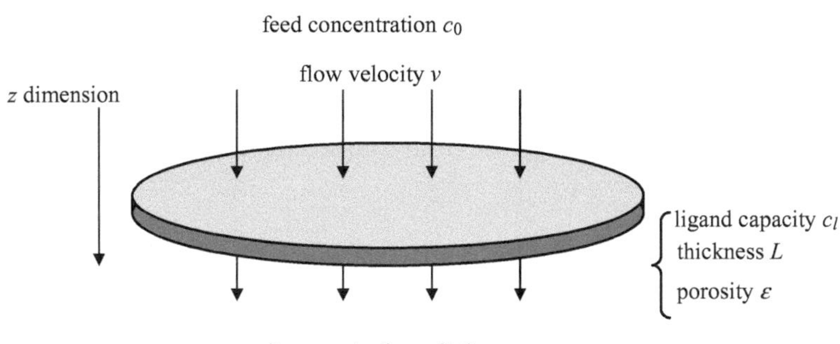

Figure 5.12 Schematic drawing of the mass transport through a flat chromatographic membrane.

A mass balance over a membrane section gives a continuity equation, which includes the variation of species concentration with time, convective flow, axial diffusion and association rate:

$$\varepsilon \frac{\partial c}{\partial t} + \varepsilon v \frac{\partial c}{\partial z} = \varepsilon D \frac{\partial^2 c}{\partial z^2} - (1-\varepsilon)\frac{\partial c_s}{\partial t} \tag{5.2}$$

Most models developed for membrane chromatography use the reversible second-order rate expression of the Langmuir isotherm type:

$$\frac{\partial c_s}{\partial t} = k_a c(c_l - c_s) - k_d c_s \tag{5.3}$$

where k_a and k_d are the association and dissociation rate constants, respectively. Analogous to packed-bed chromatography models, the ligand capacity, c_l, is based on the volume of the solid phase. It can be calculated from data reported by manufacturers: the void porosity, ε, and the binding capacity par membrane volume, $(1-\varepsilon)c_l$.

The initial conditions are set such that there is no species in the membrane:

$$c = 0 \quad \text{at } z \geq 0,\ t = 0 \tag{5.4}$$

$$c_s = 0 \quad \text{at } z \geq 0,\ t = 0 \tag{5.5}$$

In order to include axial diffusion at the entrance of the membrane and mixing at the exit, the Danckwerts' boundary conditions are used:

$$\varepsilon v c - \varepsilon D \frac{\partial c}{\partial z} = \varepsilon v c_0 \quad \text{at } z = 0,\ t > 0 \tag{5.6}$$

$$\frac{\partial c}{\partial z} = 0 \quad \text{at } z = L,\ t > 0 \tag{5.7}$$

In general, axial diffusion should be considered at the entrance of the membrane. When axial diffusion is negligible, that is, D approaches zero, equation (5.6) reduces to $c = c_0$ at the entrance (step input).

5.7.2 Dimensionless equations

Equations 5.2–5.7 can be rewritten using the following dimensionless numbers:

$$\tau = \frac{vt}{L} \quad \zeta = \frac{z}{L} \quad C = \frac{c}{c_0} \quad C_s = \frac{c_s}{c_l} \tag{5.8}$$

$$\text{Pe} = \frac{vL}{D} \quad m = \frac{(1-\varepsilon)c_l}{\varepsilon c_0} \tag{5.9}$$

$$n = \frac{(1-\varepsilon)c_l k_a \text{L}}{\varepsilon v} \quad r = 1 + \frac{c_0}{K_d} \tag{5.10}$$

where Pe is the axial Peclet number, m *is* the ratio of total ligand capacity to feed protein concentration, n is the dimensionless number of transfer units and r is the dimensionless separation factor which determines the maximum species concentration loaded onto the membrane matrix at equilibrium. K_d is the equilibrium dissociation constant and is equal to k_d/k_a.

After substitution of the dimensionless groups, equations 5.2–5.7 become:

$$\frac{\partial C}{\partial \tau} + \frac{\partial C}{\partial \zeta} = \frac{1}{\text{Pe}} \frac{\partial^2 C}{\partial \zeta^2} - m \frac{\partial C_S}{\partial \tau} \tag{5.11}$$

The initial conditions:

$$C = 0 \quad \text{at } \zeta \geq 0, \ \tau = 0 \tag{5.12}$$

$$C_s = 0 \quad \text{at } \zeta \geq 0, \ \tau = 0 \tag{5.13}$$

The boundary conditions:

$$C - \frac{1}{\text{Pe}} \frac{\partial C}{\partial \zeta} = 1 \quad \text{at } \zeta = 0, \ \tau > 0 \tag{5.14}$$

$$\frac{\partial C}{\partial \zeta} = 0 \quad \text{at } \zeta = 1, \ \tau > 0 \tag{5.15}$$

For negligible axial diffusion (Pe$\rightarrow \infty$), the analytical solution developed by Thomas [142] used to describe protein adsorption in affinity chromatography, is obtained.

5.7.3 Mixing and dead volumes

A complete model of an adsorptive membrane must consider the entire flow system and not only the membrane unit. The effects of flow non-idealities such as mixing, channelling and dead volumes (including the void volume of the membrane, the volume of flow distributors, detector flow cell and pump head) must be taken into account as they influence the breakthrough curves [143–145]. In particular, small units with layered stack membranes have often a volume smaller than the total volume of circuit, pump and detector in which dispersion is not negligible.

The experimental set-up used for chromatographic membrane operations contains mixing zones and dead volumes other than the membrane itself. A simplified method to describe this behaviour is the serial combination of the model of the membrane with a continuously stirred tank reactor (CSTR) and an ideal plug flow reactor (PFR) [143,144] as shown in Figure 5.13. The system volume is:

$$V_{\text{sys}} = V_{\text{CSTR}} + V_{\text{PFR}} \tag{5.16}$$

The CSTR model with the initial condition $c_{\text{out}}(t = 0) = 0$ is:

$$\frac{dc_{\text{out}}}{dt} = \frac{Q}{V_{\text{CSTR}}}(c_{\text{in}} - c_{\text{out}}) \tag{5.17}$$

where c_{in} and c_{out} are the input and output species concentration for CSTR, respectively, and Q is the flow rate.

The PFR model shifts the response by the delay time (t_{delay}). The first temporal moment method may be used to determine V_{CSTR} and V_{PFR} [144]. Because the PFR volume includes the void volume in the membrane, the delay time is due to the remaining PFR volume ($V_{\text{PFR}} - V_{\text{void}}$).

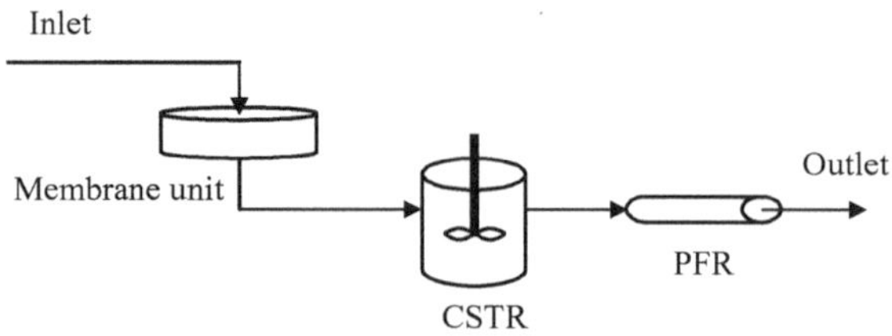

Figure 5.13 System dispersion model described by a combination of a membrane unit, a continuous stirred tank reactor (CSTR) and an ideal plug flow reactor (PFR).

5.7.4 Kinetic equations

Langmuir model

The Langmuir isotherm (equation 5.3) is the simplest and most widely used adsorption kinetic model. It assumes energetic homogeneity of the adsorption sites without steric effects. Occupation of a binding site by an adsorbed molecule is assumed not to affect adsorption onto other sites of newly adsorbed molecules. This occurs when the binding site density is so low that the average distance between adjacent sites is larger than the diameter of the adsorbed molecule.

Steric hindrance model

The Langmuir model is not valid for adsorption of large molecules, such as proteins, because occupation of a binding site may exclude some adjacent empty sites from binding because of steric hindrance effects. As binding approaches saturation, steric hindrance effects should be considered. A kinetic equation including the steric hindrance effect was proposed by Jin et al. [146]. The steric hindrance model is a modification of the Langmuir model such that the association rate constant is multiplied by a term that decreases with increasing surface coverage.

Spreading model

In both the Langmuir and steric hindrance models, proteins are not allowed to change conformation after adsorption. In reality, a protein might be captured while undergoing a conformational change such as flattening or spreading. A model accounting for adsorption and changes in conformation was proposed by Lundström [147].

5.7.5 Comparison with experimental data

Several studies have provided comparison between experimental and calculated breakthrough curves. For example, the Langmuir, steric hindrance and spreading equations associated to equations 5.11–5.15 were evaluated to fit experimental breakthrough curves of α-lactalbumin and thyroglobulin using ion-exchange membranes [144]. The breakthrough curves were sharp initially and then broadened as the membrane approached saturation. The spreading equation provided the best fit,

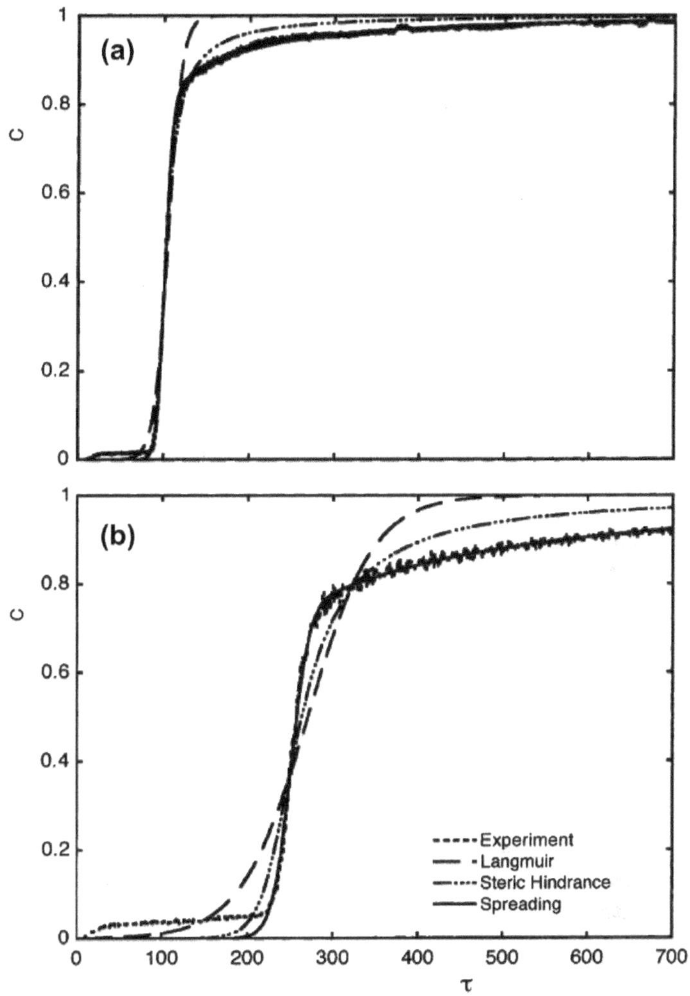

Figure 5.14 Experimental and fitted breakthrough curves for (a) α-lactalbumin and (b) thyroglobulin using an ion-exchange membrane [144].
Reprinted with permission from Ind. Eng. Chem. Res., Vol. 42, H. Yang, M.R. Etzel, Evaluation of three kinetic equations in models of protein purification using ion-exchange membranes, pp. 890–896, Copyright (2003) American Chemical Society.

followed by the steric hindrance equation, and finally by the Langmuir equation (Figure 5.14). The intrinsic rate of protein adsorption to the membrane was found to be rate-limiting, whereas effects of mass transfer and flow non-idealities were negligible.

The model was modified and used by many authors for other purposes, to study the influence of a hollow fibre membrane geometry [148,149], to explore pore-size distribution and membrane thickness variation effects on breakthrough curves [150], to include the Freundlich kinetics equation [151], to predict the adsorption of virus particles by anion- and cation-exchange membrane [119] and to investigate other dimensionless numbers [152].

References

[1] J. Thömmes, M.R. Kula, Membrane chromatography – an integrative concept in the downstream processing of proteins, Biotechnol. Prog. 11 (1995) 357–367.
[2] D.K. Roper, E.N. Lightfoot, Separation of biomolecules using adsorptive membranes, J. Chromatogr. A 702 (1995) 3–26.
[3] E. Klein, Affinity Membranes: Their Chemistry and Performance in Adsorptive Separation Processes, John Wiley & Sons, New York, 1991.
[4] X. Zeng, E. Ruckenstein, Membrane chromatography: preparation and applications to protein separation, Biotechnol. Prog. 15 (1999) 1003–1019.
[5] C. Charcosset, Purification of proteins by membrane chromatography, J. Chem. Technol. Biotechnol. 71 (1998) 95–110.
[6] E. Klein, Affinity membranes: a 10-year review, J. Membr. Sci. 179 (2000) 1–27.
[7] R. Ghosh, Protein separation using membrane chromatography: opportunities and challenges, J. Chromatogr. A 952 (2002) 13–27.
[8] C. Charcosset, Downstream processing and product recovery/membrane systems and technology, in: M.Y. Murray (Ed.), Comprehensive Biotechnology, second ed., vol. 2, Elsevier, 2011, pp. 603–618.
[9] S.Y. Suen, Y.C. Liu, C.S. Chang, Exploiting immobilized metal affinity membranes for the isolation or purification of therapeutically relevant species, J. Chromatogr. B 797 (2003) 305–319.
[10] J. Zhou, T. Tressel, Basic concepts in Q membrane chromatography for large-scale antibody production, Biotechnol. Prog. 22 (2006) 341–349.
[11] C. Boi, Membrane adsorbers as purification tools for monoclonal antibody purification, J. Chromatogr. B 848 (2007) 19–27.
[12] M.E. Avramescu, Z. Borneman, M. Wessling, Membrane chromatography, in: A.K. Pabby, S.S.H. Rizvi, A.M. Sastre (Eds.), Handbook of Membrane Separations, Chemical, Pharmaceutical, Food, and Biotechnological Applications, CRC Press, Taylor & Francis Group, 2009, pp. 25–64.
[13] D. Josić, A. Buchacher, A. Jungbauer, Monoliths as stationary phases for separation of proteins and polynucleotides and enzymatic conversion, J. Chromatogr. B 752 (2001) 191–205.
[14] H. Zhou, X. Huang, M. Ye, Q. Luo, Monolithic stationary phases for liquid chromatography and capillary electrochromatography, J. Chromatogr. A 954 (2002) 5–32.
[15] W. Guo, Z. Shang, Y. Yu, L. Zhou, Membrane affinity chromatography of alkaline phosphatise, J. Chromatogr. A 685 (1994) 344–348.
[16] T. Barroso, M. Temtem, A. Hussain, A. Aguiar-Ricardo, A.C.A. Roque, Preparation and characterization of a cellulose affinity membrane for human immunoglobulin G (IgG) purification, J. Membr. Sci. 348 (2010) 224–230.
[17] P. Lu, Y.L. Hsieh, Lipase bound cellulose nanofibrous membrane via Cibacron Blue F3GA affinity ligand, J. Membr. Sci. 330 (2009) 288–296.
[18] J. Wang, F. Dismerb, J. Hubbuch, M. Ulbricht, Detailed analysis of membrane adsorber pore structure and protein binding by advanced microscopy, J. Membr. Sci. 320 (2008) 456–467.

[19] K. Rodemann, E. Staude, Polysulfone affinity membranes for the treatment of amino acid mixtures, Biotechnol. Bioeng. 46 (1995) 503–509.

[20] W. Guo, Z. Shang, Y. Yu, Y. Guan, L. Zhou, Membrane affinity chromatography used for the separation of trypsin inhibitor, Biomed. Chromatogr. 6 (1992) 95–98.

[21] Z.W. Ma, Z.W. Lan, T. Matsuura, S. Ramakrishna, Electrospun polyethersulfone affinity membrane: Membrane preparation and performance evaluation, J. Chromatogr. B 877 (2009) 3686–3694.

[22] K. Kugel, A. Moseley, G.B. Harding, E. Klein, Microporous poly(caprolactam) hollow fibers for therapeutic affinity adsorption, J. Membr. Sci. 74 (1992) 115–129.

[23] T.C. Beeskow, W. Kusharyoto, F.B. Anspach, K.H. Kroner, W.D. Deckwer, Surface modification of microporous polyamide membranes with hydroxyethyl cellulose and their application as affinity membranes, J. Chromatogr. 715 (1995) 49–65.

[24] K.C. Hou, C.J. Hou, S. Roy, R. Zaniewski, Immobilization of protein G on nylon membrane for the removal of IgG from human plasma, Polym. Mater. Sci. Eng. 61 (1989) 670–674.

[25] K.C. Hou, R. Zaniewski, S. Roy, Protein A immobilized affinity cartridge for immunoglobulin purification, Biotechnol. Appl. Biochem. 13 (1991) 257–268.

[26] K.C. Hou, R. Zaniewski, Purification of urokinase by combined cation exchanger and affinity chromatographic cartridges, J. Chromatogr. B 525 (1990) 297–306.

[27] C. Charcosset, Z. Su, S. Karoor, G. Daun, C.K. Colton, Protein A immunoaffinity hollow fiber membranes for immunoglobulin purification: experimental characterization, Biotechnol. Bioeng. 48 (1995) 415–427.

[28] E. Klein, E. Eichhloz, D.H. Yeager, Affinity membranes prepared from hydrophilic coatings on microporous polysulfone hollow fibers, J. Membr. Sci. 90 (1994) 69–80.

[29] T.B. Tennikova, B.G. Belenkii, F. Švec, High-performance membrane chromatography – A novel method of protein separation, J. Liq. Chromatogr. 13 (1990) 63–70.

[30] T.B. Tennikova, M. Bleha, F. Svec, T.V. Almazova, B.G. Belenkii, High performance membrane chromatography of proteins, a novel method of protein separation, J. Chromatogr. 555 (1991) 97–107.

[31] H. Abou-Rebyeh, F. Korber, K. Schubert-Rehberg, Carrier membrane as a stationary phase for affinity chromatography and kinetic studies of membrane-bound enzymes, J. Chromatogr. B 566 (1991) 341–350.

[32] H. Iwata, K. Saito, S. Furusaki, T. Sugo, J. Okamato, Adsorption characteristics of an immobilized metal affinity membrane, Biotechnol. Prog. 7 (1991) 412–418.

[33] M. Kim, K. Saito, S. Furusaki, T. Sato, T. Sugo, I. Ishigaki, Adsorption and elution of bovine γ-globulin using an affinity membrane containing hydrophobic amino acids as ligands, J. Chromatogr. 585 (1991) 45–51.

[34] H. Shinano, S. Tsuneda, K. Saito, S. Furusaki, T. Sugo, Ion exchange of lysozyme during permeation across a microporous sulfopropyl-group-containing hollow fiber, Biotechnol. Prog. 9 (1993) 193–198.

[35] S. Tsuneda, K. Saito, T. Sugo, K. Makuuchi, Protein adsorption characteristics of porous and tentacle anion exchange membrane prepared by radiation-induced graft polymerization, Radiat. Phys. Chem. 46 (1995) 239–245.

[36] V. Kumar, Y.K. Bhardwaj, S.N. Jamdar, N.K. Goel, S. Sabharwal, Preparation of an anion-exchange adsorbent by the radiation-induced grafting of vinylbenzyltrimethylammonium chloride onto cotton cellulose and its application for protein adsorption, J. Appl. Polym. Sci. 102 (2006) 5512–5521.

[37] M. Ulbricht, H. Yang, Porous polypropylene membranes with different carboxyl polymer brush layers for reversible protein binding via surface initiated graft copolymerization, Chem. Mater. 17 (2005) 2622–2631.
[38] A.H.M. Yussof, M. Ulbricht, Effects of photo-initiation and monomer composition onto performance of graft-copolymer based membrane adsorbers, Desalination 200 (2006) 462–463.
[39] N. Singh, S.M. Husson, B. Zdyrko, I. Luzinov, Surface modification of microporous PVDF membranes by ATRP, J. Membr. Sci. 262 (2005) 81–90.
[40] N. Singh, J. Wang, M. Ulbricht, S.R. Wickramasinghe, S.M. Husson, Surface-initiated atom transfer radical polymerization: a new method for preparation of polymeric membrane adsorbers, J. Membr. Sci. 309 (2008) 64–72.
[41] B.V. Bhut, S.M. Husson, Dramatic performance improvement of weak anion-exchange membranes for chromatographic bioseparations, J. Membr. Sci. 337 (2009) 215–223.
[42] B.V. Bhut, S.R. Wickramasinghe, S.M. Husson, Preparation of high-capacity, weak anion-exchange membranes for protein separations using surface-initiated atom transfer radical polymerization, J. Membr. Sci. 325 (2008) 176–183.
[43] C.J.M. Nova, D. Paolucci-Jeanjean, M.P. Belleville, M. Barboiu, M. Rivallin, G. Rios, Elaboration, characterization and study of a new hybrid chitosan/ceramic membrane for affinity membrane chromatography, J. Membr. Sci. 321 (2008) 81–89.
[44] C.S. Chang, H.S. Ni, S.Y. Suen, W.C. Tseng, H.Ch. Chiu, C.P. Chou, Preparation of inorganic–organic anion-exchange membranes and their application in plasmid DNA and RNA separation, J. Membr. Sci. 311 (2008) 336–348.
[45] C. Wu, T. Xu, W. Yang, Fundamental studies of a new hybrid (inorganic–organic) positively charged membrane: membrane preparation and characterizations, J. Membr. Sci. 216 (2003) 269–278.
[46] C. Wu, T. Xu, W. Yang, A new inorganic–organic negatively charged membrane: membrane preparation and characterizations, J. Membr. Sci. 224 (2003) 117–125.
[47] W. Guo, E. Ruckenstein, Crosslinked glass fiber affinity membrane chromatography and its application to fibronectin separation, J. Chromatogr. B 795 (2003) 61–72.
[48] R. Ghosh, T. Wong, Effect of module design on the efficiency of membrane chromatographic separation processes, J. Membr. Sci. 281 (2006) 532–540.
[49] L.C.L. de Aquino, H.R.T. de Sousa, E.A. Miranda, L. Vilela, S.M.A. Bueno, Evaluation of IDA-PEVA hollow fiber membrane metal ion affinity chromatography for purification of a histidine-tagged human proinsulin, J. Chromatogr. B 834 (2006) 68–76.
[50] Y. Li, T.S. Chung, S.Y. Chan, High-affinity sulfonated materials with transition metal counterions for enhanced protein separation in dual-layer hollow fiber membrane chromatography, J. Chromatogr. A 1187 (2008) 285–288.
[51] T. Kline, Handbook of Affinity Chromatography, Chromatographic Science Series, vol. 63, Marcel Dekker Inc., New York, 1993.
[52] J. Turkova, Bioaffinity Chromatography, J. Chromatogr. Library, 51A, Elsevier, Amsterdam, 1993.
[53] C.H. Bamford, K.G. Al-Lamee, M.D. Purbrick, T.J. Wear, Studies of a novel membrane for affinity separations, I. Functionalisation and protein coupling, J. Chromatogr. 606 (1992) 19–31.
[54] T.T. Ngo, Facile activation of sepharose hydroxyl groups by 2-fluoro-1-methylpyridinium toluene-sulfonate: preparation of affinity and covalent chromatographic matrices, Biotechnology 4 (1986) 134–137.

[55] M. Unarska, P.A. Davies, M.P. Esnouf, B.J. Bellhouse, Comparative study of reaction kinetics in membrane and agarose bead affinity systems, J. Chromatogr. 519 (1990) 53–67.
[56] J. Porath, K. Asperg, H. Drevin, R. Axen, Preparation of cyanogen bromide-activated agarose gels, J. Chromatogr. 86 (1973) 53–56.
[57] X. Ma, S.F. Mohammad, S.W. Kim, Heparin removal from blood using poly(L-lysine) immobilized hollow fiber, Biotechnol. Bioeng. 40 (1992) 530–536.
[58] P.D. Weston, S. Avrameas, Proteins coupled to polyacrylamide beads using glutaraldehyde, Biochem. Biophys. Res. Commun. 45 (1971) 1574–1580.
[59] R. Mandaro, S. Roy, K.C. Hou, Filtration supports for affinity separation, Biotechnology 5 (1987) 928–932.
[60] S.M.A. Bueno, K. Haupt, M.A. Vijayalashmi, Separation of immunoglobulin G from human serum by pseudobioaffinity chromatography using immobilized Lhistidine in hollow fibre membranes, J. Chromatogr. B 667 (1995) 57–67.
[61] M. Nachman, A.R.M. Azad, P. Bailon, Efficient recovery of recombinant proteins using membrane-based immunoaffinity chromatography (MIC), Biotechnol. Bioeng. 40 (1992) 564–571.
[62] Z. Ma, S. Ramakrishna, Electrospun regenerated cellulose nanofiber affinity membrane functionalized with protein A/G for IgG purification, J. Membr. Sci. 319 (2008) 23–28.
[63] S. Brandt, R.A. Goffe, S.B. Kessler, J.L. O'Connor, S.E. Zale, Membrane-based affinity technology for commercial scale purifications, Biotechnology 6 (1988) 779–782.
[64] A. Malakian, M. Golebiowska, J. Bellefeuille, Purification of monoclonal and polyclonal IgG with affinity membrane matrix coupled with proteins A and G, Am. Lab. (1993). 40P–40V.
[65] L.R. Castilho, F.B. Anspach, W.D. Deckwer, Comparison of affinity membranes for the purification of immunoglobulins, J. Membr. Sci. 207 (2002) 253–264.
[66] C. Boi, S. Dimartino, G.C. Sarti, Performance of a new protein A affinity membrane for the primary recovery of antibodies, Biotechnol. Prog. 24 (2008) 640–647.
[67] A. Denizli, E. Piskin, Dye-ligand affinity systems, J. Biochem. Biophys. Methods 49 (2001) 391–416.
[68] B. Champluvier, M.R. Kula, Dye-ligand membranes as selective adsorbents for rapid purification of enzymes: a case study, Biotechnol. Bioeng. 40 (1992) 33–40.
[69] S. Krause, K.H. Kroner, W.D. Deckwer, Comparison of affinity membranes and conventional affinity matrices with regard to protein purification, Biotechnol. Technol. 5 (1991) 199–204.
[70] B. Champluvier, M.R. Kula, Modification of microfiltration membranes as dye-ligand adsorbents for the isolation of enzymes from crude extracts, in: D.L. Pyle (Ed.), Separations for Biotechnology, Elsevier, London, 1990, pp. 295–303.
[71] K.G. Briefs, M.R. Kula, Fast protein chromatography on analytical and preparative scale using modified microporous membranes, Chem. Eng. Sci. 47 (1992) 141–149.
[72] M.Y. Arıca, M. Yılmaz, E. Yalçın, G. Bayramoğlu, Surface properties of Reactive Yellow 2 immobilised pHEMA and HEMA/chitosan membranes: characterisation of their selectivity to different proteins, J. Membr. Sci. 240 (2004) 167–178.
[73] M.Y. Arıca, G. Bayramoğlu, Purification of lysozyme from egg white by Reactive Blue 4 and Reactive Red 120 dye-ligands immobilised composite membranes, Process Biochem. 40 (2005) 1433–1442.

[74] W. Guo, E. Ruckenstein, Separation and purification of horseradish peroxidase by membrane affinity chromatography, J. Membr. Sci. 211 (2003b) 101–111.
[75] S.N. Su, H.L. Nie, L.M. Zhu, T.X. Chen, Optimization of adsorption conditions of papain on dye affinity membrane using response surface methodology, Bioresour. Technol. 100 (2009) 2336–2340.
[76] F.J. Wolman, M. Grasselli, O. Cascone, Rapid neutral protease purification by dye-affinity membrane chromatography, Process Biochem. 41 (2006) 356–361.
[77] F.J. Wolman, D.G. Maglio, A. Grasselli, O. Cascone, One-step lactoferrin purification from bovine whey and colostrum by affinity membrane chromatography, J. Membr. Sci. 288 (2007) 132–138.
[78] A.S. Kamalanathan, M.A. Vijayalakshmi, Molecular studies of rheumatoid factor using pseudobioaffinity membrane chromatography, J. Mol. Recognit. 22 (2009) 146–153.
[79] R. Woker, B. Champluvier, M.R. Kula, Purification of S-oxynitrilase from Sorghum bicolor by immobilized metal ion affinity chromatography on different carrier materials, J. Chromatogr. B 584 (1992) 85–92.
[80] O.W. Reif, V. Nier, U. Bahr, R. Freitag, Immobilized metal affinity membrane adsorbers as stationary phases for metal interaction protein separation, J. Chromatogr. A 664 (1994) 13–25.
[81] D.S. Lee, B.M. Kim, D.W. Seol, Improved purification of recombinant adenoviral vector by metal affinity membrane chromatography, Biochem. Biophys. Res. Commun. 378 (2009) 640–644.
[82] G. Serpa, E.F.P. Augusto, W.M.S.C. Tamashiro, M.B. Ribeiro, E.A. Miranda, S.M.A. Bueno, Evaluation of immobilized metal membrane affinity chromatography for purification of an immunoglobulin G(1) monoclonal antibody, J. Chromatogr. B 816 (2005) 259–268.
[83] L. Opitz, J. Hohlweg, U. Reichl, M.W. Wolff, Purification of cell culture-derived influenza virus A/Puerto Rico/8/34 by membrane-based immobilized metal affinity chromatography, J. Virol. Methods 161 (2009) 312–316.
[84] M.W.H. Roberts, C.M. Ongkudon, G.M. Forde, M.K. Danquah, Versatility of polymethacrylate monoliths for chromatographic purification of biomolecules, J. Sep. Sci. 32 (2009) 2485–2494.
[85] D. Lütkemeyer, M. Bretschneider, H. Büntmeyer, J. Lehmann, Membrane chromatography for rapid purification of recombinant antithrombin III and monoclonal antibodies from cell culture supernatant, J. Chromatogr. 639 (1993) 57–66.
[86] W.K. Wang, S.P. Lei, H.G. Monbouquette, W.C. McGregor, Membrane adsorber process development for the isolation of a recombinant immunofusion protein, BioPharm. June (1995) 52–59.
[87] K.C. Hou, R.M. Mandaro, Bioseparation by ion exchange cartridge chromatography, BioTech 4 (1986) 358–366.
[88] M. Belanich, B. Cummings, D. Grob, J. Klein, A. O'Connor, D. Yarosh, Reduction of endotoxin in a protein mixture using strong anion-exchange membrane absorption, Pharm. Technol. 20 (1996) 142–150.
[89] I. Recio, S. Visser, Two ion-exchange chromatographic methods for the isolation of antibacterial peptides from lactoferrin. In situ enzymatic hydrolysis on an ion--exchange membrane, J. Chromatogr. A 831 (1999) 191–201.
[90] A. Karger, B. Bettin, H. Granzow, T.C. Mettenleiter, Simple and rapid purification of alphaherpesviruses by chromatography on a cation exchange membrane, J. Virol. Methods 70 (1998) 219–224.

[91] H.R. Charlton, J.M. Relton, N.K.H. Slater, Characterization of a generic monoclonal antibody harvesting system for adsorption of DNA by depth filters and various membranes, Bioseparation 8 (1999) 281–291.
[92] A.M. Ventura, H.M.F. Lahore, E.E. Smolko, M. Grasselli, High-speed protein purification by adsorptive cation-exchange hollow-fiber cartridges, J. Membr. Sci. 321 (2008) 350–355.
[93] R.H. Kutner, S. Puthli, M.P. Marino, J. Raiser, Simplified production and concentration of HIV-1-based lentiviral vectors using HYPERFlask vessels and anion exchange membrane chromatography, BMC Biotechnology 9 (2009) 10. doi:10.1186/1472-6750-9-10.
[94] T.B. Tennikova, F. Svec, High performance membrane chromatography: highly efficient separation method for proteins in ion-exchange, hydrophobic interaction and reversed-phase modes, J. Chromatogr 646 (1993) 279–288.
[95] L. Wang, R. Ghosh, Fractionation of monoclonal antibody aggregates using membrane chromatography, J. Membr. Sci. 318 (2008) 311–316.
[96] R. Ghosh, Separation of proteins using hydrophobic interaction membrane chromatography, J. Chromatogr. A 923 (2001) 59–64.
[97] R. Ghosh, L. Wang, Purification of humanized monoclonal antibody by hydrophobic interaction membrane chromatography, J. Chromatogr. A 1107 (2006) 104–109.
[98] R. Ghosh, Fractionation of human plasma proteins by hydrophobic interaction membrane chromatography, J. Membr. Sci. 260 (2005) 112–118.
[99] O.W. Reif, R. Freitag, Comparison of membrane adsorber (MA) based purification schemes for the downstream processing of recombinant h-AT III, Bioseparation 4 (1994) 369–381.
[100] D.Q. Yu, M.D. McLean, J.C. Hall, R. Ghosh, Purification of a human immunoglobulin G1 monoclonal antibody from transgenic tobacco using membrane chromatographic processes, J. Chromatogr. A 1187 (2008a) 128–137.
[101] D.Q. Yu, M.D. McLean, J.C. Hall, R. Ghosh, Purification of monoclonal antibody from tobacco extract using membrane-based bioseparation techniques, J. Membr. Sci. 323 (2008) 159–166.
[102] P. Guerrero-German, D.M.F. Prazeres, R. Guzman, R.M. Montesinos-Cisneros, A. Tejeda-Mansir, Purification of plasmid DNA using tangential flow filtration and tandem anion-exchange membrane chromatography, Bioprocess Biosystem Eng 32 (2009) 615–623.
[103] L. Wang, D.M. Kanani, R. Ghosh, Purification of humanized monoclonal antibodies by membrane-based hybrid bioseparation technique, J. Immunol. Methods 314 (2006) 1–8.
[104] M.E. Avramescu, M. Girones, Z. Borneman, M. Wessling, Preparation of mixed matrix adsorber membranes for protein recovery, J. Membr. Sci. 218 (2003) 219–233.
[105] M.E. Avramescu, W.F.C. Sager, Z. Borneman, M. Wessling, Adsorptive membranes for bilirubin removal, J. Chromatogr. B 803 (2004) 215–223.
[106] S. Saiful, Z. Borneman, M. Wessling, Enzyme capturing and concentration with mixed matrix membrane adsorbers, J. Membr. Sci. 280 (2006) 406–417.
[107] S.A. Saufi, C.J. Fee, Fractionation of beta-lactoglobulin from whey by mixed matrix membrane ion exchange chromatography, Biotechnol. Bioeng. 103 (2009) 138–147.
[108] M.E. Avramescu, Z. Borneman, M. Wessling, Particle-loaded hollow-fiber membrane adsorbers for lysozyme separation, J. Membr. Sci. 322 (2008) 306–313.
[109] O.P. Dancette, J.L. Taboureau, E. Tournier, C. Charcosset, P. Blond, Purification of immunoglobulins G by protein A/G affinity membrane chromatography, J. Chromatogr. B 723 (1999) 61–68.

[110] M.A. Teeters, S.E. Conrardy, B.L. Thomas, T.W. Root, E.N. Lightfoot, Adsorptive membrane chromatography for purification of plasmid DNA, J. Chromatogr. A 989 (2003) 165–173.
[111] H.N. Endres, J.A.C. Johnson, C.A. Ross, J.K. Welp, M.R. Etzel, Evaluation of an ion-exchange membrane for the purification of plasmid DNA, Biotechnol. Appl. Biochem. 37 (2003) 259–266.
[112] S. Zhang, A. Krivosheyeva, S. Nochumson, Large-scale capture and partial purification of plasmid DNA using anion-exchange membrane capsules, Biotechnol. Appl. Biochem. 37 (2003) 245–249.
[113] C. Haber, J. Skupsky, A. Lee, R. Lander, Membrane chromatography of DNA: Conformation-induced capacity and selectivity, Biotechnol. Bioeng. 88 (2004) 26–34.
[114] A. Arunakumari, J. Wang, G. Ferreira, Improved downstream process design for human monoclonal antibody production, BioPharm. Int. 5 (2007) 36–40.
[115] J.X. Zhou, F. Solamo, T. Hong, M. Shearer, T. Tressel, Viral clearance using disposable systems in monoclonal antibody commercial downstream processing, Biotechnol. Bioeng. 100 (2008) 488–496.
[116] T. Vicente, M.F.Q. Sousa, C. Peixoto, J.P.B. Motab, P.M. Alves, M.J.T. Carrondo, Anion-exchange membrane chromatography for purification of rotavirus-like particles, J. Membr. Sci. 311 (2008) 270–283.
[117] C. Peixoto, T.B. Ferreira, M.F.Q. Sousa, M.J.T. Carrondo, P.M. Alves, Towards purification of adenoviral vectors based on membrane technology, Biotechnol. Prog. 24 (2008) 1290–1296.
[118] R. Specht, B. Han, S.R. Wickramasinghe, J.O. Carlson, P. Czermak, A. Wolf, O.W. Reif, Densonucleosis virus purification by ion exchange membranes, Biotechnol. Bioeng. 88 (2004) 465–473.
[119] B. Han, R. Specht, S.R. Wickramasinghe, J.O. Carlson, Binding *Aedes aegypti* densonucleosis virus to ion exchange membranes, J. Chromatogr. A 1092 (2005) 114–124.
[120] B. Kalbfuss, M. Wolff, L. Geisler, A. Tappe, R. Wickramasinghe, V. Thom, U. Reichl, Direct capture of influenza A virus from cell culture supernatant with Sartobind anion-exchange membrane adsorbers, J. Membr. Sci. 299 (2007) 251–260.
[121] D. Josić, A. Buchacher, Application of monoliths as supports for affinity chromatography and fast enzymatic conversion, J. Biochem. Biophys. Methods 49 (2001) 153–174.
[122] F. Švec, T.B. Tennikova, Polymeric separation media for chromatography of biopolymers in a novel shape – macroporous membranes, J. Bioact. Compat. Polym. 6 (1991) 393–405.
[123] A. Podgornik, M. Barut, A. Štrancar, D. Josić, T. Koloini, Construction of large-volume monolithic columns, Anal. Chem. 72 (2000) 5693–5699.
[124] R. Mallik, D.S. Hage, Affinity monolith chromatography, J. Sep. Sci. 29 (2006) 1686–1704.
[125] G. Guiochon, Monolithic columns in high-performance liquid chromatography, J. Chromatogr. A 1168 (2007) 101–168.
[126] S.D. Chambers, K.M. Glenn, C.A. Lucy, Developments in ion chromatography using monolithic columns, J. Sep. Sci. 30 (2007) 1628–1645.
[127] A. Jungbauer, R. Hahn, Polymethacrylate monoliths for preparative and industrial separation of biomolecular assemblies, J. Chromatogr. A 1184 (2008) 62–79.
[128] E.G. Vlakh, T.B. Tennikova, Applications of polymethacrylate-based monoliths in high-performance liquid chromatography, J. Chromatogr. A 1216 (2009) 2637–2650.

[129] F. Plieva, B. Bober, M. Dainiak, I.Y. Galaev, B. Mattiasson, Macroporous polyacrylamide monolithic gels with immobilized metal affinity ligands: the effect of porous structure and ligand coupling chemistry on protein binding, J. Mol. Recognit. 19 (2006) 305–312.

[130] E.G. Vlakh, T.B. Tennikova, Preparation of methacrylate monoliths, J. Sep. Sci. 30 (2007) 2801–2813.

[131] A. Štrancar, P. Koselj, H. Schwinn, D. Josic, Application of compact porous disks for fast separations of biopolymers and in-process control in biotechnology, Anal. Chem. 68 (1996) 3483–3488.

[132] M. Bedair, Z. El Rassi, Affinity chromatography with monolithic capillary columns I. Polymethacrylate monoliths with immobilized mannan for the separation of mannose-binding proteins by capillary electrochromatography and nano-scale liquid chromatography, J. Chromatogr. A 1044 (2004) 177–186.

[133] M.B. Tennikov, N.V. Gazdina, T.B. Tennikova, F. Švec, Effect of porous structure of macroporous polymer supports on resolution in high-performance membrane chromatography of proteins, J. Chromatogr. A 798 (1998) 55–64.

[134] A. Mönster, O. Hiller, D. Grüger, R. Blasczyk, C. Kasper, Isolation and purification of blood group antigens using immuno-affinity chromatography on short monolithic columns, J. Chromatogr. A 1218 (2011) 706–710.

[135] M. Benčina, A. Podgornik, A. Štramcar, Characterization of methacrylate monoliths for purification of DNA molecules, J. Sep. Sci. 27 (2004) 801–810.

[136] J. Urthaler, R. Schlegl, A. Podgornik, A. Strancar, A. Jungbauer, R. Necina, Application of monoliths for plasmid DNA purification – Development and transfer to production, J. Chromatogr. A 1065 (2005) 93–106.

[137] N.L. Krajnc, F. Smrekar, J. Černe, P. Raspor, M. Modic, D. Krgovič, A. Štrancar, A. Podgornik, Purification of large plasmids with methacrylate monolithic columns, J. Sep. Sci. 32 (2009) 2682–2690.

[138] K. Branovic, D. Forcič, J. Ivancič, A. Štrancar, M. Barut, T. Kosutic-Gulija, R. Zgorelec, R. Mazuran, Application of short monolithic columns for improved detection of viruses, J. Virol. Methods 110 (2003) 163–171.

[139] P. Kramberger, N. Petrovic, A. Štrancar, M. Ravnikar, Concentration of plant viruses using monolithic chromatographic supports, J. Virol. Methods 120 (2004) 51–57.

[140] P. Kramberger, M. Peterka, J. Boben, M. Ravnikar, A. Štrancar, Short monolithic columns – A breakthrough in purification and fast quantification of tomato mosaic virus, J. Chromatogr. A 1144 (2007) 143–149.

[141] S.Y. Suen, M.R. Etzel, A mathematical analysis of affinity membrane bioseparations, Chem. Eng. Sci. 47 (1992) 1355–1364.

[142] H.C. Thomas, Heterogeneous ion exchange in a flowing system, J. Am. Chem. Soc. 66 (1944) 1664–1666.

[143] H. Yang, M. Bitzer, M.R. Etzel, Analysis of protein purification using ion-exchange membranes, Ind. Eng. Chem. Res. 38 (1999) 4044–4050.

[144] H. Yang, M.R. Etzel, Evaluation of three kinetic equations in models of protein purification using ion-exchange membranes, Ind. Eng. Chem. Res. 42 (2003) 890–896.

[145] S.E. Bower, S.R. Wickramasinghe, Elimination of non-uniform, extra-device flow effects in membrane adsorbers, J. Membr. Sci. 330 (2009) 379–387.

[146] X. Jin, J. Talbot, N.H.L. Wang, Analysis of steric hindrance effects on adsorption kinetics and equilibria, AIChE J. 40 (1994) 1685–1696.

[147] I. Lundström, Models of protein adsorption on solid surfaces, Prog. Colloid Polym. Sci. 70 (1985) 76–82.

[148] G.C. Serafica, J. Pimbley, G. Belfort, Protein fractionation using fast-flow immobilized metal chelate affinity membranes, Biotechnol. Bioeng. 43 (1994) 21–36.
[149] A. Tejeda-Mansir, R.M. Montesinos, R. Guzman, Mathematical analysis of frontal affinity chromatography in particle and membrane configurations, J. Biochem. Biophys. Methods 49 (2001) 1–28.
[150] H. Liu, J.R. Fried, Breakthrough of lysozyme through an affinity membrane of cellulose-Cibacron Blue, AIChE J. 40 (1994) 40–49.
[151] W. Shi, F. Zhang, G. Zhang, Mathematical analysis of affinity membrane chromatography, J. Chromatogr. A 1081 (2005) 156–162.
[152] J. Labandaa, J. Sabatéb, J. Llorensa, Modeling of the dynamic adsorption of an anionic dye through ion-exchange membrane adsorber, J. Membr. Sci. 340 (2009) 234–240.

6 Membranes for the preparation of emulsions and particles

Chapter Outline

Membrane Processes in Biotechnology and Pharmaceutics. DOI: 10.1016/B978-0-444-56334-7.00006-X

6.1 Principles

6.1.1 Introduction

Emulsions are dispersed systems of two (or more) insoluble liquids, for example water and oil. Depending on which is the dispersed phase and which is the continuous phase, there are various types of emulsions: oil-in-water (o/w), water-in-oil (w/o) emulsions and water-in-oil-in-water (w/o/w) emulsions. Emulsions play an important role in the formulation of cosmetics, pharmaceuticals, paints and foods. Emulsions are usually prepared using high-pressure homogenizers, ultrasound homogenizers and rotor/stator systems, such as stirred vessels, colloid mills or toothed disc dispersing machines [1]. In the dispersing zone of these machines high shear stresses are applied to deform and disrupt large droplets. Therefore, high-energy inputs are required and shear-sensitive ingredients such as proteins or starches may lose functional properties.

The production of monodisperse emulsions and microparticles has been investigated by several new techniques such as emulsification with microchannels [2], microcapillaries [3] and other microfluidic devices [4,5]. Scaling up to industrial volumes is a major limitation. Unlike these methods, membrane emulsification has the potential for scaling up the production of emulsions with droplets of well-defined size.

Membrane emulsification has received increasing attention over the last 20 years as an alternative to other methods of emulsification. The dispersed phase is pressed through the pores of a microporous membrane, while the continuous phase flows along the membrane surface (Figure 6.1). Droplets grow at pore openings until they detach when having reached a certain size. Surfactant molecules in the continuous phase stabilize the newly formed interface, to prevent droplet coalescence immediately after formation. The distinguishing feature is that the resulting droplet size is controlled

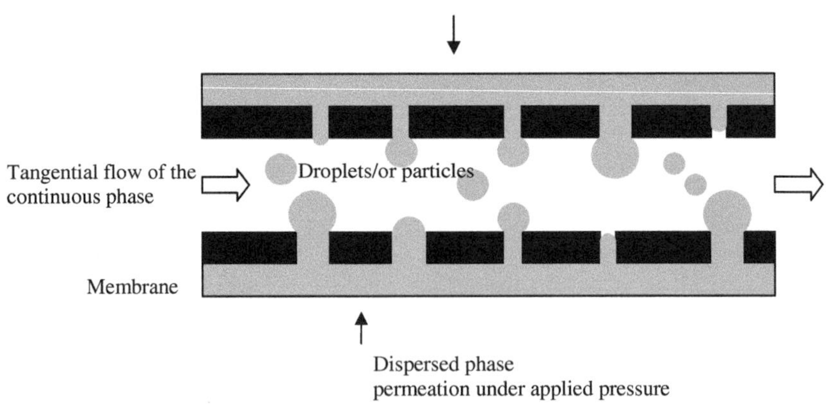

Figure 6.1 Schematic diagram of the membrane emulsification process [6].
This article was published in J. Food Eng., Vol. 92, C. Charcosset, Preparation of emulsions and particles by membrane emulsification for the food processing industry, pp. 241–249, Copyright Elsevier (2009).

primarily by the choice of the membrane and not by the generation of turbulent droplet break-up. The apparent shear stress is lower than in classical emulsification systems, because small droplets are directly formed by permeation of the dispersed phase through the micropores, instead of disruption of large droplets in zones of high energy density. Besides the possibility of using shear-sensitive ingredients, emulsions with narrow droplet size distributions can be produced. Furthermore, membrane emulsification processes allow the production of emulsions at lower energy input (10^4–10^6 J/m^3) compared to conventional mechanical methods (10^6–10^8 J/m^3) [7].

6.1.2 Previous reviews

Previous reviews on membrane emulsification [8–11] focused on principles, influence of process parameters, and comparison with other methods and applications. Nakashima et al. [12] provided a review recalling that membrane emulsification was introduced by these authors at the annual Meeting of the Society of Chemical Engineers, Japan, in 1988. The fundamentals of membrane emulsification were presented and the applications – food emulsions, synthesis of monodispersed microspheres and drug delivery systems – were described. In a book chapter, Piacentini et al. [13] presented a review on membrane emulsification, including the different configurations and the large range of dispersions prepared, including o/w, w/o, multiple emulsions and particles. A special attention was devoted to patents, with 60 patents referring to membranes, apparatus, methods and a broad range of applications. The patents came mainly from Japan (60%).

Other reviews focused on specific aspects. For example, Gijsbertsen et al. [10] presented a state of the art on membrane emulsification, as well as an analysis of an industrial scale production of culinary cream, for which a microsieve membrane with a low porosity was found the best suitable. Vladisavljević and Williams [11] provided a very complete review on manufacturing emulsions and particulate products using membranes, ranging from the production of simple o/w and w/o emulsions to multiple emulsions of different types, solid-in-oil-in-water (s/o/w) dispersions, coherent solids (silica particles, solid lipid microspheres, solder metal powder) and structured solids (solid lipid microcarriers, gel microbeads, polymeric microspheres, core-shell microcapsules and hollow polymeric microparticles). Liu et al. [14] focused on membrane emulsification for the preparation of multiple emulsions and micro/nano particulates for drug delivery. The membrane emulsification technique can be directly used to prepare monodisperse multiple emulsions for drug delivery. Also, a variety of uniform-sized colloidal suspensions can be obtained by means of sequential secondary processes/reactions after primary emulsification, such as solidification, crystallization, freeze-drying, evaporation, droplet swelling, gelation, polymerization and so on.

6.1.3 Basic parameters

Pressures

Membrane emulsification involves using a TMP to force the dispersed phase to permeate through the membrane into the continuous phase. The TMP is defined as the difference between the pressure of the dispersed phase, P_d, and the mean pressure of the continuous phase:

$$\mathrm{TMP} = P_d - \frac{(P_{c,in} + P_{c,out})}{2} \tag{6.1}$$

where $P_{c,in}$ and $P_{c,out}$ are the pressure of the flowing continuous phase at the inlet and at the outlet of the membrane module, respectively. The applied TMP required to make the discontinuous phase (i.e. oil) flow can be estimated from the capillary pressure, assuming that the pores are ideal cylinders:

$$P_c = \frac{4\gamma \cos \theta}{\overline{d}_p} \tag{6.2}$$

where P_c is the critical pressure, γ the o/w interfacial tension, θ the contact angle of the oil droplet against the membrane surface well wetted with the continuous phase and $\overline{d}_p$ the average pore diameter. The actual TMP required to make the discontinuous phase flow may be greater than predicted by Equation (6.2), due to tortuosities in the pores, irregular pore openings at the membrane surface and the significant effects of surface wettability [15].

Dispersed phase flux

The emulsification result is expressed in terms of the dispersed phase flux, J_d, through the membrane calculated as [16]:

$$J_d = \frac{M_d}{\rho_d A} \tag{6.3}$$

where M_d is the mass flow rate of the dispersed phase, A the membrane surface area and ρ_d the dispersed phase density. The dispersed phase flux allows the comparison of results from different type or size of membranes.

Droplet size distribution

The droplet size distribution and the parameters deriving from this distribution can be measured using a laser light scattering system (such as the Mastersizer, Malvern Instruments, USA). This method allows the detection of droplets with a minimum diameter of 0.1 μm. The mean droplet diameter is expressed as the mean Sauter diameter, $d_{3,2}$, which is the diameter of a spherical droplet having the same area per unit volume, S_v, as that of the total collection of droplets in the emulsion:

$$d_{3,2} = \frac{6}{S_v} = \left(\sum_{1}^{k_s} \frac{v_i}{d_i}\right)^{-1} \tag{6.4}$$

where v_i is the volume fraction of droplets in the ith range of sizes, the mean diameter of which is d_i, and k_s is the number of size ranges.

The droplet uniformity is expressed as the span of droplet size distribution:

$$\text{span} = \frac{d_{90} - d_{10}}{d_{50}} \tag{6.5}$$

where d_x is the diameter corresponding to x vol.% on a relative cumulative droplet size distribution curve. This parameter is also termed particle size dispersion coefficient.

For droplet with a mean diameter below 5 μm, the size distribution can be determined by photon correlation spectroscopy (e.g. using a Zetasizer Nano-series, Malvern, USA). The droplet size distribution is then characterized by the z-average size (mean size based on the intensity data) and the polydispersity index.

6.2 Configurations and membranes

6.2.1 Configurations

Experimental set-up

A schematic picture of a typical membrane emulsification set-up is shown in Figure 6.2. The system incorporates a tubular microfiltration membrane, a pump, a feed vessel and a pressurized (N_2) oil container. The dispersed phase is pumped under gas pressure through the pores of the membrane into the continuous phase

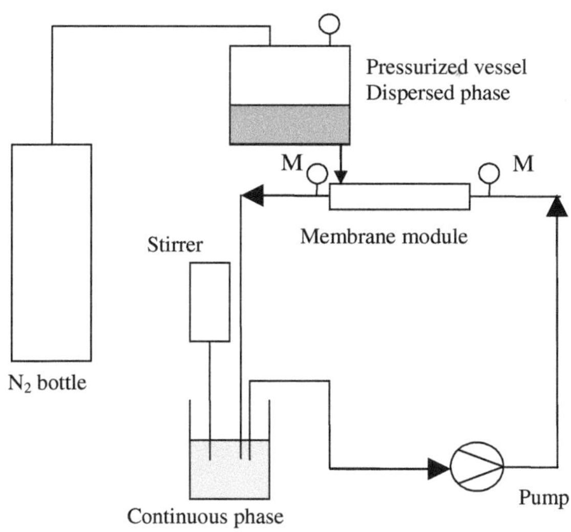

Figure 6.2 Typical experimental set-up for the membrane emulsification process. M: manometer [6]. This article was published in J. Food Eng., Vol. 92, C. Charcosset, Preparation of emulsions and particles by membrane emulsification for the food processing industry, pp. 241–249, Copyright Elsevier (2009).

which circulates through the membrane device. The membrane should not be wetted with the dispersed phase. Therefore, at the beginning of the experiment, the membrane is wetted with the continuous phase, that is a hydrophilic membrane for o/w emulsions is wetted with the water phase and a hydrophobic membranes for w/o emulsions is wetted with the oil phase. At the end of the experiment, the membrane is cleaned, using an appropriate solution, until the pure water flux is restored to its original value.

Flow configurations

In conventional direct membrane emulsification, fine droplets are formed at the membrane/continuous phase interface by pressing the disperse phase through the membrane. In order to ensure a regular droplet detachment from the pore outlets, shear stress is generated at the membrane/continuous phase interface by recirculating the continuous phase using a pump or by agitation in a stirring vessel [11]. The rate of mixing should be high enough to provide the required tangential shear on the membrane surface, but not too excessive to induce further droplet break-up.

The membrane emulsification process has also been carried out in a dead-end mode without tangential flow of the continuous phase [17], or in a stirred cell configuration [18] (Figure 6.3). A stirred cell is not a common device for membrane emulsification, because it is usually believed that a uniform shear field at the

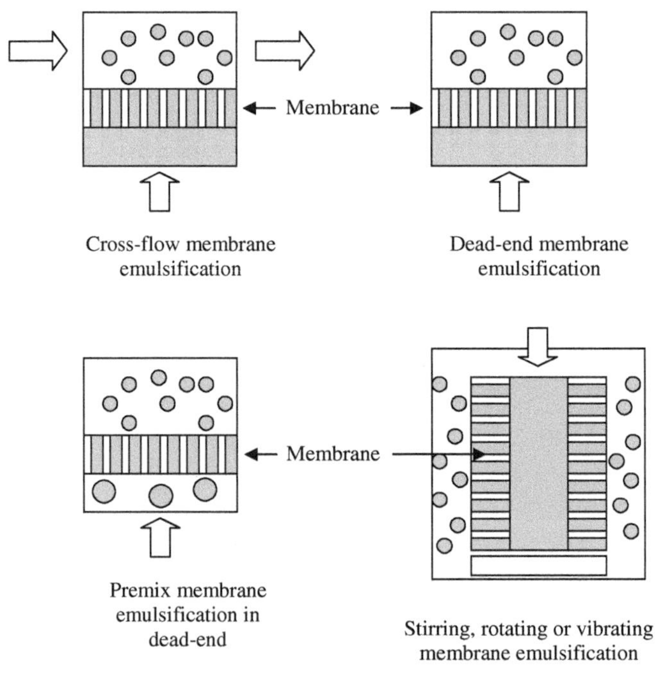

Figure 6.3 Various configurations for membrane emulsification.

membrane surface is required for the generation of uniform droplets. However, some studies showed that a stirred cell with a varying radial shear field at the surface of a flat disc membrane could produce uniform droplets of paraffin wax and refined sunflower oil [18,19]. In these studies, micro-engineered flat disc membranes were used on top of which a paddle blade stirrer was operated to induce surface shear. These configurations are particularly suited for the preparation of small amounts of emulsions, microcapsules and microparticles loaded with high values chemicals.

Other systems use a moving membrane, in which the droplet detachment from the pore outlets is obtained by rotation or vibration of the membrane within the stationary continuous phase. Droplets can be spontaneously detached from the pore outlets at small disperse phase fluxes, particularly in the presence of fast-adsorbing emulsifiers in the continuous phase and for a pronounced non-circular cross-section of the pores. These configurations have the advantage of eliminating external pumps to circulate the continuous phase. This is particularly attractive in case of coarse emulsion droplets or fragile particulate products, as their structure can be easily destroyed during their circulation inside the pump.

Rotating membrane devices were tested to increase the performances of the membrane emulsification process, especially to increase the flux of the dispersed phase through the membrane [20–22]. The centrifugation force increases with the diameter of the tubes (or column) [22]. With the rotation speed (higher wall shear stress), the particle size decreases. For example, Aryanti et al. [23] used a 100 μm pore-size rotating stainless steel tubular membrane for the preparation of o/w emulsions with a low viscosity paraffin wax as the dispersed phase, Tween 20 or sodium dodecyl sulphate (SDS) as the emulsifier and carbomer as the stabilizer, respectively. Coarse monodisperse emulsions were obtained with droplets of average diameter 80–570 μm and coefficient of variation (CV) ranging from 9.8 to 33.6%. The droplet size was found to be primarily a function of the membrane pore size, membrane rotational speed and interactions of surfactants at the oil/water interface.

A recent configuration was proposed using turbulence static promoters to lower the aqueous phase flow rate compared to other cross-flow membrane emulsification systems [24]. Two kinds of self-fabricated cross-section reducers were installed within the lumen side of a tubular Shirasu porous glass (SPG) membrane: a cylindrical rod or a double helix-shaped-ribbon reducer. Both promoters had a length equal to the membrane. Emulsions up to 38% of soybean oil in aqueous Tween 80 solution were prepared at various experimental conditions. Good emulsion quality in terms of droplet size distribution and stability were obtained even operating at high dispersed phase flux values.

Premix membrane emulsification is another configuration of membrane emulsification. A pre-emulsion with a large droplet size is passed through the porous membrane into the continuous phase, instead of directly pushing the oil or water through the membrane. The droplets of the pre-emulsion are disrupted into fine droplets during their permeation through the membrane. For similar mean pore sizes, the mean droplet size resulting from premix membrane emulsification is smaller than in direct membrane emulsification, which is often an advantage [7,25]. Repeating the processes with the same membrane results in smaller mean droplet size, narrower droplet size distribution and long-term physical stability. Premix membrane emulsification has been reviewed recently by focussing on droplet formation mechanisms (localized shear, interfacial tension, steric hindrance between droplets) and process parameters (membrane properties, TMP, continuous phase viscosity and number of homogenization cycles) [26]. Premix membrane emulsification is increasingly reported for the preparation of single emulsions, multiple emulsions, gel microbeads and polymer microspheres.

6.2.2 Membranes

Some commercial membranes used for emulsion and particle preparation are presented in Table 6.1. Scanning electron micrograph of these membranes is presented in Figure 6.4.

SPG membranes

The most commonly used membrane for the preparation of emulsions is the SPG membrane (Ise Chemical Co., Japan), because of its narrow pore size distribution and tubular shape [27]. The SPG membrane is synthesized from $CaO\text{-}Al_2O_3\text{-}B_2O_3\text{-}SiO_2$ type glass which is made from 'Shirasu', a Japanese volcanic ash. The SPG membrane has uniform cylindrical interconnected micropores, a wide spectrum of available mean pore sizes (0.05–30 μm) and a high porosity (50–60%). The surface wettability can be changed by reaction with organic silanes, such as octadecyltrichlorosilane [28,29]. The SPG membrane has been characterized by liquid permeability measurements, SEM, Hg porosimetry [30], high-resolution X-ray microtomography and microscopic observation of droplet formation in real time [31]. A new type of asymmetric SPG membrane was prepared from two types of primary glass in the $NaO\text{-}CaO\text{-}Al_2O_3\text{-}B_2O_3\text{-}ZrO_2\text{-}SiO_2$

Table 6.1 Properties of some commercialized membranes for emulsion and particle preparation

Filter	Manufacturer	Geometry, material, pore size	Effective filtration area
SPG Type 1	SPG Technology, Ltd Japan	Tubular, Shirasu porous glass, 0.1 to 20 μm	0.2 to 125 cm^2
SPG Type 2		Disc square or circle, Shirasu porous glass, 1 (smaller on request) to 20 μm	
Micropore	Micropore Technologies Ltd, UK	Flat or tubular, metal sieve, uniform pores >2 μm	Not available
Aquamarijn	Aquamarijn Microfiltration BV, The Netherlands	Flat, silicon nitride, uniform pores 0.1 to 100 μm	0.1 cm^2 to 225 cm^2

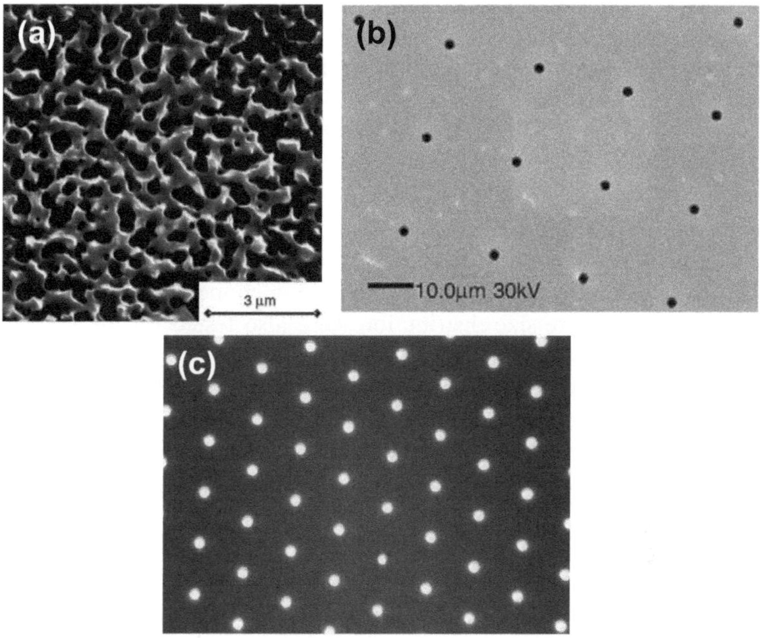

Figure 6.4 Scanning electron micrograph of (a) the surface of a hydrophilic SPG membrane with 0.262 μm pore size (SPG Technology Co.) [30]; (b) the surface of a micromachined silicon nitride membrane (Aquamarijn Microfiltration) [32] and (c) the surface of a membrane with a 30 μm regular pore size (Micropore Technologies Ltd) [18].
Figure 4a: This article was published in J. Membr. Sci., Vol. 250, G.T. Vladisavljević, M. Shimizu, T. Nakashima, Permeability of hydrophilic and hydrophobic Shirasu-porous-glass (SPG) membranes to pure liquids and its microstructure, pp. 69–77, Copyright Elsevier (2005).
Figure 4b: This article was published in J. Membr. Sci., Vol. 261, J. Zhu, D. Barrow, Analysis of droplet size during crossflow membrane emulsification using stationary and vibrating micromachined silicon nitride membranes, pp. 136–144, Copyright Elsevier (2005).
Figure 4c: Reprinted with permission from Ind. Eng. Chem. Res., Vol. 46, M.T. Stillwell, R.G. Holdich, S.R. Kosvintsev, G. Gasparini, I.W. Cumming, Stirred cell membrane emulsification and factors influencing dispersion drop size and uniformity, pp. 965–972, Copyright (2007) American Chemical Society.

system with different growth rates of phase separation [17]. The dispersed phase flux through the asymmetric membrane increased by a factor of approximately 20 compared with that through the symmetric membrane, due to the much smaller resistance of the asymmetric membrane.

Other membranes

In addition to SPG membranes, o/w emulsions were successfully prepared using silicon and silicon nitride microsieves membranes (Aquamarijn Microfiltration BV,

The Netherlands) [32,33]. These are made by photolithographic treatment of a silicon wafer and subsequent etching, or by electrochemical metal deposition on a skeleton in an electrolysis bath, respectively. These membranes have interesting properties, such as a smooth and flat surface, a very low membrane resistance and narrow pore size distribution. Different pore geometries (circular, square, slit shaped), pore size, pore edges and membrane porosities are available.

Flat or tubular metal membranes with monosized circular pores distributed in a highly regular array are also available (Micropore Technologies Ltd, UK) [18,34]. The membranes are chemically treated to make the surface hydrophilic for the preparation of o/w emulsions. Polycarbonate track-etch membranes (Millipore, Inc.) having a very narrow pore-size distribution were also tested for the preparation of particles [35].

Other commercial microfiltration membranes are attractive because of their availability in very large surface area, and their high flux through the membrane pores, such as ceramic aluminium oxide (α-Al_2O_3) membranes (Membraflow, Germany) [16,36], α-alumina- and zirconia-coated membranes (SCT, France) [37] and PTFE membranes (Advantec Tokyo Ltd., Japan [38,39] and Goretex Co. Ltd., Japan [40]). W/o emulsions were also successfully prepared using microporous polypropylene hollow fibres (Microdyn module, Wuppertal, Germany) [41], macroporous silica glass membranes [42], PTFE membranes [38,43] and polyamide hollow fibres membrane [44].

6.2.3 *Influence of parameters*

The major factors influencing membrane emulsification include membrane parameters, phase parameters and process parameters. Their influence of the emulsifying process (droplet size distribution, dispersed phase flux, dispersed phase percentage) is summarized in Figure 6.5.

Membrane parameters

Several authors have shown that the average droplet diameter, $\overline{d}_d$, increases with the average membrane pore diameter, $\overline{d}_p$, by a linear relationship, for given operating conditions:

$$\overline{d}_d = c\overline{d}_p \tag{6.6}$$

where c is a constant. For SPG membranes, values of c range typically from 2 to 10. This range was explained by differences in operating conditions, and by the type of SPG membrane used [45]. For membranes other than SPG, the values reported for c are higher, typically 3–50.

Monodispersed emulsions can be produced if the membrane pore-size distribution is sufficiently narrow. Using SPG membranes, Omi et al. [46] stated that fairly uniform droplets were obtained with a CV around 10%, due to the uniformity of the membrane pore size. Typical droplet size distributions are shown in Figure 6.4 [47]. The SPG membrane enabled to produce o/w emulsions with a very narrow droplet size distribution over a wide range of mean pore sizes. The span of the droplet size

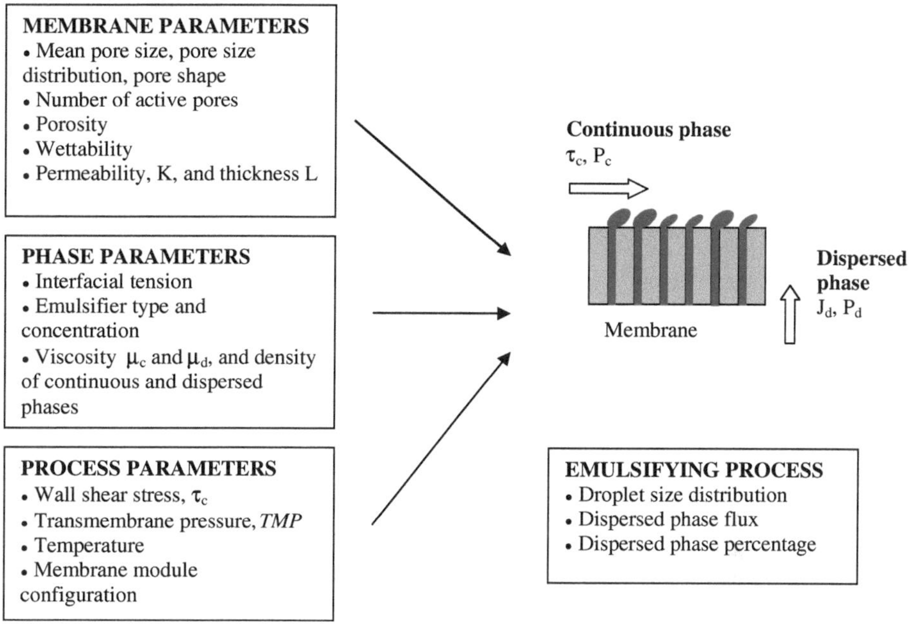

Figure 6.5 Parameters influencing the result of membrane emulsification [6].
This article was published in J. Food Eng., Vol. 92, C. Charcosset, Preparation of emulsions and particles by membrane emulsification for the food processing industry, pp. 241–249, Copyright Elsevier (2009).

distribution of 0.26–0.45 for the SPG emulsification was much lower than 0.94–2.7 for the microfluidization technique. At the same pore size and under the same experimental conditions, the oil droplets produced by utilizing the SPG membrane were more uniform than droplets prepared by using the α-Al_2O_3 membrane. However, the obtained span values of 0.51–0.59 for the α-Al_2O_3 membrane were much smaller than those usually reported. For example, Williams et al. [15] obtained the span of 0.83 at the oil flux of $8\ l\ m^{-2}\ h^{-1}$ using an α-Al_2O_3 membrane with the mean pore size of 0.5 μm. The narrower droplet size distribution obtained by Vladisavljević et al. [47] was explained by a careful membrane cleaning by ultrasonication. Moreover, the span of droplet size distribution for the emulsions prepared by microchannel emulsification was close to the upper limit for SPG emulsification (Figure 6.6).

The porosity of the membrane surface is also an important parameter for the emulsification membrane process because it determines the distance between two adjacent pores [15]. This distance is critical to ensure that two adjacent droplets do not come sufficiently close to allow contact with each other, which may lead to coalescence. Abrahamse et al. [48] calculated the maximum membrane porosity to be 1.5% to prevent coalescence of droplets growing on neighbouring pores of 5 μm diameter. However, a low porosity had the negative effect of a low dispersed flux.

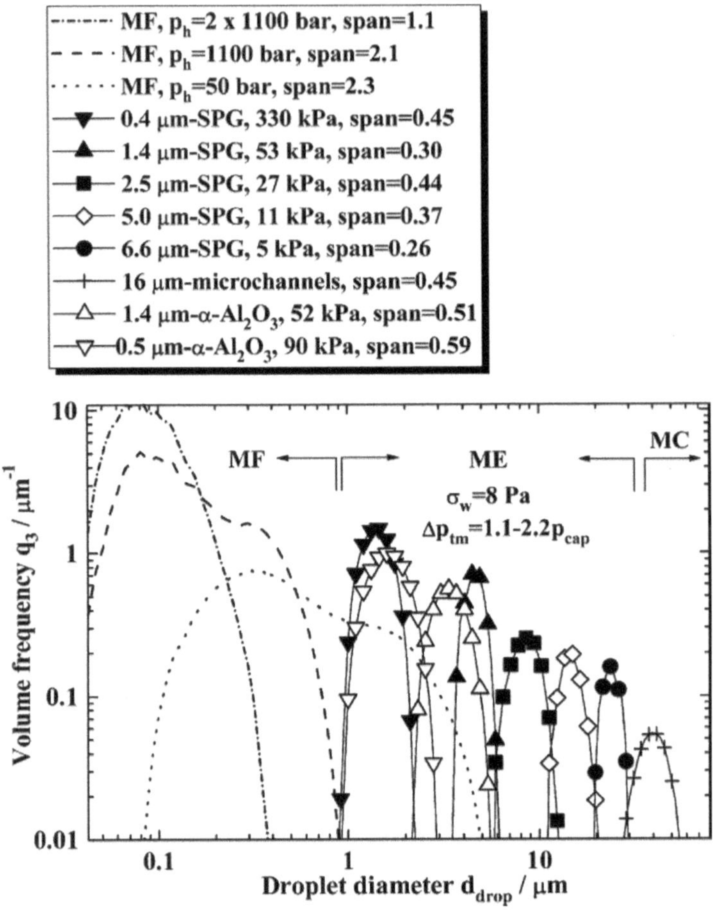

Figure 6.6 Influence of emulsification method on droplet size distribution (dispersed phase content $\varphi \approx 1$ vol.%) [47].
This article was published in Colloids Surf. A, Vol. 232, G.T. Vladisavljević, U. Lambrich, M. Nakajima, H. Schubert, Production of O/W emulsions using SPG membranes, ceramic α-aluminium oxide membranes, microfluidizer and a silicon microchannel plate – a comparative study, pp. 199–207, Copyright Elsevier (2004).

Transmembrane pressure and cross-flow velocity

The dispersed phase flux is an essential parameter for the membrane emulsification process. Increasing TMP increases the flux of dispersed phase through the membrane, according to Darcy's law. At high fluxes, the average droplet size and the size distribution tend to increase because of increased droplet coalescence at the membrane surface. Therefore, an increase in flux may be at the expense of droplet size distribution. The effect of TMP is dependent on operating conditions, as cross-flow velocity and type of surfactant [36,49–52].

Droplets formed at the membrane/continuous phase interface detach under the shear stress of the continuous phase. The characteristic parameter of the flowing continuous phase is the cross-flow velocity or the wall shear stress. It is shown that the droplet size becomes smaller as the wall shear stress increases and that the influence is greater for small wall shear stresses [15,37,50,51,53]. The effect of the wall shear stress on reducing droplet size is dependent on the membrane pore size, being more effective for smaller membrane pore size [36].

Surfactants

The influence of the type of surfactant in the membrane emulsification process has been studied by several authors [28,36,49,42,52–54]. Surfactants played two main roles in the formation of an emulsion. Firstly, they lowered the interfacial tension between oil and water. This facilitated droplet distribution and in case of membranes lowers the minimum emulsification pressure. Secondly, surfactants stabilize the droplets against coalescence and/or aggregation.

Schröder et al. [49] and Schröder and Schubert [36] showed that the type of surfactant used greatly influenced the droplet size. Droplet diameters obtained with Tween 20 were about twice the size of the droplets stabilized with SDS, in agreement with the ratio of equilibrium interfacial tensions. These authors suggested that the interfacial tension force was one of the key forces governing droplet formation during the membrane emulsification process. Van der Graaf et al. [55] carried out droplet formation experiments with a micro-engineered membrane by measuring the droplet diameter and droplet formation time as a function of the surfactant concentration in the continuous phase. Their experiments confirmed that the interfacial tension influenced the process of droplet formation: higher surfactant concentrations lead to smaller droplets and shorter droplet formation times.

The rules applying to conventional emulsification methods are suitable to the membrane emulsification process. The classical Bancroft rule states: 'The phase in which a surfactant is more soluble constitutes the continuous phase.' As a consequence, the type of emulsion (o/w or w/o) is dictated by the emulsifier and the emulsifier should be soluble in the continuous phase. The hydrophile–lipophile balance (HLB) number is a measure of the ratio of the hydrophilic and lipophilic groups of the surfactants molecule. Non-ionic surfactants have HLB numbers ranging from 0 to 20. HLB numbers >10 have an affinity for water (hydrophilic) and HLB number <10 have an affinity for oil (lipophilic). Ionic surfactants have recently been given relative HLB values, extending the range of HLB numbers to 60. The HLB number can be used in order to select the surfactant for a given emulsion. A water continuous phase system benefits from a surfactant with a high HLB number and an oil continuous phase system from a surfactant with a low HLB number.

A particularity of cross-flow membrane emulsification is that a fresh oil–water interface is continuously generated with a high rate at the membrane surface [56]. This fact implies that the surfactant is continuously depleted from the liquid near the membrane. To ensure sufficient stabilization of the droplets that are forming, high surfactant concentrations are usually used in the continuous phase ensuring high enough transport towards the interfaces. Wagdare et al. [56] have shown that a surfactant supplied in the dispersed phase will typically have a HLB number that is not appropriate for stabilization; however, it does ensure that the surfactant is available at the forming interfaces, as it is transferred with the dispersed phase itself. Emulsification with high porosity micro-engineered membranes lead to stable emulsions with a low droplet span when, besides a surfactant in the continuous phase, an additional, suitable surfactant was used in the dispersed phase. This surfactant should exhibit relatively fast adsorption dynamics, which is more critical when the surfactant in the continuous phase has slower dynamics.

Like in conventional emulsification, a large range of surfactants can be applied to the membrane emulsification process. Trentin et al. [57] studied premix membrane emulsification with organic microfiltration membranes to produce BSA-stabilized o/w emulsions (10%, v/v oil fraction). The effect of two different surface surfactants, a protein (BSA) and a non-ionic emulsifier (Tween 20), used as a single emulsifier or in mixtures, was investigated on the dispersed phase flux and the droplet size distribution. It was shown that protein fouling caused by the presence of BSA reduced the dispersed phase flux, but did not affect the droplet size diameter of the resulting emulsion. The droplet size of the final emulsion was controlled by the interfacial tension and the pore size of the membrane (pore size close to 1 μm).

Viscosity

The viscosity of the dispersed phase has also an important effect on the membrane emulsification performance. According to Darcy's law, the dispersed flux is inversely proportional to the dispersed phase viscosity. If the dispersed phase viscosity is higher, then the dispersed flux will be lower, and as a consequence the droplet diameter will be large compared to the mean pore diameter. Kukizaki and Goto [58] showed that in a system composed of decane containing liquid paraffin and SDS solution containing polyethylene glycol, the resulting droplet diameter increased with increasing water-phase viscosity, while droplet diameter decreased with increasing oil-phase viscosity. Droplet diameter decreased as the ratio of oil-phase to water-phase viscosity increased. However, droplet diameter did not change in the case of a constant viscosity ratio.

6.3 Applications

A very large range of emulsions and particles were prepared using the membrane emulsification technique. Some of these emulsions and particles are presented in Figure 6.7 and include w/o and o/w simple emulsions, $w_1/o/w_2$, $o_1/w/o_2$ multiple emulsions and particles such as polymeric, lipid nano and microparticles, and nano and microcapsules.

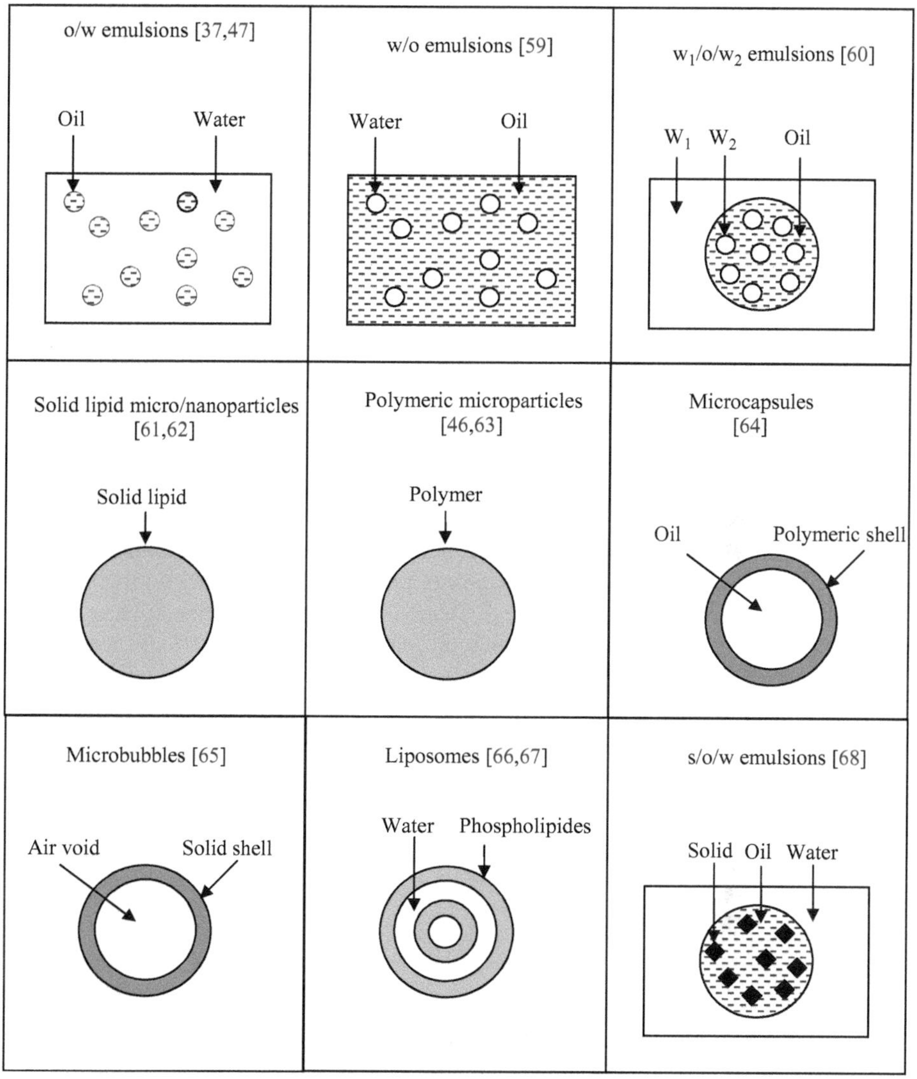

Figure 6.7 Examples of emulsions and particles prepared by membrane emulsification.

6.3.1 *Emulsions*

Emulsions, microemulsions and nanoemulsions are of particular interest in the biotechnology and pharmaceutical fields. Each of these colloidal delivery systems has specific physicochemical properties [69]. Emulsions are thermodynamically unstable systems that consist of oil, surfactant and water. Their particle size is greater than 100 nm, so they appear either turbid or opaque. Microemulsions are thermodynamically stable systems that typically consist of oil, surfactant (sometimes co-surfactant/co-solvent) and water. They have relatively small particles sizes ($r < 50$ nm) compared to the wavelength of light ($\lambda = 390$–750 nm). They scatter light weakly and appear either transparent or slightly turbid. Nanoemulsions are thermodynamically unstable systems that consist of oil, surfactant and water. They have small particle sizes ($r < 100$ nm). They are either transparent or only slightly turbid.

Numerous studies have been realized on the preparation of emulsions using membrane emulsification as reported in Sections 6.1 and 6.2. Some specific applications in the biotechnology and pharmaceutical fields are presented below. O/w emulsions prepared by membrane emulsification were used in a multiphase reactor developed by immobilizing the lipase from *Candida rugosa* in a polymeric membrane [44,70]. The configuration was a two separate phase membrane reactor constituted by an emulsion plus enzyme-loaded membrane and an organic and an aqueous phase recycled along the two separate sides of the membrane. It was shown that the presence of the emulsion within the membrane improved the catalytic activity and the enantioselectivity of the immobilized enzyme as well as the transport rate of the hydrophobic reagent through the hydrophilic membrane.

A recent study reported the preparation of nanoemulsion of flurbiprofen consisting of methylene chloride as the dispersed phase, polyvinyl alcohol (PVA) as the stabilizer and a mixture of Tween 20 and Tween 80 in demineralized water as the continuous phase [71]. SPG membranes with a mean pore size of 2.5 μm were used. Emulsion droplets with a mean droplet size of 25 times smaller than the mean pore size and a narrow droplet size distribution were produced using 5% emulsifier at a feed pressure of 15 kPa. In addition, the droplets in the emulsion containing the drug were not smaller but were more uniform with a narrower distribution compared to those without the drug.

6.3.2 Multiple emulsions

Multiple emulsions are complex system, termed 'emulsions of emulsions', that is the droplets of the dispersed phase contain even smaller dispersed droplets themselves [72]. W/o/w emulsions consist of water droplets dispersed within larger oil droplets, which are dispersed in an aqueous continuous phase. O/w/o emulsion consists of larger water droplets enclosing smaller oil droplets, which are dispersed in a continuous oil phase. Multiple emulsions can be produced by forcing a primary emulsion through a microporous membrane into a continuous phase liquid [73–76]. The shear is lower than in a conventional emulsification process so that the droplets are intact and high entrapment efficiency and monodispersity can be obtained [76]. In addition, the mean size of the outer drops can be controlled by changing experimental parameters such as membrane pore size.

The first application of membrane emulsification was a w/o/w multiple emulsion for the treatment of liver cancer by arterial injection chemotherapy [60,74]. The aqueous anti-cancer drug solution (epirubicin or carboplatin) was mixed with the oil (iodinated poppy-seed oil or lipiodol). This mixture was then sonicated to form submicron sized w/o emulsion, and permeated through a SPG membrane into a glucose solution forming w/o/w emulsions. The efficacy of the w/o/w emulsions was proved clinically, for patients bearing hepatocellular carcinoma nodules recurrent after hepatectomy.

Vladisavljević et al. [76] prepared w/o/w emulsions for drug delivery systems by extruding a coarse w/o/w emulsion through SPG membranes with mean pore size from 5.4 to 20.3 μm. The encapsulation efficiency of a hydrophilic marker ($CaNa_2$-EDTA) was 83–85% in the emulsion containing 30 vol. % of inner droplets and 30–50 vol. % of outer drops. At low continuous phase viscosity, uniform droplets with a span values of 0.28–0.34 were produced at high transmembrane fluxes exceeding 200 $m^3\ m^{-2}\ h^{-1}$. The authors concluded to the suitability of repeated SPG membrane homogenization for controlled production of w/o/w emulsions with high encapsulation efficiency and high dispersed phase fluxes.

Using membrane emulsification, Piacentini et al. [77] prepared glucose-sensitive multiple emulsions with potential use as insulin-controlled delivering systems were able to mimic the pancreatic activity in the diabetes treatment. Concanavalin A, a glucose-binding protein, was selected as the biomolecule sensor. In the preparation of water-in-oil emulsion (w_1/o), concanavalin A was used as the emulsifier. In addition, concanavalin A was used, also together with other emulsifiers, in the preparation of water-in-oil-in-water (w_1/o/w_2) emulsions. The influence of emulsion interfacial composition and operating conditions on droplets stability and functional properties was investigated. The authors also studied the release of concanavalin A as a function of glucose stimulus and its concentration. The authors concluded to the potentiality of the membrane emulsification process for preparation of drug delivery w/o/w emulsions which could mimic and/or replace important physiological functions.

6.3.3 *Food emulsions*

Food emulsions can provide interesting applications in biotechnologies and pharmaceutics. Emulsions play an important role in the formulation of foods for production of o/w emulsions, such as dressings, artificial milks, cream liqueurs, as well as for preparation of some w/o emulsions, such as margarines and low fat spreads. Various preparation of food emulsions by membrane emulsification are reported [37,50,51,54,59,78]. When preparing w/o food emulsions, Katoh et al. [50] showed that the dispersed phase flux was increased 100 times using a hydrophilic membrane pre-treated by immersion in the oil phase. They prepared a low fat spread with a fat content of 25% (v/v), and concluded that the membrane emulsification process was suitable for preparation of large-scale w/o food emulsions. Using SPG membranes, Scherze et al. [51] and Muschiolik et al. [54] prepared o/w emulsions with liquid butter fat or sunflower oil as the dispersed phase and a continuous phase containing milk proteins. The emulsions so obtained were characterized by particle size distribution, creaming behaviour and protein adsorption at the dispersed phase. The advantage of membrane emulsification was pointed out to be the low shear forces on the physicochemical and molecular properties of the proteins.

Aerated food gels were produced recently by membrane foaming [65]. The method is based on pressing the dispersed phase (gas) through the pores of a tubular membrane into the continuous phase. The bubbles formed are covered with surface-active substances of the continuous phase and they are detached from the membrane surface by the shear forces exerted by the phase flowing along the membrane surface. Bals and Kulozik [65] investigated the influence of pore size, foaming temperature and viscosity of the continuous phase on the properties of foams produced by membrane foaming. An important factor is that the added amount of gas must be stabilized as completely as possible in the foam. Raising the foaming temperature increased the quantity of stabilized gas. The whey proteins then diffuse faster to the bubble surfaces and stabilize these by unfolding and networking reactions to prevent the coalescence of the bubbles. Recently, Müller-Fischer et al. [79] proposed a dynamically enhanced membrane foaming process consisting of two cylindrical cylinders: the inner cylinder is rotated, the outer cylinder being fixed. The membrane was either mounted to the inner or outer cylinder. The gas was pressed through the membrane and was detached as small bubbles by the acting flow shear stresses. The authors showed that by applying this technique, the foam microstructure was significantly improved, that is smaller mean bubble sizes and narrower size distributions were achieved.

6.3.4 Microspheres and microcapsules

Microspheres and microcapsules have widespread applications in various fields ranging from cosmetics, chromatography column packings, coatings, pesticides, inks, electronic photocopying, catalysis, biomedical and pharmaceutics [80–84]. Microcapsules prepared from biodegradable polymer such as poly(glycolide), poly(ε-caprolactone), poly(saccharides) or polylactide have been prepared to deliver several types of drugs, such as antibiotics, anti-cancer, anti-microbial drugs, vaccines and proteins [85–88]. Among other applications of biodegradable microspheres are hollow microcapsules as ultrasound contrast agents. Gas microbubbles are encapsulated in a thin polymer or protein shell, which enable the gas core to oscillate in an acoustic field and reflect the ultrasound signal [89].

Membrane emulsification has been extensively used to prepare microspheres and microcapsules. Omi et al. [46] synthesized various microspheres with diameters from 2 to 100 μm in aqueous solution, by combining the membrane emulsification process and subsequent polymerization. Monomer droplets containing solvents, cross-linkers and an initiator, as well as a water-insoluble oligomer, are formed by permeation through the membrane pores, suspended in the aqueous solution of stabilizing agents, and transferred to a reactor for polymerization. These preparations include polystyrene [90,91], polymethacrylate [92], poly(styrene-*co*-methyl methacrylate) [45], poly(methyl methacrylate) [93], poly(styrene-*co*-*N*-dimethylaminoethyl methacrylate) [94], poly(glycidyl methacrylate) microspheres [95] and titanium dioxide polystyrene particles [96].

Microspheres were also prepared by the same authors, by combining the membrane emulsification process and subsequent solvent evaporation, to prepare poly(styrene)-poly(methyl methacrylate) [97], polyurethane urea [52], magnetite [93] and titanium dioxide microcapsules [98]. Other techniques reported were the droplet swelling method, to prepare hydrophilic poly(methyl methacrylate) [99] and p(HEMA-*co*-EDMA) microspheres [100], and the thermal-gelation method [63].

Ma and her co-workers prepared a large range of particles, such as uniform polyurethane urea–vinyl polymer microspheres of about 20 μm, using a SPG membrane emulsification technique and subsequent radical suspension polymerization [101]. For this, a mixture of a 40% urethane prepolymer solution of xylene and a vinyl monomer containing an initiator was permeated through the uniform pores of the SPG membrane into a continuous phase containing a stabilizer to form uniform droplets. The droplets were then allowed to stand for chain extension at room temperature with di- or triamines for 2 h in the absence or presence of ethyl acetate, followed by suspension polymerization at 70 °C for 24 h.

In another example, uniform-sized pH-sensitive quaternized chitosan microspheres were prepared by combining membrane emulsification technique and a thermal-gelation method (Figure 6.8) [63]. In this preparation process, the mixture of quaternized chitosan solution and alpha-beta-glycerophosphate (alpha-beta-GP) was used as the water phase and dispersed in the oil phase to form uniform w/o emulsion by SPG membrane emulsification technique. The droplets solidified into microspheres at 37 °C by the thermal-gelation method. The obtained microspheres had a porous structure and showed apparent pH-sensitivity which affected its drug release behaviour. BSA as a drug model was encapsulated in these microspheres. BSA released rapidly in acid solution and slowly in neutral medium, demonstrating their potential as drug delivery systems.

Another application is the synthesis and characterization of polymeric microcapsules for ultrasound-triggered delivery of lipophilic drugs [102]. Microcapsules (mean diameter between 1.22 and 1.31 μm) with a shell of fluorinated end-capped poly(l-lactic acid) were prepared through premix emulsification. In addition to a gaseous phase, they contained different amounts of hexadecane oil as a drug-carrier reservoir. These partially oil-filled microcapsules were reported to have a great potential for ultrasound-triggered local delivery of lipophilic drugs under ultrasound image-guidance.

Preparations from other authors include poly(acrylamide-*co*-acrylic acid) [103], poly(*N*-isopropylacrylamide-*co*-acrylic acid) [104], agarose [105] and chitosan microspheres [106]. For size-exclusion chromatography applications, Yoshizawa et al. [107] prepared polyethylene glycol monomethacrylate microspheres by the combination of the SPG membrane emulsification technique and the emulsion swelling method.

Several drug delivery systems were obtained. The treatment of gastric ulcerogenicity with diclofenac sodium suspension prepared via the membrane emulsification technique was reported by Piao et al. [108]. Uniform-sized microspheres were prepared for the controlled release of other drugs such as the anti-cancer drug anthracycline [109], carotenoid astaxanthin [110], haemoglobin [111], insulin [68,112] and spironolactone for paediatric use [113]. The encapsulation efficiency and/or drug release properties were measured with model drugs, for example,

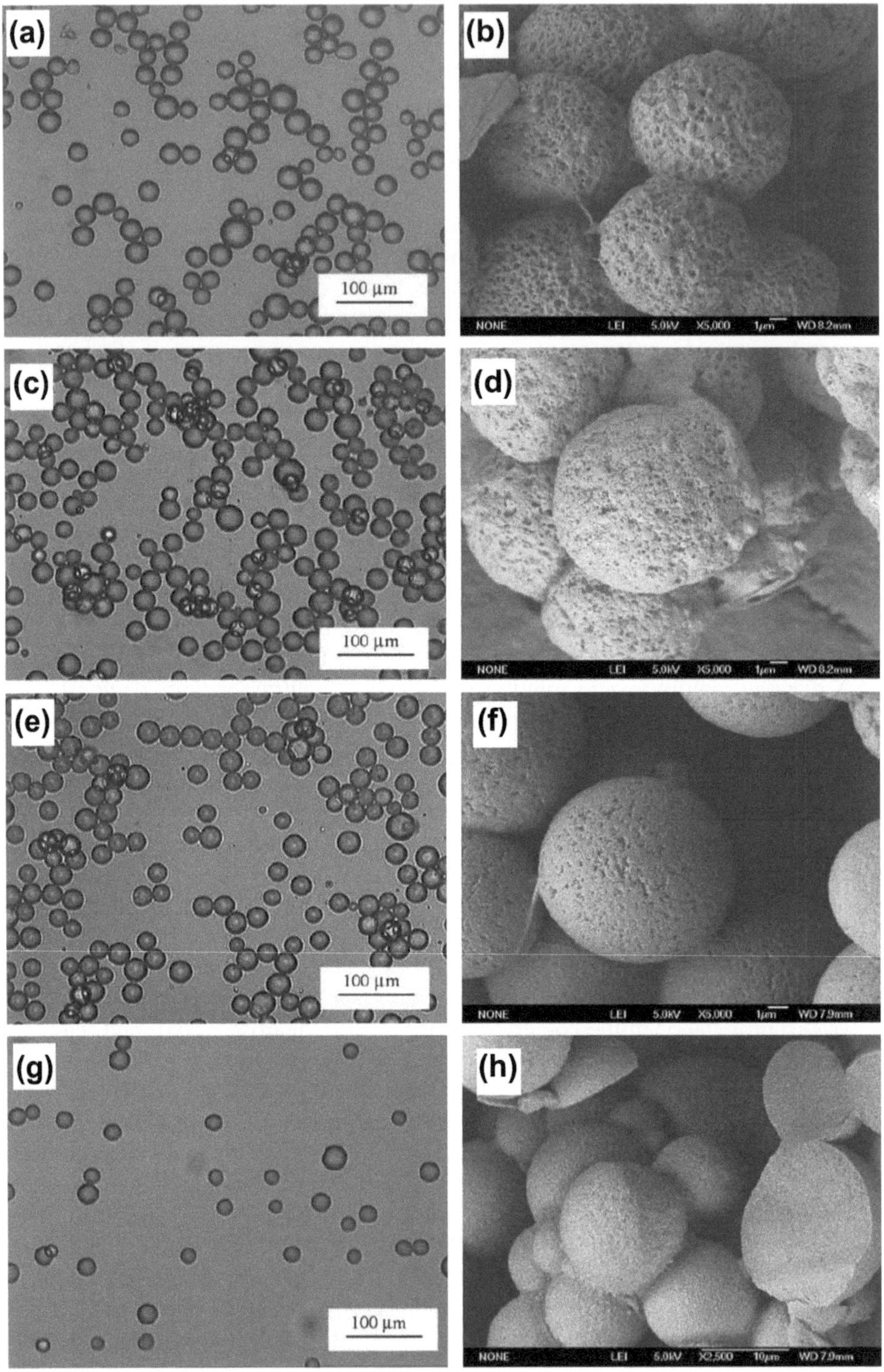
(a)
100 μm
(b)
NONE LEI 5.0kV X5,000 1μm WD 8.2mm
(c)
100 μm
(d)
NONE LEI 5.0kV X5,000 1μm WD 8.2mm
(e)
100 μm
(f)
NONE LEI 5.0kV X5,000 1μm WD 7.9mm
(g)
100 μm
(h)
NONE LEI 5.0kV X2,500 10μm WD 7.9mm

recombinant human insulin (rhI) [112], blue dextran [34], lidocaine–hydrochloride, sodium salicylate, and 4-acetaminophen [114], BSA [115] and haloperidol [116]. The results suggested that the release behaviour could be adjusted by changing precisely the diameters of microcapsules, prepared with the membrane emulsification technique.

6.3.5 Other particles

A large range of other particles has been prepared with the membrane emulsification technique. Solid lipid particles and microcapsules were prepared using a high melting point lipid as the encapsulating material, followed by cooling of the preparation at room temperature [61,62,117]. Polymeric nanoparticles were obtained successfully by the nanoprecipitation or the interfacial polymerization methods [118].

A membrane contactor for the preparation of polymeric nanoparticles was proposed by Charcosset and Fessi [118] and Limayem et al. [113]. The organic phase was pressed through the membrane pores via the filtrate side. The aqueous phase circulated inside the membrane module, and swept away the nanoparticles forming at the pore outlets. Unloaded nanoparticles were obtained as well as nanoparticles loaded with vitamin E and spironolactone (Figure 6.9). It was shown that nanoparticles as small as 260 nm were obtained with the 1000 Da nanofiltration membrane, a TMP of 3 bar and a cross-flow velocity of 1.7 m s^{-1}. Very high fluxes were obtained with the 0.1 μm pore-size microfiltration membrane (1.6 $m^3/h\ m^2$), leading to the preparation of 1.8 L nanoparticles, with an average diameter of 360 nm, in 4 min. The advantages of this membrane contactor compared to other processes for nanoparticle preparation are its scale-up ability and the possibility to control nanoparticle size by an appropriate choice of the membrane.

Magnetic polymer microspheres have attracted extensive attention in the past decades because of their rapid and low-cost separation properties in external magnetic field [119,120]. Various applications of magnetic polymer microspheres are found such as cell isolation, enzyme immobilization, immunoassay, protein separation, biochemical assays and drug targeting. Magnetic polymer particles at nano- and micro-scales were prepared by encapsulating magnetic components with dissolved or in situ formed polymers and subsequent membrane emulsification [119]. Yang et al. [120] proposed a novel method to prepare uniform magnetic

Figure 6.8 Microphotographs and SEM images of formed HTCC microspheres with 2.5 wt.% (a and b), 3.0 wt.% (c and d), 3.5 wt.% (e and f), 4.0 wt.% (g and h) of HTCC at 8.0 wt.% α-β-GP. (HTCC: *N*-[(2-hydroxy-3-trimethylammonium) propyl] chitosan chloride; α-β-GP: α-β-glycerophosphate) [63].

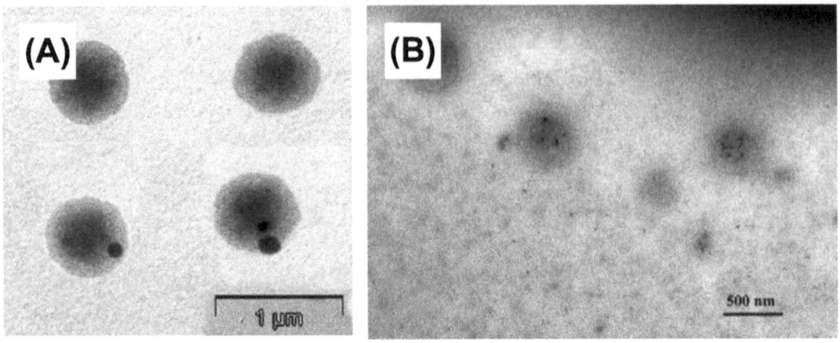

Figure 6.9 TEM of (A) spironolactone-loaded nanocapsules obtained in a membrane device [113], (B) vitamin E-loaded nanocapsules obtained in a membrane device [118].
Figure 9A: This article was published in Int. J. Pharm., Vol. 325, I. Limayem, C. Charcosset, S. Sfar, H. Fessi, Preparation and characterization of spironolactone-loaded nanocapsules for paediatric use, pp. 124–131, Copyright Elsevier (2006).
Figure 9B: This article was published in J. Membr. Sci., Vol. 266, C. Charcosset, H. Fessi, Preparation of nanoparticles with a membrane contactor, pp. 115–120, Copyright Elsevier (2005).

microspheres including the preparation of coarse double emulsions with ferrous ferric solution as the internal water phase and screening it into a uniform emulsion by premix membrane emulsification, then adding the ammonia solution into the external water phase to in situ magnetization of the ferrous ferric solution into nanoparticles, and finally solidifying the emulsion. The method was shown to facilitate the enhancement of encapsulation efficiency with a high magnetization up to 14.23 emu g^{-1}.

Stirred cell membrane emulsification has been used recently to prepare Pickering emulsions [121]. Pickering emulsions are emulsion droplets that are stabilized by solid particles such as silica sols and polystyrene latexes [122,123]. Membrane emulsification studies were conducted using a stirred cell (Micropore Technologies Ltd, UK). The device used a motor to drive a paddle stirrer above a stationary nickel membrane to create surface shear. The Pickering emulsions (44–269 μm in size) had narrower droplet size distributions than those prepared by conventional homogenization (25% vs. 74% CV). Finally, the addition of a suitable oil-soluble polymeric cross-linker enabled low polydispersity colloidosomes to be prepared in situ from the corresponding Pickering emulsions. Yuan et al. [124] also reported the preparation of silica-stabilized emulsions using both cross-flow and rotating membrane emulsification. In each case, stable Pickering emulsions were obtained with narrower droplet size distributions compared to those produced via rotor-stator homogenization using otherwise identical conditions.

Wei et al. [125] have shown that chitosan chemically cross-linked with glutaraldehyde display autofluorescent properties without the need to conjugate any external fluorochromes. This property was used in the preparation of fluorescent chitosan microspheres by the SPG membrane emulsification technique. The fluorescence colour varied with different cross-linkers and was modulated by further chemical reduction, whereas the fluorescence intensity was controlled by tuning the particle size and degree of cross-linking. In addition, the autofluorescence of the microspheres was applied to study the phagocytosis of HepG2 cells using the microspheres as tracers.

Other particles prepared via membrane emulsification included silica hydrogel particles [78], microcapsules of *Lactobacillus casei* [126], core-shell enzyme-loaded microcapsules with environment-responsive membranes [64], monodisperse thermo-sensitive poly(*N*-isopropylacrylamide) hollow microcapsules [127] and pH-sensitive particles [128].

6.3.6 Liposomes

Liposomes are spherical particles with size in the nanometre to micrometre range formed by polar lipids. The vesicular particles may consist of one or several bilayer membranes. Because of their structure, liposomes can entrap hydrophilic drugs in their internal aqueous compartment or lipophilic drugs within the lipid membrane [129]. Due to their biocompatibility, biodegradability and low toxicity, potential applications of liposomes as pharmaceutical carriers for efficacy enhancement and toxicity reduction are well recognized [130]. Since the pioneering discovery of Bangham, several techniques have been reported in the literature for liposome preparation: the thin-film hydration method [131], reversed phase evaporation [132], solvent injection techniques [133,134] and membrane extrusion [66,135].

For the membrane extrusion method, the dispersed phase containing large liposomes is forced to pass through a membrane with a uniform pore-size distribution generating a homogenous population of smaller vesicles. Membrane extrusion for the preparation of liposomes is reported to be an easy, reproducible method, producing no detectable degradation of the phospholipids, and a high encapsulation efficiency of the liposome preparation [66,135,136]. Multilamellar vesicles are shown to have mean diameters approaching the pore diameter of the polycarbonate membrane through which they are extruded. Other parameters which influence the process are the temperature and the applied pressure across the membrane. The 'membrane extrusion' method for the preparation of liposomes may be seen as a precursor process of membrane emulsification. The productivity of the membrane method for liposome preparation could probably be improved using recent findings on membrane emulsification (new materials and new configurations).

Recently, Jaafar-Maalej et al. [67] applied the ethanol injection technique using a membrane device for large-scale liposomes production. In this method, a lipid phase (ethanol, phospholipid and cholesterol) was pressed through a tubular SPG membrane. Pressures below 2 bar were sufficient for passing the organic phase through the membrane. At the same time, the aqueous phase flew tangentially to the membrane surface and swept away the liposomes formed within the membrane device. Drug-loaded liposomes were prepared and filled with two hydrophobic drug models. High entrapment efficiency values were successfully achieved for

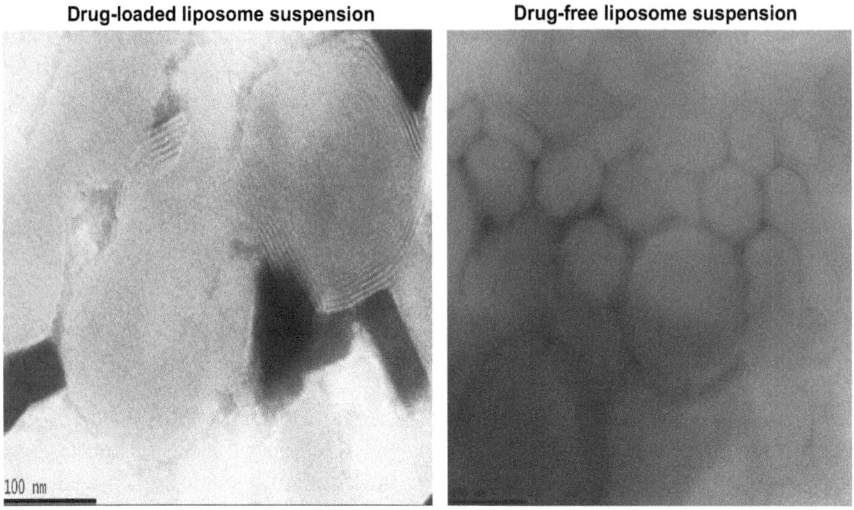

Figure 6.10 TEM micrographs of drug-free and drug-loaded liposome suspension. (Experimental conditions of liposome suspension preparation: 500 mL of water, 250 mL of ethanol containing 5 g of lipoid® E80 and 750 mg of spironolactone), feed pressure 1.8 bar [137]. This article was published in Int. J. Pharm., Vol. 415, A. Laouini, C. Jaafar-Maalej, S. Sfar, C. Charcosset, H. Fessi, Liposome preparation using a hollow fiber membrane contactor–Application to spironolactone encapsulation, pp. 53–61, Copyright Elsevier (2011).

indomethacin (63%) and beclomethasone dipropionate (98%). The advantages of this new process are its simplicity, the control of the liposome size by changing the process parameters and its scaling-up abilities.

The method was also effective using a hollow fibre membrane device (Liqui-cel® Mini Module X50, Alting, France) [137]. Liposomes were filled with a hydrophobic drug model, spironolactone, that could be used for a paediatric medication. Figure 6.10 shows TEM images of the liposomes obtained. The mean size of drug-free and drug-loaded liposomes was 113 and 123 nm, respectively. High entrapment efficiency values were successfully achieved (93%) (Table 6.2). The release profile showed a rapid and complete release within about 5 h. In addition, special attention was paid on process reproducibility and long-term lipid vesicles stability. Results confirmed the suitability of the membrane-based ethanol injection technique for liposome preparation.

6.3.7 Industrial applications

Membrane systems are particularly suitable for large-scale production because they are easy to scale-up, by adding more membranes to a module. Williams et al. [15] have shown that pilot-scale membrane emulsification can be operated successfully in both batch and semi-continuous modes. In Table 6.3, the advantages and drawbacks of membrane emulsification compared to traditional processes are

Table 6.2 Reproducibility data of drug-free and drug-loaded liposome suspensions [137]

Batch N°	Mean size ± SD* (nm)	Zeta potential ± SD* (mV)	Processing time (s)	Encapsulation efficiency ± SD* (%)
R1	114 ± 5	−44 ± 1.0	85	
R2	110 ± 3	−44 ± 0.6	84	
R3	115 ± 3	−41 ± 0.6	84	
S1	123 ± 2	−23 ± 1.8		94.2 ± 1.51
S2	120 ± 4	−21 ± 0.8		92.7 ± 1.62
S3	126 ± 5	−24 ± 1.0		92.0 ± 1.60

(Experimental conditions of liposome suspension preparation: 500 mL of water, 250 mL of ethanol containing 5 g of lipoid® E80 and 750 mg of spironolactone), feed pressure 1.8 bar.
* SD: standard deviation ($n = 3$).

Table 6.3 Advantages and drawbacks of membrane emulsification compared to conventional methods [6]

Advantages	Drawbacks
Membrane contactors (general)	
High and constant specific interfacial area [143]	Additional resistance to mass transfer created by the membrane
Independence of the fluid phases in contact	
No moving parts, use of plastic equipments (absence of corrosion)	Relatively low membrane lifetime; high replacement costs
High modularity and compatibility, straight-forward scale-up, easy control	
Membrane emulsification	
Lower energy input with respect to conventional emulsifiers [7]	Low permeation rates associated to narrow droplet size distribution and reduced productivity
No foaming, reduced coalescence phenomena [143]	
Narrow droplet size distributions	
Various configurations: premix emulsification [142], rotating devices [20] to reduce fouling and to increase productivity	Fouling phenomena on the membrane surface and/or in the pores

summarized. However, to date, there is only one documented product produced using membrane emulsification ('Yes light', a very low fat spread, Moringa Milk Industry, Japan) [12].

Some authors have compared the emulsification result using membranes and stirring methods [138,139]. Okochi and Nakano [138] provided a comparative study of properties of w/o/w emulsions prepared by membrane emulsification and a two-stage stirring method. The w/o/w emulsion prepared by membrane emulsification was found to be more homogeneous and provided slower release of low molecular–weight drugs. Dowding et al. [139] found that the droplet size distribution obtained by membrane emulsification was narrower than that obtained by stirring.

The main disadvantage of the membrane emulsification process may be a low dispersed phase flux through the membrane. Using SPG membranes, oil fluxes for preparation of o/w emulsions typically range from 5 to 15 $dm^3\ m^{-2}\ h^{-1}$ with a 0.2 μm membrane and 10 to 45 $dm^3\ m^{-2}\ h^{-1}$ with a 0.5 μm membrane [51]. Higher fluxes are obtained with ceramic membranes, that is 50–250 $dm^3\ m^{-2}\ h^{-1}$ using a 0.2 μm α-alumina membrane [50]. For preparation of w/o emulsions, water fluxes of 200 $dm^3\ m^{-2}\ h^{-1}$ are achieved using a 0.5 μm oil-pretreated hydrophilic SPG membrane and 2300 $dm^3\ m^{-2}\ h^{-1}$ using a 1 μm membrane [50]. The flux through the membrane is determined by the applied TMP, the permeability of the membrane and the number of active pores (Darcy's law). The SPG membranes have a low permeability, because they are quite thick (0.45–0.75 mm) and are homogeneous in structure. The number of active pores is reported to be very low, that is 0.3–0.5% [140] and 2% [1]. For ceramic membranes, the permeability is expected to be higher, because the top layer with the smallest pores is very thin; this layer being mechanically supported by layers with larger pore diameters [141].

The dispersed phase flux can be increased using repeated premix membrane emulsification [142,7]. As reported previously, the preliminary emulsified o/w or w/o emulsion is dispersed into the continuous phase via the membrane. Oil fluxes as high as 20,000 $dm^3\ m^{-2}\ h^{-1}$ were obtained by single premix emulsification using a 1 μm hydrophilic PTFE membrane [7].

6.4 Theory

6.4.1 Forces on a spherical droplet

Theoretical data for droplet formation are obtained by calculating the overall forces, assuming a rigid and spherical droplet [49,144–146].

The following are the main forces that act on the forming droplets (Figure 6.11):

- The interfacial tension force, F_γ, which represents the effects of dispersed phase adhesion around the edge of the pore opening:

$$F_\gamma = \pi \bar{d}_P \gamma \qquad (6.7)$$

- The static pressure difference force, F_{SP}, due to the pressure difference between the dispersed phase and the continuous phase at the membrane surface:

$$F_{SP} = \frac{\gamma}{\bar{d}_d} \pi \bar{d}_p^2 \qquad (6.8)$$

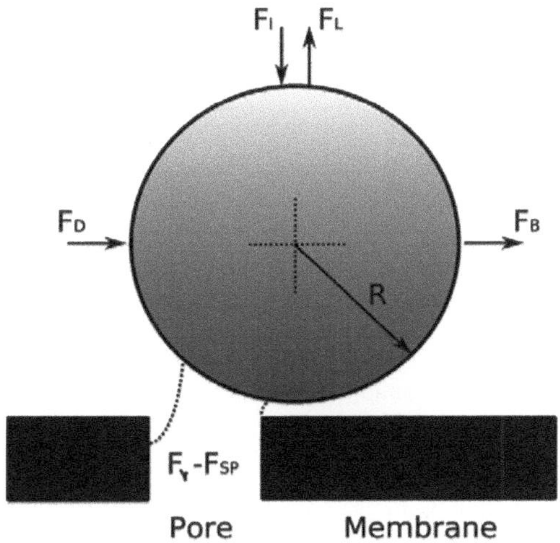

Figure 6.11 Forces acting on a droplet coming out of a pore.

- The drag force, F_{D}, created by the continuous phase flowing past the droplet parallel to the membrane surface:

$$F_{\mathrm{D}} = \frac{3}{2} k_{\mathrm{x}} \pi \tau_{\mathrm{c,s}} \overline{d}_{\mathrm{d}}^{2} \tag{6.9}$$

- The dynamic lift force, F_{L}, which results from the asymmetric velocity profile of the continuous phase near the droplet:

$$F_{\mathrm{L}} = 0.761 \frac{\tau_{\mathrm{c,s}}^{1.5} \rho_{\mathrm{c}}^{0.5}}{\mu_{\mathrm{c}}} \overline{d}_{\mathrm{d}}^{3} \tag{6.10}$$

- The buoyancy force, F_{B}, due to the density difference between the continuous phase and the dispersed phase:

$$F_{\mathrm{B}} = \frac{1}{6} \pi g \Delta \rho \overline{d}_{\mathrm{d}}^{3} \tag{6.11}$$

- The inertial force, F_{I}, is the linear momentum force associated with a mass of the fluid flowing out from the opening of the pore:

$$F_{\mathrm{I}} = \int_{A_{\mathrm{N}}} \rho_{\mathrm{d}} v_{\mathrm{p}}^{2} dA = \rho_{\mathrm{d}} \left(\frac{J_{\mathrm{d}}}{\varepsilon} \right)^{2} A_{\mathrm{N}} \tag{6.12}$$

The following equation is used to obtain the shear stress (at the membrane surface) from the average cross-flow velocity v_{c} of continuous phase, flowing in the membrane tube, with D_{i} the inner diameter:

$$\tau_{\mathrm{c,s}} = 0.0395 \frac{\mu_{\mathrm{c}}^{0,25} \rho_{\mathrm{c}}^{0,75} v_{\mathrm{c}}^{1,75}}{D_{i}^{0,25}} \tag{6.13}$$

with

A_N: cross-sectional area of the droplet neck,
k_x: equal to 1.7 and takes into account the wall correction factor for a single sphere touching an impermeable wall,
v_p: velocity of the dispersed phase inside the pore,
ε: membrane porosity,
$\tau_{c,s}$: wall shear stress,
γ: dynamic interfacial tension,
ρ_c: density of the continuous phase,
ρ_d: density of the dispersed phase,
μ_c: viscosity of the continuous phase,
$\Delta\rho$: the difference between continuous and dispersed phase density.

The relative magnitude of these forces changes as the droplet increases in size. Peng and Williams [144] proposed a torque balance between the drag, buoyancy and the interfacial tension force. Schröder et al. [49] stated that over the range of their experimental conditions, only the interfacial tension force, the dynamic effect of the pressure difference between the phases and the drag of the continuous phase had to be taken into account. However, they did not compare experimental and calculated data.

The force balance model was applied to other configurations, such as vibrating membranes by adding the additional forces induced by the transversal membrane movement [147], and drop detachment from a pore by adding the hydrodynamic force due to the liquid flux outgoing from a capillary [148]. Recently, De Luca et al. [149] compared the results obtained with two approaches: a model based on an algebraic torque balance equation (TBE) and one on a force balance equation (FBE) against experimental data available in literature. The analysis showed that the FBE model yielded better results than the TBE model under conditions of wall shear stress equal or larger than 7 Pa and membrane pore diameters below 1.5 μm.

6.4.2 Numerical studies

Numerical studies with various calculation methods have been reported for droplet formation. Abrahamse et al. [150] simulated the process of droplet break up at a 5 μm pore in cross-flow membrane emulsification using computational fluid dynamics (CFD). The maximum membrane porosity was calculated to be 1.5% to prevent coalescence of droplets growing on neighbouring pores. Both Kobayashi et al. [151] and Rayner et al. [152,153] simulated droplet formation from straight-through microchannels using a CFD software package with the so-called piecewise linear interface construction method and the Surface Evolver, respectively. Recently, the formation of oil drops from a single capillary with a diameter of 200 μm into a cross-flowing continuous water phase has been studied experimentally with the particle image velocimetry technique and numerically with the CFD software Fluent [154]. The drop formation time and the volume of the detached drop were used as validation parameters. The results from the two methods corresponded well, with a difference of less than 5% for the drop formation time and 10% for the drop volume.

Droplet formation was also simulated with the lattice Boltzmann method, a method suitable for modelling on the mesoscale [155]. Droplet detachment in a T-junction was investigated experimentally and theoretically, and both the shape of the droplet and the final droplet size were found comparable in the range of parameters investigated.

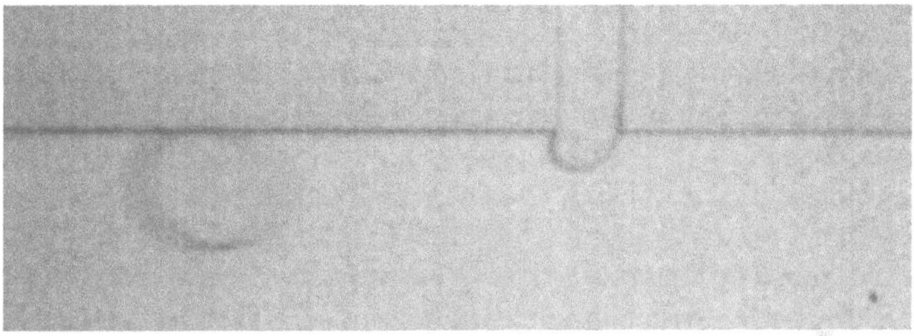

(a)

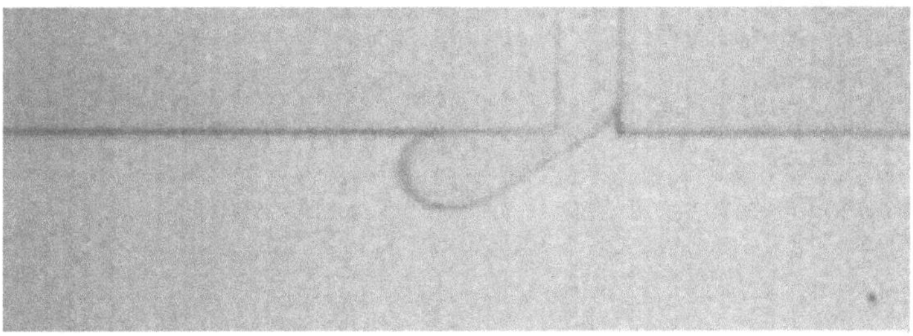

(b)

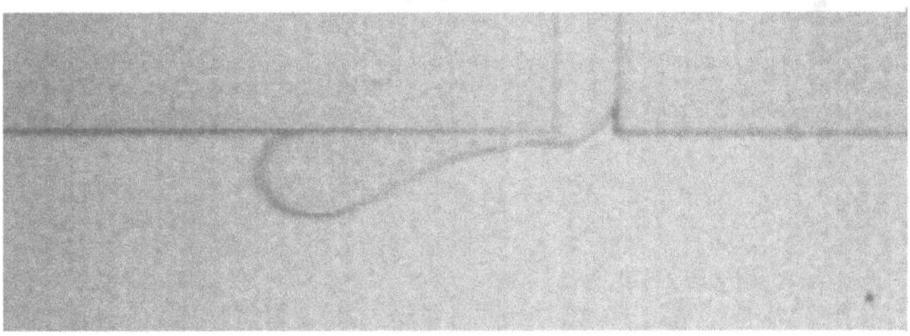

(c)

Figure 6.12 Hexadecane droplets formed in de-ionized water, continuous phase flow rate 10 μL min^{-1} and oil flow rate 0.04 μL min^{-1}. The previous droplet has just detached at (a) $t = 0$ s; (b) the droplet grows at $t = 0.006$ s and (c) the new droplet is going to detach at $t = 0.01$ s [157].
This article was published in Colloids Surf. A, Vol. 266, S. Van der Graaf, M.L.J. Steegmans, R.G.M. van de Sman, C.G.P.H. Schröen, R.M. Boom, Droplet formation in a T-shaped microchannel junction: A model system for membrane emulsification, pp. 106–116, Copyright Elsevier (2005).

6.4.3 Visualization of droplet formation

Visualization of droplet formation using single pores, micro-engineered membranes and microporous membranes were performed to better understand the membrane emulsification process using high-speed camera systems or optical microscopy [48,53,144,156,157]. Peng and Williams [144] used a high-speed video camera to record the formation of droplets from single pores (glass capillaries of diameter ranging from 5 to 200 μm, embedded in epoxy). The droplet growth and the detachment processes were measured as a function of TMP, membrane pore size and cross-flow velocity. Abrahamse et al. [48] visualized by microscopy the droplet formation at a micro-engineered membrane with uniform pores. Although the membrane pores were uniform in size, the obtained emulsions were polydispersed. Observations showed this to be due to steric hindrance of droplets forming simultaneously. Kobayashi et al. [53] provided a real-time microscopic observation of emulsion droplet formation from a polycarbonate membrane. Scherze et al. [156] developed an approach that determines the distribution of the inner aqueous phase in multiple emulsions by means of light microscopy in conjunction with image processing.

Van der Graaf et al. [157] studied droplet formation in a glass microchip with a small channel containing the dispersed phase perpendicular to a large channel with a cross-flow continuous phase. In this model system of membrane emulsification, droplets were formed at a T-junction of these two rectangular channels. The droplet formation and detachment were studied with a microscope connected to a high-speed camera. Monodispersed hexadecane droplets were formed in aqueous solutions with various concentrations of ethanol, SDS and Tween 20. As an example, Figure 6.12 shows hexadecane droplets formed in water. Droplet formation started with a half disc of hexadecane formed at the pore opening. The droplet grew further and was deformed in the direction of the aqueous phase flow. A neck was formed which holded the droplet connected to the pore while the continuous water phase flew through the pore. Finally, the droplet detached at the left side of the pore opening.

References

[1] G. Vladisavljević, H. Schubert, Production of Emulsions with a Narrow Droplet Size Distribution Using a Crossflow Shirasu Porous Glass (SPG) Membrane, 3ième Congrès Mondial de l'Emulsion, 24–27 September 2002, Lyon, France.

[2] M. Saito, L.J. Yin,˙I. Kobayashi, M. Nakajima, Preparation characteristics of monodispersed oil-in-water emulsions with large particles stabilized by proteins in straight-through microchannel emulsification, Food Hydrocolloid 19 (2005) 745–751.

[3] A.S. Utada, E. Lorenceau, D.R. Link, P.D. Kaplan, H.A. Stone, D.A. Weitz, Monodisperse double emulsions generated from a microcapillary device, Science 308 (2005) 537–541.

[4] S. Abraham, E.H. Jeong, T. Arakawa, S. Shoji, K.C. Kim, I. Kim, J.S. Go, Microfluidics assisted synthesis of well-defined spherical polymeric microcapsules and their utilization as potential encapsulants, Lab Chip 6 (2006) 752–756.

[5] P. Garstecki, M.J. Fuertsman, H.A. Stone, G.M. Whitesides, Formation of droplets and bubbles in a microfluidic T-junction—scaling and mechanism of break up, Lab Chip 6 (2006) 437–446.
[6] C. Charcosset, Preparation of emulsions and particles by membrane emulsification for the food processing industry, J. Food Eng. 92 (2009) 241–249.
[7] J. Altenbach-Rehm, K. Suzuki, H. Schubert, Production of O/W-Emulsions with Narrow Droplet Size Distribution by Repeated Premix Membrane Emulsification, 3ième Congrès Mondial de l'Emulsion, 24–27 September 2002, Lyon, France.
[8] S.M. Joscelyne, G. Trägårdh, Membrane emulsification-a literature review, J. Memb. Sci. 169 (2000) 107–117.
[9] C. Charcosset, I. Limayem, H. Fessi, The membrane emulsification process – a review, J. Chem. Biochem. Technol. 79 (2004) 209–218.
[10] A.J. Gijsbersten-Abrahamse, A. van der Padt, R.M. Boom, Status of cross-flow membrane emulsification and outlook for industrial application, J. Memb. Sci. 230 (2004) 149–159.
[11] G.T. Vladisavljević, R.A. Williams, Recent developments in manufacturing emulsions and particulate products using membranes, Adv. Colloid Interface Sci. 113 (2005) 1–20.
[12] T. Nakashima, M. Shimizu, M. Kukizaki, Particle control of emulsion by membrane emulsification and its application, Adv. Drug Deliv. Rev. 45 (2000) 47–56.
[13] E. Piacentini, A. Figoli, L. Giorno, E. Drioli, Membrane emulsification, in: E. Drioli, L. Giorno (Eds.), Comprehensive Membrane Science and Engineering, Elsevier, Kidlington, UK, 2010, pp. 47–75.
[14] W. Liu, X.L. Yang, W.S.W. Ho, Preparation of uniform-sized multiple emulsions and micro/nano particulates for drug delivery by membrane emulsification, J. Pharm. Sci. 100 (2011) 75–93.
[15] R.A. Williams, S.J. Peng, D.A. Wheeler, N.C. Morley, D. Taylor, M. Whalley, D.W. Houldsworth, Controlled production of emulsions using a crossflow membrane, part II: industrial scale manufacture, Chem. Eng. Res. Des. 76 (1998) 902–910.
[16] V. Schröder, H. Schubert, Production of emulsions using microporous, ceramic membranes, Colloids Surf. A 152 (1999a) 103–109.
[17] M. Kukizaki, M. Goto, Preparation and characterization of a new asymmetric type of Shirasu porous glass (SPG) membrane used for membrane emulsification, J. Memb. Sci. 299 (2007a) 190–199.
[18] M.T. Stillwell, R.G. Holdich, S.R. Kosvintsev, G. Gasparini, I.W. Cumming, Stirred cell membrane emulsification and factors influencing dispersion drop size and uniformity, Ind. Eng. Chem. Res. 46 (2007) 965–972.
[19] S.R. Kosvintsev, G. Gasparini, R.G. Holdich, I.W. Cumming, M.T. Stillwell, Liquid–liquid membrane dispersion in a stirred cell with and without controlled shear, Ind. Eng. Chem. Res. 44 (2005) 9323–9330.
[20] N. Aryanti, R.A. Williams, R. Hou, G.T. Vladisavljevic, Performance of rotating membrane emulsification for o/w production, Desalination 200 (2006) 572–574.
[21] V. Schadler, E.J. Windhab, Continuous membrane emulsification by using a membrane system with controlled pore distance, Desalination 189 (2006) 130–135.
[22] G. Muschiolik, H. Bunjes, Multiple Emulsionen, Behr's Verlag, 2007, pp. 61 and 312.
[23] N. Aryanti, R. Hou, R.A. Williams, Performance of a rotating membrane emulsifier for production of coarse droplets, J. Memb. Sci. 326 (2009) 9–18.
[24] A. Koris, E. Piacentini, G. Vatai, E. Bekassy-Molnar, E. Drioli, L. Giorno, Investigation on the effects of a mechanical shear-stress modification method during cross-flow membrane emulsification, J. Memb. Sci. 371 (2011) 28–36.

[25] G.T. Vladisavljević, M. Shimizu, T. Nakashima, Preparation of monodisperse multiple emulsions at high production rates by multi-stage premix membrane emulsification, J. Memb. Sci. 244 (2004a) 97–106.
[26] A. Nazir, K. Schroën, R. Boom, Premix emulsification: a review, J. Memb. Sci. 362 (2010) 1–11.
[27] T. Nakashima, M. Shimizu, M. Kukizaki, Membrane emulsification by microporous glass, Key Eng. Mater. 61–62 (1991) 513–516.
[28] K. Kandori, K. Kishi, T. Ishikawa, Formation mechanisms of monodispersed W/O emulsions by SPG filter emulsification method, Colloids Surf. 61 (1991) 269–270.
[29] C.J. Cheng, L.Y. Chu, R. Xie, Preparation of highly monodisperse W/O emulsions with hydrophobically modified SPG membranes, J. Colloid Interface Sci. 300 (2006) 375–382.
[30] G.T. Vladisavljević, M. Shimizu, T. Nakashima, Permeability of hydrophilic and hydrophobic Shirasu-porous-glass (SPG) membranes to pure liquids and its microstructure, J. Memb. Sci. 250 (2005) 69–77.
[31] G.T. Vladisavljević, I. Kobayashi, M. Nakajima, R.A. Williams, M. Shimizu, T. Nakashima, Shirasu porous glass membrane emulsification: characterization of membrane structure by high-resolution X-ray microtomography and microscopic observation of droplet formation in real time, J. Memb. Sci. 302 (2007) 243–253.
[32] J. Zhu, D. Barrow, Analysis of droplet size during crossflow membrane emulsification using stationary and vibrating micromachined silicon nitride membranes, J. Memb. Sci. 261 (2005) 136–144.
[33] M.J. Geerken, R.G.H. Lammertink, M. Wessling, Tailoring surface properties for controlling droplet formation at microsieve membranes, Colloids Surf. A 292 (2007) 224–235.
[34] G. Gasparini, S.R. Kosvintsev, M.T. Stillwell, R.G. Holdich, Preparation and characterization of PLGA particles for subcutaneous controlled drug release by membrane emulsification, Colloids Surf.B 61 (2008) 199–207.
[35] R. Tangirala, R. Revanur, T.P. Russell, T. Emrick, Sizing nanoparticle-covered droplets by extrusion through track-etch membranes, Langmuir 23 (2007) 965–969.
[36] V. Schröder, H. Schubert, Influence of emulsifier and pore size on membrane emulsification, Spec. Publ – R Soc. Chem. 227 (1999) 70–80.
[37] S.M. Joscelyne, G. Trägårdh, Food emulsions using membrane emulsification: conditions for producing small droplets, J. Food Eng. 39 (1999) 59–64.
[38] K. Suzuki, I. Fujiki, Y. Hagura, Preparation of corn oil/water and water/corn oil emulsions using PTFE membranes, Food Sci. Technol. 4 (1998) 164–167.
[39] S. Kanichi, O. Yuko, H. Yoshio, Properties of Solid Fat O/W Emulsions Prepared by Membrane Emulsification Method Combined with Pre-emulsification. 3ième Congrès Mondial de l'Emulsion, 24–27 September 2002, Lyon, France.
[40] N. Yamazaki, H. Yuyama, M. Nagai, G.H. Ma, S. Omi, A comparison of membrane emulsification obtained using SPG (Shirasu Porous Glass) and PTFE [poly(tetrafluoroethylene)] membranes, J. Dispersion Sci. Technol. 23 (2002) 279–292.
[41] K. Sotoyama, Y. Asano, K. Ihara, K. Takahashi, K. Doi, Water/Oil emulsions prepared by the membrane emulsification method and their stability, J. Food Sci. 64 (1999) 211–215.
[42] T. Fuchigami, M. Toki, K. Nakanishi, Membrane emulsification using sol-gel derived macroporous silica glass, J. Sol-Gel Sci. Technol. 19 (2000) 337–341.
[43] N. Yamazaki, K. Naganuma, M. Nagai, G.-H. Ma, S. Omi, Preparation of w/o (water-in-oil) emulsions using a PTFE (polytetrafluoroethylene) membrane – a new emulsification device, J. Dispersion Sci. Technol. 24 (2003) 249–257.

[44] L. Giorno, N. Li, E. Drioli, Preparation of oil-in-water emulsions using polyamide 10 kDa hollow fiber membrane, J. Memb. Sci. 217 (2003) 173–180.
[45] S. Omi, Preparation of monodisperse microspheres using the Shirasu porous glass emulsification technique, Colloids Surf. A 109 (1996) 97–107.
[46] S. Omi, G.H. Ma, M. Nagai, Membrane emulsification – a versatile tool for the synthesis of polymeric microspheres, Macromol. Symp. 151 (2000) 319–330.
[47] G.T. Vladisavljević, U. Lambrich, M. Nakajima, H. Schubert, Production of O/W emulsions using SPG membranes, ceramic α-aluminium oxide membranes, microfluidizer and a silicon microchannel plate – a comparative study, Colloids Surf. A 232 (2004b) 199–207.
[48] A.J. Abrahamse, R. van Lierop, R.G.M. van der Sman, A. van der Padt, R.M. Boom, Analysis of droplet formation and interactions during cross-flow membrane emulsification, J. Memb. Sci. 204 (2002) 125–137.
[49] V. Schröder, O. Behrend, H. Schubert, Effect of dynamic interfacial tension on the emulsification process using microporous, ceramic membranes, J. Colloid Interface Sci. 202 (1998) 334–340.
[50] R. Katoh, Y. Asano, A. Furuya, K. Sotoyama, M. Tomita, Preparation of food emulsions using a membrane emulsification system, J. Memb. Sci. 113 (1996) 131–135.
[51] I. Scherze, K. Marzilger, G. Muschiolik, Emulsification using micro porous glass (MPG): surface behaviour of milk proteins, Colloids Surf. B 12 (1999) 213–221.
[52] H. Yuyama, T. Watanabe, G.-H. Ma, M. Nagai, S. Omi, Preparation and analysis of uniform emulsion droplets using SPG membrane emulsification technique, Colloids Surf. A 168 (2000) 159–174.
[53] I. Kobayashi, M. Yasuno, S. Iwamoto, A. Shono, K. Satoh, M. Nakajima, Microscopic observation of emulsion droplet formation from a polycarbonate membrane, Colloids Surf. A 207 (2002) 185–196.
[54] G. Muschiolik, S. Dräger, I. Scherke, H.M. Rawel, M. Stang, Protein-stabilized emulsions prepared by the micro-porous glass method, in: E. Dickinson, B. Bergenstähl (Eds.), Food Colloids: Proteins, Lipids and Polysaccharides, Royal Society of Chemistry, Cambridge, 1997, pp. 393–400.
[55] S. van der Graaf, C.G.P.H. Schroën, R.G.M. van der Sman, R.M. Boom, Influence of dynamic interfacial tension on droplet formation during membrane emulsification, J. Colloid Interface Sci. 277 (2004) 456–463.
[56] N.A. Wagdare, A.T.M. Marcelis, O.B. Ho, R.M. Boom, C.J.M. van Rijn, High throughput vegetable oil-in-water emulsification with a high porosity micro-engineered membrane, J. Memb. Sci. 347 (2010) 1–7.
[57] A. Trentin, C. Güell, F. López, M. Ferrando, Microfiltration membranes to produce BSA-stabilized O/W emulsions by premix membrane emulsification, J. Memb. Sci. 356 (2010) 22–32.
[58] M. Kukizaki, M. Goto, Effects of interfacial tension and viscosities of oil and water phases on monodispersed droplet formation using a Shirasu-porous-glass (SPG) membrane, Membrane 31 (2006) 215–220.
[59] G.T. Vladisavljević, S. Tesch, H. Schubert, Preparation of water-in-oil emulsions using microporous polypropylene hollow fibers: influence of some operating parameters on droplet size distribution, Chem. Eng. Process. 41 (2002) 231–238.
[60] S. Higashi, M. Shimizu, T. Setoguchi, Preparation of new lipiodol-emulsion containing water soluble anticancer agent by membrane emulsification technique, Drug Deliv. Sys. 8 (1993) 59–61.

[61] C. Charcosset, A.A. El-Harati, H. Fessi, Preparation of solid lipid nanoparticles for controlled drug delivery using a membrane contactor, J. Control. Release 108 (2005) 112–120.
[62] M. Kukizaki, M. Goto, Preparation and evaluation of uniformly sized solid lipid microcapsules using membrane emulsification, Colloids Surf. A 293 (2007b) 87–94.
[63] J. Wu, W. Wei, L.Y. Wang, Z.G. Su, G.H. Ma, Preparation of uniform-sized pH-sensitive quaternized chitosan microsphere by combining membrane emulsification technique and thermal-gelation method, Colloids Surf. B 63 (2008) 164–175.
[64] K. Akamatsu, T. Yamaguchi, Novel preparation method for obtaining pH-responsive core-shell microcapsule reactors, Ind. Eng. Chem. Res. 46 (2007) 124–130.
[65] A. Bals, U. Kulozik, The influence of pore size, the foaming temperature and the viscosity of the continuous phase on the properties of foams produced by membrane foaming, J. Memb. Sci. 220 (2003) 5–11.
[66] F. Olson, C.A. Hunt, F.C. Szoka, W.J. Vail, D. Papahadjopoulos, Preparation of liposomes of defined size distribution by extrusion through polycarbonate membranes, Biochim. Biophys. Acta 557 (1979) 9–23.
[67] C. Jaafar-Maalej, C. Charcosset, H. Fessi, A new method for liposome preparation using a membrane contactor, J. Liposome Res. 21 (2011) 213–220.
[68] E. Toorisaka, H. Ono, K. Arimori, N. Kamiya, M. Goto, Hypoglycemic effect of surfactant-coated insulin solubilized in a novel solid-in-oil-in-water (S/O/W) emulsion, Int. J. Pharm. 252 (2003) 271–274.
[69] J. Rao, D.J. McClements, Food-grade microemulsions, nanoemulsions and emulsions: fabrication from sucrose monopalmitate & lemon oil, Food Hydrocolloid. 25 (2011) 1413–1423.
[70] L. Giorno, E. Piacentini, R. Mazzei, E. Drioli, Membrane emulsification as a novel method to distribute phase-transfer biocatalysts at the oil/water interface in bioorganic reactions, J. Memb. Sci. 317 (2008) 19–25.
[71] D.H. Oh, P. Balakrishnan, Y.K. Ohc, D.D. Kim, C.S. Yong, H.G. Choi, Effect of process parameters on nanoemulsion droplet size and distribution in SPG membrane emulsification, Int. J. Pharm. 404 (2011) 191–197.
[72] N. Garti, Double emulsions scope, limitations and new achievements, Colloid Surf. A 123/124 (2000) 233–246.
[73] Y. Mine, M. Shimizu, T. Nakashima, Preparation and stabilization of simple and multiple emulsions using a microporous glass membrane, Colloid Surf. B 6 (1996) 261–268.
[74] S. Higashi, T. Setoguchi, Hepatic arterial injection chemotherapy for hepatocellular carcinoma with epirubicin aqueous solution as numerous vesicles in iodinated poppy-seed oil microdroplets: clinical application of water-in-oil-in-water emulsion prepared using a membrane emulsification technique, Adv. Drug Deliv. Rev. 45 (2000) 57–64.
[75] S. van der Graaf, C.G.P.H. Schroöen, R.M. Boom, Preparation of double emulsions by membrane emulsification—a review, J. Memb. Sci. 251 (2005) 7–15.
[76] G.T. Vladisavljević, M. Shimizu, T. Nakashima, Production of multiple emulsions for drug delivery systems by repeated SPG membrane homogenization: influence of mean pore size, interfacial tension and continuous phase viscosity, J. Memb. Sci. 284 (2006) 373–383.
[77] E. Piacentini, E. Drioli, L. Giorno, Preparation of stimulus responsive multiple emulsions by membrane emulsification using Con A as biochemical sensor, Biotechnol. Bioeng. 108 (2011) 913–923.

[78] K. Kandori, K. Kishi, T. Ishikawa, Preparation of uniform silica hydrogel particles by SPG filter emulsification method, Colloids Surf. 62 (1992) 259–262.
[79] N. Müller-Fischer, H. Bleuler, E.J. Windhab, Dynamically enhanced membrane foaming, Chem. Eng. Sci. 62 (2007) 4409–4419.
[80] P. Perugini, I. Genta, F. Pavanetto, B. Conti, S. Scalia, A. Baruffini, Study on glycolic acid delivery by liposomes and microspheres, Int. J. Pharm. 196 (2000) 51–61.
[81] H.L. Guo, X.P. Zhao, Preparation of a kind of red encapsulated electrophoretic ink, Opt. Mater. 26 (2004) 297–300.
[82] P.G. Shukla, B. Kalidhass, A. Shah, D.V. Palaskar, Preparation and characterization of microcapsules of water-soluble pesticide monocrotophos using polyurethane as carrier material, J. Microencapsul. 19 (2002) 293–304.
[83] M.M. Maciejewska, J. Osypiuk, B. Gawdzik, Preparation and characterization of the chromatographic properties of ethylene glycol dimethacrylate/divinylbenzene polymeric microspheres, J. Polym. Sci. 43 (2005) 3049–3058.
[84] S. Freiberg, X.X. Zhu, Polymer microspheres for controlled drug release, Int. J. Pharm. 282 (2004) 1–18.
[85] J. Hanes, J.L. Cleland, R. Langer, New advances in microsphere-based single-dose vaccines, Adv. Drug Deliv. Rev. 28 (1997) 97–119.
[86] H. Chen, R. Langer, Oral particulate delivery: status and future trends, Adv. Drug Deliv. Rev. 34 (1998) 339–350.
[87] S. Prior, C. Gamazo, J.M. Irache, H.P. Merkle, B. Gander, Gentamicin encapsulation in PLA/PLGA microspheres in view of treating Brucella infections, Int. J. Pharm. 196 (2000) 115–125.
[88] M.H. Perez, C. Zinutti, A. Lamprecht, N. Ubrich, A. Astier, M. Hoffman, R. Bodmeier, P. Maincent, The preparation and evaluation of poly([epsilon]-caprolactone) microparticles containing both a lipophilic and a hydrophilic drug, J. Control. Release 65 (2000) 429–438.
[89] P.A. Dayton, J.E. Chomas, A.F.H. Lum, J.S. Allen, J.R. Lindner, S.I. Simon, K.W. Ferrara, Optical and acoustical dynamics of microbubble contrast agents inside neutrophils, Biophys. J. 80 (2001) 1547–1556.
[90] S. Omi, K. Katami, A. Yamamoto, M. Iso, Synthesis of polymeric microspheres employing SPG emulsification technique, J. Appl. Polym. Sci. 51 (1994) 1–11.
[91] G.H. Ma, A.Y. Chen, Z.G. Su, S. Omi, Preparation of uniform hollow polystyrene particles with large voids by a glass-membrane emulsification technique and a subsequent suspension polymerization, J. Appl. Polym. Sci. 87 (2003) 244–251.
[92] S. Tawonsree, S. Omi, S. Kiatkamjornwong, Control of various morphological changes of poly(meth)acrylate microspheres and their swelling degrees by SPG emulsification, J. Polym. Sci. 38 (2000) 4038–4056.
[93] S. Omi, T. Senba, M. Nagai, G.H. Ma, Morphology development of 10-μm scale polymer particles prepared by SPG emulsification and suspension polymerization, J. Appl. Polym. Sci. 79 (2001) 2200–2220.
[94] G.H. Ma, S. Omi, Mechanism of formation of monodisperse polystyrene hollow particles prepared by membrane emulsification technique. Effect of hexadecane amount on the formation of hollow particles, Macromol. Symp. 179 (2002) 223–240.
[95] R. Wang, Y. Zhang, G. Ma, Z. Su, Preparation of uniform poly(glycidyl methacrylate) porous microspheres by membrane emulsification – polymerization technology, J. Appl. Polym. Sci. 102 (2006) 5018–5027.
[96] A. Supsakulchai, G.H. Ma, M. Nagai, S. Omi, Preparation of uniform titanium dioxide (TiO_2) polystyrene-based composite particles using the glass membrane emulsification

process with a subsequent suspension polymerization, J. Microencapsul. 20 (2003) 1–18.
[97] G.H. Ma, M. Nagai, S. Omi, Effect of lauryl alcohol on morphology of uniform polystyrene-poly(methyl methacrylate) composite microspheres prepared by porous glass membrane emulsification technique, J. Colloid Interface Sci. 219 (1999) 110–128.
[98] A. Supsakulchai, G.H. Ma, M. Nagai, S. Omi, Uniform titanium dioxide (TiO_2) microcapsules prepared by glass membrane emulsification with subsequent solvent evaporation, J. Microencapsul. 19 (2002) 425–449.
[99] S. Omi, K. Katami, T. Taguchi, K. Kaneko, M. Iso, Synthesis of uniform PMMA microspheres employing modified SPG (Shirasu Porous Glass) emulsification technique, J. Appl. Polym. Sci. 57 (1995) 1013–1024.
[100] H. Qu, F. Gong, G. Ma, Z. Su, Preparation and characterization of large porous poly(HEMA-co-EDMA) microspheres with narrow size distribution by modified membrane emulsification method, Appl. Polym. Sci. 105 (2007) 1632–1641.
[101] G.H. Ma, C.J. An, H. Yuyama, Z.G. Su, S. Omi, Synthesis and characterization of polyurethaneurea-vinyl polymer (PUU-VP) uniform hybrid microspheres by SPG emulsification technique and subsequent suspension polymerization, J. Appl. Polym. Sci. 89 (2003) 163–178.
[102] K. Kooiman, M.R. Böhmer, M. Emmer, H.J. Vos, C. Chlon, W.T. Shi, C.S. Hall, S.H.P.M. de Winter, K. Schroën, M. Versluis, N. de Jong, A. van Wamel, Oil-filled polymer microcapsules for ultrasound-mediated delivery of lipophilic drugs, J. Control. Release 133 (2009) 109–118.
[103] S. Nagashima, M. Koide, S. Ando, K. Makino, T. Tsukamoto, H. Ohshima, Surface properties of monodisperse poly(acrylamide-co-acrylic acid) hydrogel microspheres prepared by a membrane emulsification technique, Colloids Surf. A 153 (1999) 221–227.
[104] K. Makino, H. Agata, H. Ohshima, Dependence of temperature-sensitivity of poly(N-isopropylacrylamide-co-acryl acid) hydrogel microspheres upon their sizes, J. Colloid Interface Sci. 230 (2000) 128–134.
[105] Q.Z. Zhou, L.Y. Wang, G.H. Ma, Z.G. Su, Preparation of uniform-sized agarose beads by microporous membrane emulsification technique, J. Colloid Interface Sci. 311 (2007) 118–127.
[106] S.B. Park, Y.J. Jeon, S. Haam, H.Y. Park, W.S. Kim, Preparation of chitosan microspheres using membrane emulsification and its size modelling, J. Microencapsul. 21 (2004) 539–552.
[107] H. Yoshizawa, M. Maruta, S. Ikeda, Y. Hatate, Y. Kitamura, Preparation and pore-size control of hydrophilic monodispersed polymer microspheres for size-exclusive separation of biomolecules by the SPG membrane emulsification technique, Colloid Polym. Sci. 282 (2004) 965–971.
[108] H. Piao, A. Hirata, H. Yokoyama, T. Fujii, I. Shimizu, S. Ito, N. Kamiya, M. Goto, Reduction of gastric ulcerogenicity during multiple administration of diclofenac sodium by a novel solid-in-oil suspension, Pharm. Dev. Technol. 12 (2007) 321–325.
[109] M.S. Costa, M.M. Cardoso, Effect of uniform sized polymeric microspheres prepared by membrane emulsification technique on controlled release of anthracycline anticancer drugs, Desalination 200 (2006) 498–500.
[110] H.S. Ribeiro, L.G. Rico, G.G. Badolato, H. Schubert, Production of O/W emulsions containing astaxanthin by repeated premix membrane emulsification, J. Food Sci. 70 (2005) E117–E123.

[111] A.J. Ribeiro, C. Silva, D. Ferreira, F. Veiga, Chitosan-reinforced alginate microspheres obtained through the emulsification/internal gelation technique, Eur. J. Pharm. Sci. 25 (2005) 31–40.
[112] R. Liu, G.H. Ma, Y.H. Wan, Z.G. Su, Influence of process parameters on the size distribution of PLA microcapsules prepared by combining membrane emulsification technique and double emulsion-solvent evaporation method, Colloids Surf. B 45 (2005) 144–153.
[113] I. Limayem, C. Charcosset, S. Sfar, H. Fessi, Preparation and characterization of spironolactone-loaded nanocapsules for paediatric use, Int. J. Pharm. 325 (2006) 124–131.
[114] T.H. Yun, K.S. Kim, S.H. Cho, K.H. Youm, Preparation of polycaprolactone microcapsules by membrane emulsification method and its drug release properties, Membrane J. 17 (2007) 67–69.
[115] W. Wei, L. Yuan, G. Hu, L.Y. Wang, J. Wu, X. Hu, Z.G. Su, G.H. Ma, Monodisperse chitosan microspheres with interesting structures for protein drug delivery, Adv. Mater. 20 (2008) 2292–2296.
[116] R.F. Meyer, W.B. Rogers, M.T. McClendon, J.C. Crocker, Producing monodisperse drug-loaded polymer microspheres via cross-flow membrane emulsification: the effects of polymers and surfactants, Langmuir 26 (2010) 14479–14487.
[117] C. D'Oria, C. Charcosset, A. Barresi, H. Fessi, Preparation of solid lipid particles by membrane emulsification: influence of process parameters, Colloids Surf. A 338 (2009) 114–118.
[118] C. Charcosset, H. Fessi, Preparation of nanoparticles with a membrane contactor, J. Memb. Sci. 266 (2005) 115–120.
[119] Q. Yuan, R.A. Williams, Large scale manufacture of magnetic polymer particles using membranes and microfluidic devices, China Particuology 5 (2007) 26–42.
[120] J. Yang, D.X. Hao, C.X. Bi, Z.G. Su, L.Y. Wang, G.H. Ma, Rapid synthesis of uniform magnetic microspheres by combing premix membrane emulsification and in situ formation techniques, Ind. Eng. Chem. Res. 49 (2010) 6047–6053.
[121] K.L. Thompson, S.P. Armes, D.W. York, Preparation of Pickering emulsions and colloidosoes with relatively narrow size distributions by stirred cell membrane emulsification, Langmuir 27 (2011) 2357–2363.
[122] B.P. Binks, S.O. Lumsdon, Stability of oil-in-water emulsions stabilised by silica particles, Phys. Chem. Chem. Phys. 1 (1999) 3007–3016.
[123] J.I. Amalvy, S.P. Armes, B.P. Binks, J.A. Rodrigues, G.F. Unali, Use of sterically-stabilised polystyrene latex particles as a pH-responsive particulate emulsifier to prepare surfactant-free oil-in-water emulsions, Chem. Commun. 15 (2003) 1826–1827.
[124] Q.C. Yuan, O.J. Cayre, M. Manga, R.A. Williams, S. Biggs, Preparation of particle-stabilized emulsions using membrane emulsification, Soft Matter 6 (2010) 1580–1588.
[125] W. Wei, L.Y. Wang, L. Yuan, Q. Wei, X.D. Yang, Z.G. Su, G.H. Ma, Preparation and application of novel microspheres possessing autofluorescent properties, Adv. Funct. Mater. 17 (2007) 3153–3158.
[126] S.H. Song, Y.H. Cho, J. Park, Microencapsulation of Lactobacillus casei YIT 9018 using a microporous glass membrane emulsification system, J. Food Sci. 68 (2003) 195–200.
[127] C.J. Cheng, L.Y. Chu, P.W. Ren, J. Zhang, L. Hu, Preparation of monodisperse thermo-sensitive poly(N-isopropylacrylamide) hollow microcapsules, J. Colloid Interface Sci. 313 (2007) 383–388.

[128] N. Sheibat-Othman, T. Burne, C. Charcosset, H. Fessi, Preparation of pH-sensitive particles by membrane emulsification, Colloids Surf. A 315 (2008) 13–22.
[129] V.P. Torchilin, Recent advances with liposomes as pharmaceutical carriers, Drug Discov. 4 (2005) 145–160.
[130] T. Lian, R.J. Ho, Trends and developments in liposome drug delivery systems, J. Pharm. Sci. 90 (2005) 667–680.
[131] A.D. Bangham, Properties and uses of lipid vesicles: an overview, Ann. N. Y. Acad. Sci. 308 (1978) 2–7.
[132] F.J. Szoka, D. Papahadjopoulos, Procedure for preparation of liposomes with large internal aqueous space and high capture by reverse-phase evaporation, Proc. Natl. Acad. Sci. U S A 75 (1978) 4194–4198.
[133] D.W. Deamer, Preparation and properties of ether-injection liposomes, Ann. N. Y. Acad. Sci. 308 (1978) 250–258.
[134] P. Stano, S. Bufali, C. Pisano, F. Bucci, M. Barbarino, M. Santaniello, P. Carminati, P.L. Luisi, Novel camptothecin analogue (gimatecan)-containing liposomes prepared by the ethanol injection method, J. Liposome Res. 14 (2004) 87–109.
[135] B. Mui, L. Chow, M.J. Hope, Extrusion technique to generate liposomes of defined size, Methods Enzymol. 367 (2003) 3–14.
[136] M.J. Hope, M.B. Bally, G. Webb, P.R. Cullis, Production of large unilamellar vesicles by a rapid extrusion procedure. Characterization of size distribution, trapped volume and ability to maintain a membrane potential, Biochim. Biophys. Acta 812 (1985) 55–65.
[137] A. Laouini, C. Jaafar-Maalej, S. Sfar, C. Charcosset, H. Fessi, Liposome preparation using a hollow fiber membrane contactor – application to spironolactone encapsulation, Int. J. Pharm. 415 (2011) 53–61.
[138] H. Okochi, M. Nakano, Comparative study of two preparation methods of w/o/w emulsions: stirring and membrane emulsification, Chem. Pharm. Bull. 45 (1997) 1323–1326.
[139] P.J. Dowding, J.W. Goodwin, B. Vincent, Production of porous suspension polymer beads with a narrow size distribution using a cross-flow membrane and a continuous tubular reactor, Colloids Surf. A 180 (2001) 301–309.
[140] M. Yasuno, M. Nakajima, S. Iwamoto, T. Maruyama, S. Sugiura, I. Kobayashi, A. Shono, K. Satoh, Visualization and characterization of SPG membrane emulsification, J. Memb. Sci. 210 (2002) 29–37.
[141] A.J. Gijsbersten-Abrahamse, A. van der Padt, R.M. Boom, Membrane Emulsification: The Influence of Pore Geometry and Wall Contact Angle on Process Performance, 3ième Congrès Mondial de l'Emulsion, 24–27 September 2002, Lyon, France.
[142] S.H. Park, T. Yamaguchi, S.I. Nakao, Transport mechanism of deformable droplets in microfiltration of emulsions, Chem. Eng. Sci. 56 (2001) 3539–3548.
[143] E. Drioli, E. Curcio, G. Di Profio, State of the art and recent progresses in membrane contactors, Trans. IChemE A 83 (2005) 223–233.
[144] S.J. Peng, R.A. Williams, Controlled production of emulsions using a crossflow membrane, part I: droplet formation from a single pore, Trans. IChemE 76 (1998) 894–901.
[145] M. Rayner, G. Trägårdh, Membrane emulsification modelling: how can we get from characterisation to design? Desalination 145 (2002) 65–172.
[146] G. De Luca, E. Drioli, Force balance conditions for droplet formation in cross-flow membrane emulsifications, J. Colloid Interface Sci. 294 (2006) 436–448.
[147] J.D.H. Kelder, J.J.M. Janssen, R.M. Boom, Membrane emulsification with vibrating membranes: a numerical study, J. Memb. Sci. 304 (2007) 50–59.

[148] K.D. Danov, D.K. Danova, P.A. Kralchevsky, Hydrodynamic forces acting on a microscopic emulsion drop growing at a capillary tip in relation to the process of membrane emulsification, J. Colloid Interface Sci. 316 (2007) 844–857.

[149] G. De Luca, F.P. Di Maio, A. Di Renzo, Droplet detachment in cross-flow membrane emulsification: comparison among torque- and force-based models, Chem. Eng. Process. 47 (2008) 1150–1158.

[150] A.J. Abrahamse, A. van der Padt, R.M. Boom, W.B.C. Heij, Process fundamentals of membrane emulsification: simulation with CFD, AIChE J. 47 (2001) 1285–1291.

[151] I. Kobayashi, S. Mukataka, M. Nakajima, CFD simulation and analysis of emulsion droplet formation from straight-through microchannels, Langmuir 20 (2004) 9868–9877.

[152] M. Rayner, G. Trägårdh, C. Trägårdh, P. Dejmek, Using the surface evolver to model droplet formation processes in membrane emulsification, J. Colloid Interface Sci. 279 (2004) 175–185.

[153] M. Rayner, G. Trägårdh, C. Trägårdh, The impact of mass transfer and interfacial expansion rate on droplet size in membrane emulsification processes, Colloid Surf. A 266 (2005) 1–17.

[154] A. Timgren, G. Trägårdh, C. Trägårdh, Effects of pore spacing on drop size during cross-flow membrane emulsification—a numerical study, J. Memb. Sci. 337 (2009) 232–239.

[155] S. van der Graaf, T. Nisisako, C.G.P.H. Schröen, R.G.M. van der Sman, R.M. Boom, Lattice Boltzmann simulations of droplet formation in a T-shaped microchannel, Langmuir 22 (2006) 4144–4152.

[156] I. Scherze, R. Knöfel, G. Muschiolik, Automated image analysis as a control tool for multiple emulsions, Food Hydrocolloid. 19 (2005) 617–624.

[157] S. van der Graaf, M.L.J. Steegmans, R.G.M. van de Sman, C.G.P.H. Schröen, R.M. Boom, Droplet formation in a T-shaped microchannel junction: a model system for membrane emulsification, Colloids Surf. A 266 (2005) 106–116.

7 Other membrane processes

Chapter Outline

In this chapter, two other membrane processes related to biotechnology and pharmaceutical applications are described. Liquids that are immiscible with the source (feed) and receiving (product) phases can also be used as membrane materials. These membranes, called liquid membranes, have been used for purification of antibiotics,

Membrane Processes in Biotechnology and Pharmaceutics. DOI: 10.1016/B978-0-444-56334-7.00007-1

organic acids and amino acids. Another field of application of liquid membranes concerns chiral resolution. Some principles, applications and theories of liquid membranes are presented in the first section of this chapter while the second section focuses on membrane bioreactors. Membrane bioreactors are alternative methods to classical techniques of immobilizing enzymes, microorganisms and antibodies. The biocatalysts are suspended in solution and compartmentalized by a membrane in a reaction vessel or immobilized within the membrane matrix itself. In the second section of this chapter, membrane bioreactors are presented including membrane bioreactor configurations, applications and modelling.

7.1 Liquid membranes

7.1.1 Introduction

A broad definition of a membrane is that it is an immiscible barrier between two phases. Therefore, liquids that are immiscible with the source (feed) and receiving (product) phases are considered as liquid membranes. Differences in the permeability coefficient produce a separation between solutes at constant driving force. Because the diffusion coefficients in liquids are typically orders of magnitude higher than in polymers, a larger flux can be obtained with liquid membranes [1].

Liquid membranes are mainly used for recovering cationic metals (cadmium, copper, zinc, nickel, etc.) and anions (nitrates, chromates, etc.) from wastewater [2,3]. For biological applications, they have been evaluated for concentration and separation of biochemical compounds like amino acids, organic acids and penicillin. Another field of application of liquid membranes concerns chiral resolution. At present, the technology has found industrial applications in the treatment of relatively low concentration waste streams where other technologies are not economical.

Liquid membranes are known for many years and are well presented in books and book chapters [1,4,5]. A recent book chapter discussed emulsion liquid membranes (ELMs) and supported liquid membranes (SLMs) with a specific attention to SLMs with strip dispersion [6]. For each of the processes, the following sections were presented: general description, theory and applications. The key feature of SLM with strip dispersion is the constant supply of organic membrane solution, which may ensure long-term stability as detailed further. In another book chapter, Perera and Stevens [7] provided a very clear picture of ELM systems in chemical and biotechnological separations. Their review focused on the ELM process, which can be divided into three stages: emulsification, permeation and settling and de-emulsification. Each stage and associated critical parameters, such as surfactants, diluents and carriers, were described. In addition, modelling was summarized as well as applications and pilot plant studies. Tables presented the main results in this field. A compete book was recently devoted to liquid membranes focusing on principles and applications in chemical separations and wastewater treatment [8]. The book is divided into four sections: general theory of the liquid membrane transport, reviews of the three basic liquid membrane configurations with theories and applications, specific applications including separation of gases and wastewater treatment, and perspectives in liquid membrane technology development. A very clear and complete picture of the topic was provided.

Several review papers have discussed general or specific aspects of liquid membranes. Pellegrino and Noble [9] have discussed the principles of liquid membranes, with and without a complexing agent (carrier), and the configurations that might be used in bioseparations. Patnaik [10] has given an overview of liquid emulsion membranes, focusing on principles, problems and application in fermentation processes. Sastre et al. [11] have reviewed improvements in liquid membrane separation techniques, including modified bulk, supported and emulsion liquid membranes, as well as hollow fibre contained liquid membranes, electrostatic pseudo liquid membranes and hybrid configurations. Their discussion also included design of ion-specific carriers, stability and modelling of liquid membranes and their applications in the separation/removal of metal cations. A detailed review on SLM has been published recently by Kocherginsky et al. [12], including mechanism and kinetic studies of SLM-based separations, stability of SLM and new applications of SLM. Recent developments in liquid membrane technology have been presented by San Román et al. [13] focusing on transport mechanisms, membrane contactors and liquid membrane separations using hollow fibre contactors. In a recent chapter, Kislik [14] discussed hybrid liquid membrane processes with organic water-immiscible carriers and their application in chemical and biochemical separations. The term organic hybrid liquid membrane includes all bulk liquid membrane (BLM) processes incorporating liquid–liquid extraction and membrane separation in one continuously working module. The chapter focused on mass transfer mechanisms and kinetics, module design for separation systems and selected applications.

7.1.2 Basic definitions

Enhanced transport

The principle of solute extraction using liquid membranes is shown in Figure 7.1. The membrane phase is immiscible with both the feed and receiving phases. For the extraction of polar solutes, phase transfer or ‘carrier’ species are added to the liquid membrane phase to facilitate extraction and to increase the selectivity of the extraction process. Stripping of the solute from the organic phase is usually induced by a change in pH or salt concentration, although other means such as photometric and thermally induced back-extraction have been reported [9].

The process of liquid membrane extraction is analogous to the unit operation of solvent extraction. While the extraction step is identical for both processes, the stripping step of the liquid membrane occurs simultaneously with extraction, instead of in a different contacting unit for the solvent extraction.

There are two basic mechanisms for enhanced transport: coupled and facilitated. In coupled transport, the reversible reaction is an exchange reaction (usually ion exchange), and the solute flux is coupled to the exchange species flux (Figure 7.2a). The reaction normally occurs at a liquid interface since the ions are not soluble in the

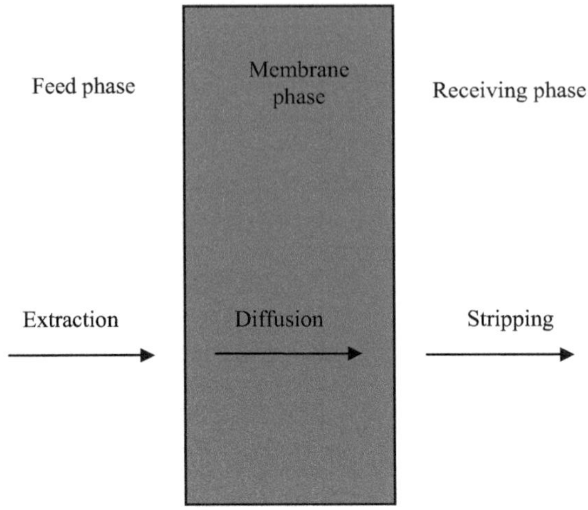

Figure 7.1 Principle of liquid membrane extraction.

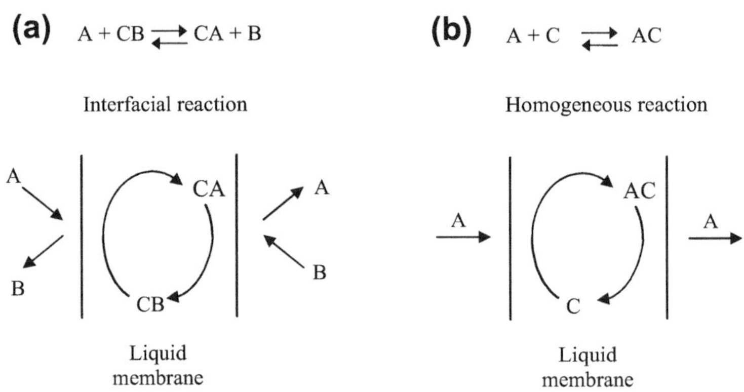

Figure 7.2 Schematic of (a) coupled transport and (b) facilitated transport. A is the solute of interest, B is the counter transporting species and C is the carrier species [from 9].

organic phase. Coupled transport provides a means of pumping solutes against their concentration gradient. Facilitated transport involves the reversible reaction between the solute and carrier and is not coupled to other components (Figure 7.2b). This reaction normally takes place throughout the liquid membrane phase. In this case, the driving force for separation is the concentration gradient across the membrane. Enhanced transport may be viewed as a form of extraction (or adsorption) that is carried out in a membrane.

Kinetic profiles

During solute extraction, the kinetic profile represents the variation of solute mass or solute concentration in both the feed and receiving phases versus time. Typical kinetic

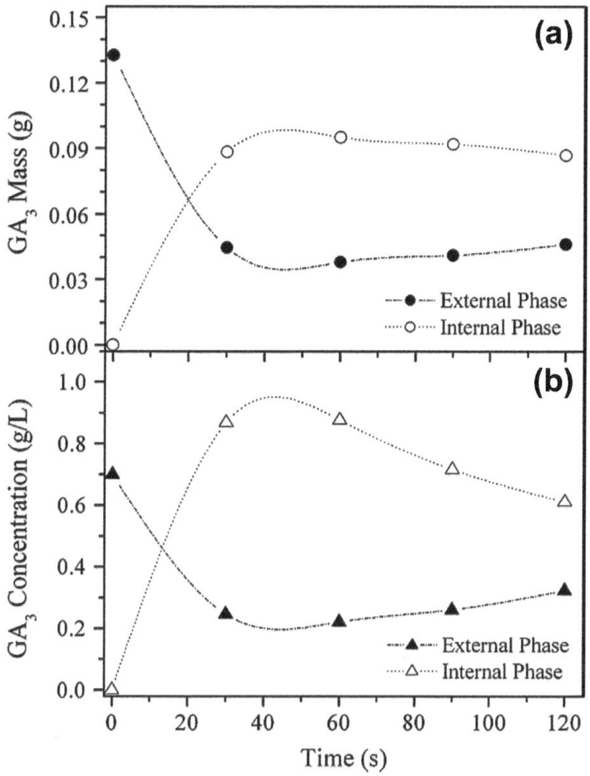

Figure 7.3 Kinetics of GA_3 extraction in external and internal phases: (a) GA_3 mass evolution in each phase; (b) GA_3 concentration in the same experiment. Operation conditions: Aliquat 336 336.1% (v/v); SPAN 80.1% (v/v); KCl 2 M in the internal phase; phosphate buffer 0.1 M, pH = 5 in both aqueous phases; GA_3 0.7 g L^{-1} initial concentration in the external phase [15].
This article was published in J. Membr. Sci., Vol. 348, J. Berrios, D.L. Pyle, G. Aroca, Gibberellic acid extraction from aqueous solutions and fermentation broths by using emulsion liquid membrane, pp. 91–98, Copyright Elsevier (2010).

profiles are shown in Figure 7.3 for the extraction of gibberellic acid (GA_3) from aqueous solutions and fermentation broths by using ELMs [15]. The extraction was carried out by placing the emulsion in the external phase that consisted of either aqueous solution or fermentation broth containing GA_3. Figure 7.3A shows that the mass variation of GA_3 in the internal phase reached a maximum value. The opposite phenomenon was observed at the same time in the external phase, indicating that a small amount of the GA_3 extracted returned to the external phase. During the extraction process by ELMs, the leakage of the internal phase towards the external phase may occur, due to the emulsion coalescence and/or the diminution of the emulsion stability when the volume of the internal phases increases [16,17]. In addition, the GA_3 concentration profiles are different from the mass profiles (Figure 7.3B), since the decrease of GA_3 concentration in the internal phase after reaching the maximum was larger than the equivalent increase in the external phase. This behaviour may be explained by the internal phase swelling, due to transfer of water from the external aqueous phase to the internal water droplets [15,18]. Swelling often causes rapid increase in the volume of the internal phase. The degree of emulsion swelling (E_s) is defined as the ratio of the volume increment at any time (V_{em}) to the initial volume of the emulsion phase ($V_{em,0}$) [19]:

$$E_s(\%) = \frac{V_{em} - V_{em,0}}{V_{em,0}} \times 100 \tag{7.1}$$

7.1.3 Liquid membrane configurations

Liquid membranes are broadly classified into three types: bulk, supported and ELMs (Figure 7.4). In a BLM, a relatively thick layer of immiscible liquid is used to separate the feed and receiving phases. To provide a thinner liquid membrane, the liquid can be impregnated in the pore structure of a microporous solid, such as a polymeric filter. This configuration is known as an immobilized or supported liquid membrane (ILM or SLM). In the third configuration, the receiving phase is emulsified in an immiscible liquid membrane. The liquid surfactant membrane or ELM is then dispersed in the feed solution and mass transfer takes place from the feed phase to the internal receiving phase. The classification and definitions presented below are adopted by several authors [12,13].

Bulk liquid membranes (BLMs)

BLMs usually consist of an aqueous feed and stripping phase, separated by a water-immiscible liquid membrane phase in a U-tube (Figure 7.4a). The experimental set-up for conducting BLM transport experiments is very simple. However, due to the thickness of the membrane, the amount of species transported is very low. Therefore, BLM transport systems are usually utilized in studies of transport mechanisms.

Supported liquid membranes (SLMs)

In SLMs, the organic liquid is usually imbedded in small pores of a polymer support and is kept there by capillary forces (Figure 7.4b) [12]. If the organic liquid is immiscible with the aqueous feed and strip streams, SLM can be used to separate the two aqueous phases. Relatively small volume of organic components in the membrane and simultaneous extraction and re-extraction in one step offers the advantages of possible usage of expensive carriers, high separation factors, easy scale-up, low energy requirements, low capital and operating costs.

An SLM can be fabricated in at least three different geometries. Planar or flat sheet SLM are very useful for laboratory research and development purposes, but the surface area to volume ratio of flat sheets may be too low for industrial applications. Spiral wound and hollow fibre geometries can be used to increase the surface area of liquid membrane modules. Surface area to volume ratios for hollow fibre modules can approach 10^4 m^2/m^3 and 10^3 for spiral wound modules [1].

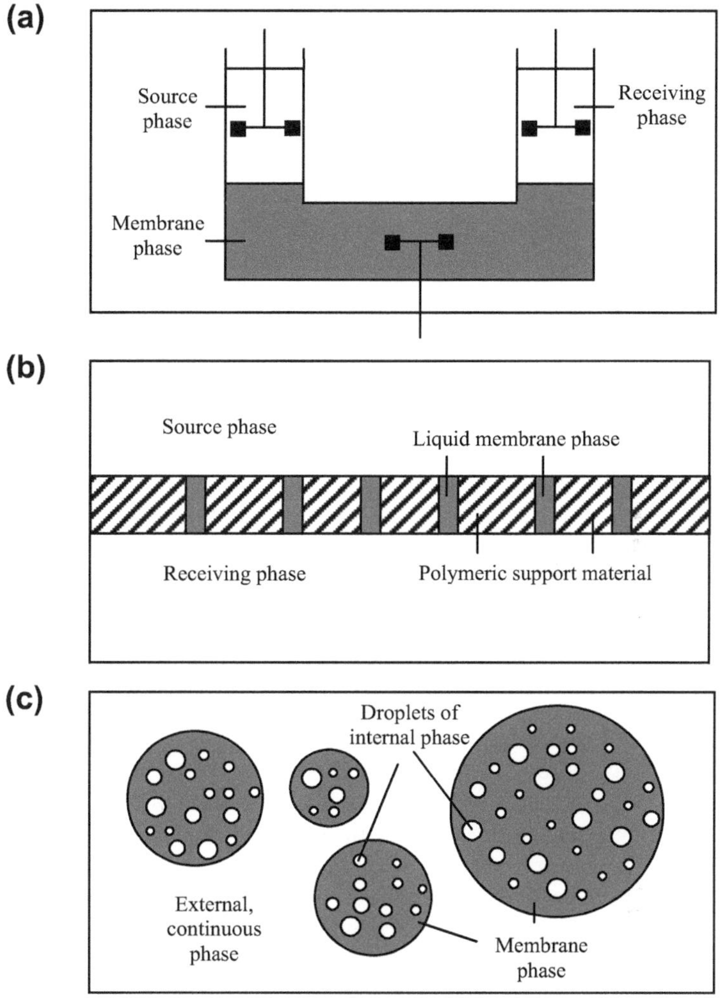

Figure 7.4 Comparison of U-tube BLM, supported liquid membrane, and ELM systems.

Although SLMs have been widely studied for the separation and concentration of a variety of compounds and present many potential advantages over other separation methods, there have been very few large-scale applications of SLM due to insufficient membrane stability. This problem can be due to the loss of the carrier and/or solvent from the membrane, which has an influence on both flux and selectivity. The degradation mechanisms include progressive wetting of the pores in the membrane support by the aqueous phase pressure difference over the membrane, mutual solubility of species from the aqueous phase and liquid membrane phase, emulsion formation in the liquid membrane phase and blockage of membrane pores by precipitation of a carrier complex at the surface [12].

Several studies have been carried out to enhance the SLM stability, by choosing optimum membrane materials and/or operating conditions. Once these factors have been properly chosen, SLM lifetime can be enhanced by various means as the pre-saturation of aqueous phases, re-impregnation of the support with the liquid membrane phase, total or partial gelation of organic phase, interfacial polymerization top layers and development of new configurations [20].

SLMs with strip dispersion were first reported by Ho et al. [21,22] to overcome problems of ELMs and SLMs. An aqueous strip solution is dispersed in an organic membrane solution by a mixer, and passed on to one side of a membrane support [6]. The organic phase of the dispersion is therefore imbedded in the pores of the support, typically hollow fibre hydrophobic membrane modules, forming a stable SLM. The organic phase extends from the pores to the strip side. Stability is maintained by having a constant supply of organic membrane solution to the pores. An aqueous feed solution is passed on to the other side of the membrane support and the target species is extracted into the organic solution by a selective carrier. The target species is then stripped by the aqueous strip solution. For final recovery of the target species, the mixer is turned off or a settler is used for the loaded strip dispersions, and the dispersion quickly separates.

Hollow fibre and spiral wound modules provide high surface area to volume ratio [11,13]. They have led to different configurations of liquid membranes creating new forms of separation processes such as emulsion pertraction (EPT) [23,24], pseudo-emulsion-based hollow fibre strip dispersion (PEHFSD) [25] and hollow fibre renewal liquid membrane (HFRLM) [26]. In the EPT process, the aqueous phase with the target species is kept apart from the emulsion phase by a hydrophobic microporous membrane [23,24]. The emulsion phase consists of an organic phase with a dissolved extractant where an aqueous strip solution is dispersed forming aqueous droplets. In the PEHFSD method, a pseudo-emulsion (organic and strip) is formed by a mixer and then passed through the shell side [25]. Since the membrane is hydrophobic, the pore of the membrane is immobilized by the extractant. The interface is maintained inside the pore mouth of the hollow fibre membrane, in which a feed solution comes into contact with the organic solution. The species is extracted into the pseudo-emulsion and subsequently stripped in the stripping solution present in the pseudo-emulsion. In the HFRLM process, hydrophobic hollow fibres are used, whose pores are filled with the organic solution (Figure 7.5) [26]. The stirred mixture of organic solution and stripping solution (or feed) circulates through the lumen side of the module. The organic droplets are homogeneously dispersed in the mixture by stirring. The feed phase (or the stripping phase) flows through the shell side, usually in counter-current mode.

A double-SLM device was also proposed to replace the single organic immersed organic membrane [27]. The organic solvent is sealed between two membranes; therefore, the extraction can be sustained long enough even though the organic solvent is evaporating during the extraction. This system was successfully applied to the extraction of nicotine from tobacco using chloroform, an easy volatile solvent with high solubility in alkaloids, as the organic membrane liquid.

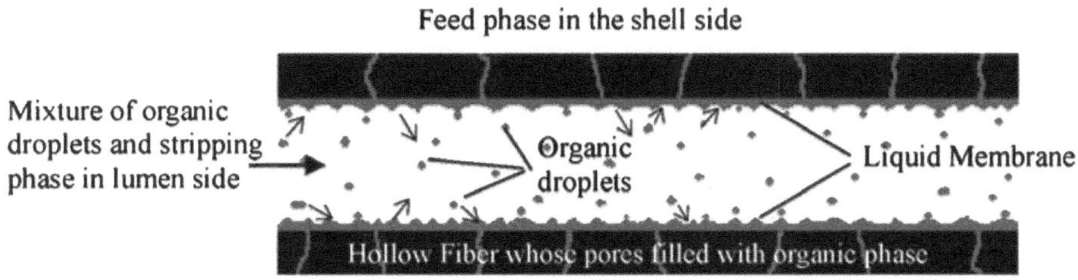

Figure 7.5 Principle of the HFRLM technique [26].
This article was published in Chem. Eng. J., Vol. 146, Z. Ren, W. Zhang, H. Li, W. Lin, Mass transfer characteristics of citric acid extraction by hollow fiber renewal liquid membrane, pp. 220–226, Copyright Elsevier (2009).

Emulsion liquid membranes (ELM)

ELMs, discovered by Li [28], are usually prepared by first forming an emulsion between two immiscible phases, and then dispersing the emulsion in a third (continuous) phase. The membrane is the liquid phase that separates the encapsulated, internal droplets in the emulsion from the external, continuous phase (Figure 7.4c). In general, the internal, encapsulated phase, and the external, continuous phase are miscible. However, the membrane phase must not be miscible with either of these two phases in order to be stable. The emulsion is of the W/O type if the external phase is water, and it is of the O/W type if the external, continuous phase is oil. To maintain the integrity of the emulsion during the extraction process, the membrane phase generally contains some surfactants and additives as stabilizing agents. In general, the internal phase droplets are small, with diameters in the order of 1–10 μm. The emulsion globules are much larger, usually 0.1–2 mm in diameter.

The ELM process can be divided into three stages: emulsification, permeation and settling and de-emulsification [7]. Emulsification is usually achieved by high speed or ultrasonic stirrers for batch operations and high pressure static dispersion or colloid mills for continuous mode. The emulsion is then dispersed by mechanical agitation into the external feed phase containing the solute to be extracted. The efficiency of the extraction is dependent on several parameters, such as surfactant concentration, volume fraction of the internal phase, carrier concentration, continuous phase and internal phase composition. After extraction, the membrane must be broken and the solute recovered. The breaking or demulsification of a loaded emulsion is an important step in the process as the membrane phase must be available for recycling. The de-emulsification process can be divided into three stages: droplet coalescence and growth, droplet settling and coalescence of the large water and oil droplets with their respective continuous phases in the coalescer. De-emulsification is usually performed by chemical or physical treatments.

In the recovery of fermentation products, ELMs are reported to have a number of advantages, especially for very low final concentration in the culture broth, since ELMs have the capacity to perform the separation even against the concentration gradient [15]. ELMs are claimed to be more efficient than solvent extraction, requiring smaller working volumes of solvents and solutions. They have large specific areas of extraction that allow high extraction rates. Other advantages are that the separation and concentration occur in only one single operation, and design and scaling of the process can be based on existing well-defined systems of liquid–liquid extraction. The limitations are that the emulsions have to be produced before the process. They have to be stable enough so that leakage is reduced to a minimum, but still not very stable so that they could be destroyed after the separation, thus allowing the removal of the transported species.

7.1.4 Applications

Separation by using liquid membranes has been studied in the recovery of fermentation products having an industrial and biotechnological interest: antibiotics, organic acids and amino acids. Another field of application is chiral resolution. These various applications are discussed in the following sections of this chapter. Other components separated using liquid membranes may be found in the literature, such as proteins [29], sugars [30] and gibberellic acid, which is a vegetal growth promoter [15].

Amino acids

In the last decade, the production of amino acids has increased enormously [31]. In the fermentation broth media, the concentration of amino acids is very low and they are present in a mixture of different compounds. So, separation and purification steps are necessary during the downstream production of amino acids. Some of the separation techniques used include ion exchange, chromatography, filtration, evaporation, reverse osmosis and electrodialysis. The use of liquid membranes is an alternative to these separation techniques. Many studies have been carried out using liquid membranes for separation and concentration of amino acids from fermentation broths.

With ELM, cation exchange and anion exchange carriers have been used [32]. Both types are associated with counter-ions to maintain electroneutrality in the apolar membrane phase. These counter-ions can be exchanged with the amino acid at the donor-membrane interface to form a carrier–amino acid complex. This complex diffuses across the membrane phase until it reaches the acceptor–membrane interface, where the same or another counter-ion replaces the amino acid, and the former carrier-counter-ion associate diffuses back across the membrane to repeat the extraction cycle.

The cationic carrier is a positively charged lipophilic compound such as Aliquat 336, a quaternary ammonium salt [33]. An example of a negatively charged anionic carrier is the lipophilic acid di-2-ethylhexyl phosphoric acid (DEHPA) [34–37]. With DEHPA as an extractant, a proton gradient between the acceptor and donor side is the driving force for the mass transfer. Recently, Calzado et al. [31] studied the facilitated transport and separation of three kinds of aromatic amino acids through activated composite membranes (ACMs) with DEHPA as a carrier, and found that the amino acid transport selectivity through these ACMs followed a sequence of tryptophan (Trp) > phenylalanine (Phe) > tyrosine (Tyr), regardless of a different amino acid-DEHPA affinity sequence, Phe > Trp > Tyr.

In solid–liquid membranes, studies concerning the transport of amino acids have been performed with neutral cyclic compounds such as crown ethers as extractants [38,39]. Other carriers include chloride of the quaternary ammonium salt (Aliquat 336) [40] and DEHPA [32]. In this last study, the transport efficiency of five kinds of amino acids, namely, arginine (Arg), glutamic acid (Glu), phenylalanine (Phe), tyrosine (Tyr) and tryptophan (Trp), through the di-(2-ethylhexyl) phosphoric acid-tri- (2-ethylhexyl) phosphate was investigated and a 56.2% of extraction efficiency for Trp was then obtained.

BLMs have also been used for separation of amino acids. With DEHPA as a carrier, Ma et al. [41] studied the transport selectivity and kinetics of some amino acids such as L-phenylalanine, L-tryptophan, L-leucine and L-isoleucine, showing high transport selectivity. The transport behaviour was mainly affected by the lipophilicity constant of amino acid side chain and the fraction of the protonated amino acid in the feed solution. Calixarene and its derivatives were also used as a carrier in BLMs [42,43]. The cavity of the calixarene can recognize the size of a guest molecule, and the framework is available for the arrangement of the functional groups to create a specific affinity for a target guest molecule. Chang et al. [42] introduced a calix[6]arene ethyl ester derivative as a mobile carrier for the transport of amino acid esters in a BLM system. The transport was considered to proceed by the inclusion of an ammonium group of the guest amino acid with the carrier. The calix[6]arene forms a stable complex with an amino acid ester and a quantitative extraction was achieved. The structure of the complex was confirmed by spectroscopic studies for the calix[6]arene including a guest molecule in the cavity.

Antibiotics

Penicillin G is an important intermediate for many other medical substances in the antibiotic industry. It is obtained in a typical biotechnological fermentation process. After incubation and fermentation periods of 40 and 190 h, respectively, the fermentation broth contains a maximum of 2% of penicillin G. An alternative to conventional liquid–liquid extraction is the use of liquid membranes.

For the extraction and subsequent derivatization of penicillin G from a fermentation broth, Scheper et al. [44] utilized an ELM system containing the enzyme penicillin acylase in the internal phase to convert the extracted penicillin G to the products 6-aminopenicillinic acid and phenylacetic acid. Lee et al. performed extraction of penicillin G using batch ELM systems and amines by varying surfactant concentration, polymer concentration and w/o ratio (i.e. volume ratio of internal aqueous phase to organic membrane phase) [19,45–47]. Figure 7.6 shows the mechanism of facilitated transport of penicillin G using the amines [46]. The general reaction of an amine (A), such as Amberlite LA-2 or PARABAR 9551, with hydrogen ion (H^+) and penicillin acid anion (P^-) occurs at the external and internal interfaces between the aqueous and membrane phases as follows:

$$mA(org.) + nH^+(aq.) + np^-(aq.) \leftrightarrow A_m(HP)_n(org.) \qquad (7.2)$$

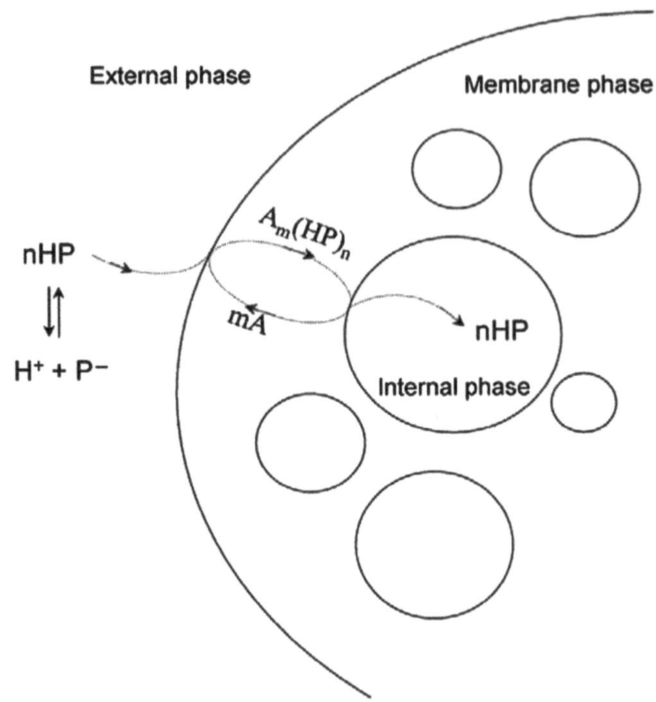

Figure 7.6 Schematic diagram for transport of penicillin G from external phase into an emulsion drop [47].
This article was published in J. Ind. Eng. Chem., Vol. 14, S.C. Lee, Development of a more efficient emulsion liquid membrane system with a dilute polymer solution for extraction of penicillin G, pp. 207–212, Copyright Elsevier (2008).

n moles of (H^+) and (P^-) react with *m* moles of amine A at the external interface to form 1 mol of complex [$A_m(HP)_n$]. The complex then diffuses across the membrane phase until it reaches an internal droplet. At the internal interface, (H^+) and (P^-) are released to the internal phase by the stripping reaction because of a very high pH of the internal phase. The uncharged amine diffuses back across the membrane to repeat the separation cycle. Table 7.1 gives typical experimental conditions for continuous extraction of penicillin G. Since the volume ratio of the internal phase to the external phase was very small, separation and concentration of penicillin G occurred simultaneously during its transport from the external phase to the internal phase. Good extraction efficiency was obtained with a few combinations of membrane composition and w/o ratio. However, emulsion swelling easily diluted penicillin G accumulated in the internal phase at high surfactant concentration despite high degree of extraction, while the degree of extraction was quite low at low surfactant concentration despite low emulsion swelling.

Ho [48] recovered penicillin G with an efficiency of 93% using the secondary amine, Amberlite LA-2 in Isopar L. With a feed solution concentration of 8840 ppm

Table 7.1 Typical experimental conditions for continuous extraction of penicillin G [47]

Parameter	Condition
Concentration of penicillin G in the feed solution	20 mmol/dm^3
pH of feed solution	5.0
Concentration of extractant (Amberlite LA-2) in the membrane phase	20 mmol/dm^3
Concentration of surfactant (PARANOX 147) in the membrane phase	2 vol.%
Concentration of polymer in the membrane phase	2 wt%
Concentration of Na_2CO_3 in the internal phase	0.35 M for w/o ratio of 1/3, 0.175 M for w/o ratio of 1/1
Flow rate of feed solution	120 mL/min
Flow rate of emulsion phase	20 mL/min
Stirring speed	330 rpm

penicillin G, this antibiotic was concentrated to 41000 ppm in the aqueous strip solution containing sodium carbonate. The removal and stripping of penicillin G by was recently performed using hollow fibre non-dispersive solvent extraction (HFNDSX) [49]. These authors checked the viability of the simultaneous extraction/stripping of penicillin G using two large scale hollow fibre modules. The recovery process of penicillin G was realized at pilot scale using HF modules with an effective surface area of 7.7 m^2.

Other studies showed that erythromycin, a macrolide antibiotic, can be recovered and concentrated by an SLM without a carrier. The antibiotic distribution coefficient between the membrane used and the feed was found to be 122, providing relatively high solute fluxes across the membrane, although the transfer resistance remains in the filled membrane pores [50]. To improve the separation performance, Habaki et al. [51] conducted erythromycin permeation experiments by both batch and continuous operations with a spray column. With this new contacting method, the separation performance and the stability of the emulsion membrane were improved.

Organic acids

Citric acid is one of the important organic acids successfully developed by fermentation process [26]. The principal uses of citric acid are in food, pharmaceutical and chemical industries. The recovery of citric acid from fermentation broth is one of the important steps in a commercial process of citric acid production, especially from dilute solutions. The only commercially viable technique on large-scale processes is actually the precipitation technique using calcium salt followed by filtration and subsequently treated with sulphuric acid.

Liquid membrane techniques are an alternative method for recovery of citric acid from fermentation broth [52–55]. Boey and Garcia Del Cerro [52] studied the recovery of citric acid using the ELM technique. Alamine 336 was used as a carrier, *n*-heptane as a diluent, Span-80 and Lubrizol as addition agents and Na_2CO_3 solution as a stripping phase. More than 80% of citric acid was extracted from aqueous solutions. Other researchers have also investigated this recovery process using ELM. Juang and Chen [54] and Guo and Zhang [55] studied the transport process of citric acid across an SLM. However, due to the lack of their long-time stability, difficult operation of the emulsification and de-emulsification steps in ELM processes, liquid membrane techniques have not been adopted for large-scale industrial processes. In order to overcome some of the difficulties arising in the conventional liquid membrane systems, Ren et al. [26] applied the HFRLM for simultaneous extraction and concentration of citric acid from dilute aqueous solutions. The organic phase of 30% N235 + 20% *n*-octanol + 50% kerosene was used as the liquid membrane phase, with NaOH as the stripping phase. The authors concluded that HFRLM was a valuable recovery method for separation and concentration of compounds from dilute solutions.

Succinic acid is another organic acid used in chemical, food and pharmaceutical industries. In conventional processes, it is manufactured by chemical synthesis from petroleum feed stocks. Recently, its biological production from renewable biomass by fermentation was proposed as an alternative to the petrochemical-based processes [56]. In the fermentation-based processes, downstream purification processing is essential to remove impurities such as protein, carbon source and acetic acid from the culture broth. The traditional purification methods, such as precipitation by calcium hydroxide or calcium carbonate and physical extraction by organic solvents, recover succinic acid directly from a fermentation broth.

Liquid membranes were recently evaluated for the recovery of succinic acid. Hong et al. [57] and Hong and Hong [58] separated acetic acid from succinic acid by reactive extraction with tri-n-octylamine (TOA) in 1-octanol, where selective separation of acetic acid from succinic acid was possible only near neutral pH and multistage extraction steps were used in order to raise purity of succinic acid. However, these authors obtained only a percentage of removal of acetic acid of about 70%, and a purity of succinic acid of 86%. Recently, Lee and Hyun [56] developed an ELM system for removal of acetic acid from succinic acid by changing various experimental parameters. Acetic acid is produced as a major by-product in the fermentation of succinic acid and the most serious inhibitor during its microbial production. Removal of acetic acid from succinic acid in the fermentation broth can therefore be a preliminary step for purification of succinic acid. For optimized operating conditions, the degree of extraction of acetic acid reached 99% and its maximal enrichment ratio was higher than 7, while the degree of extraction of succinic acid and its enrichment ratio were very low.

Chiral resolution

Chirality of chemical substances is ubiquitous in nature. The essential substances that form living organisms, such as amino acids and sugars, are chiral and usually exist as pure enantiomeric forms. In the last two decades, enantioseparation technology has developed in response to demand for optically pure compounds in a wide variety of applications.

A very complete review on membranes and membrane processes for chiral resolution has been provided by Xie et al. [59]. To achieve the chiral resolution by liquid membranes, two key considerations are taken into account. First, an appropriate chiral selector is chosen that preferentially binds one of the two isomers, and second, the solubility of the chiral selector and its complex as well as the free isomers in the feed, membrane and stripping phases is considered. To minimize passive transport, the chiral selector and its complex must be soluble only in the liquid membrane, and the uncomplexed enantiomer molecules soluble only in the feeding and receiving phases.

In general, the selectivity of enantioselective membranes can be described by the separation factor, α, and the enantiomeric purity defined by the enantiomeric excess, *ee*. The separation factor, α, is defined as the concentration ratio of the two isomers in the stripping phase divided by that in the feed phase:

$$\alpha = \frac{[X]_{\mathrm{s}}/[\theta]_{\mathrm{s}}}{[X]_{\mathrm{f}}/[Y]_{\mathrm{f}}} \tag{7.3}$$

where $[X]$ and $[Y]$ denote the concentrations of the enantiomer preferentially transported through the membrane and retained in the feed solution, respectively. The subscripts s and f denote the stripping and feed phases, respectively. The chiral selectivity of membranes is also frequently calculated in terms of the enantiomeric excess, *ee*, of permeates. The *ee* value is defined as the ratio of the concentration difference and the concentration summation of both enantiomers in the stripping phase:

$$ee = \frac{[\theta]_{\mathrm{s}}-[Y]_{\mathrm{s}}}{[X]_{\mathrm{s}}+[Y]_{\mathrm{s}}} \times 100 \tag{7.4}$$

The main three types of liquid membranes have been used for chiral resolution. In the SLM system, the membrane phase is fixed in place by non-enantioselective materials on both sides, or it fills the pores of a non-enantioselective porous membrane. SLMs offer several advantages compared with ELMs and BLMs. One of the most important is that only a small amount of expensive chiral carrier is required to achieve enantiomeric resolution. In addition, the mechanical strength of the membrane phase of SLMs is higher than those of ELMs and BLMs.

For example, an SLM system using a polypropylene hollow fibre membrane module was reported by Hadik et al. [60,61] for the separation of the racemic amino acids, D/L-lactic acid and D/L-alanine (D/L-Ala). In this system, toluene solution containing the chiral selector N-3,5-dinitrobenzoyl-L-phenylalanine-octylester was used as the liquid membrane phase, and it separated hydrophilic lumen and shell-side aqueous phases. This allowed the preferential transport of the D-enantiomer of the analyte and gave α and *ee* values of 2.0 and 33.5%, respectively, for D/L-lactic acid and 1.75 and 27.17%, respectively, for D/L-Ala.

In SLM systems, sandwich membranes have been the most commonly used configurations. These systems are composed of a porous membrane impregnated with a chiral selector-containing liquid, which is then surrounded by two relatively dense films. A similar strategy for enantioselective SLMs containing physically entrapped enzymes as chiral carriers was proposed [62,63]. Lakshmi and Martin [62] incorporated D-amino acid oxidase apoenzyme into the sandwich membrane as a molecular recognition agent. Using a 30-nm-pore membrane loaded with D-amino acid oxidase apoenzyme, the selective transport of D-phenylalanine (D-Phe) from a racemic mixture with a maximal a value of 4.9 was achieved.

BLMs have been reported by several authors for chiral resolution. Krieg et al. [64] performed the enantiomeric enrichment of a racemic drug chlorthalidone using BLMs containing β-cyclodextrin (β-CD) as a chiral mobile carrier in an aqueous membrane phase. During the initial stages of separation (after 28 h), at a low carrier concentration (1:4 β-CD/chlorthalidone) and a pH of 5, the maximal α value was 1.05 for a single BLM and 1.41 with a multiple BLM containing three membranes and three stripping phases. A BLM has also been reported to achieve enrichment of enantiomers of mandelic acid and phenylglycine [65]. This system employed the chiral carrier cinchonidine dissolved in a mixture of dodecane/decanol (v/v = 1:1) as the membrane phase. When the initial concentration ratio of the carrier to analyte in the feed phase was 0.5 and the pH values of the feed and receiving phase were 4–5 and 8, respectively, a maximal α value of 1.5 was reached for L-mandelic acid.

ELMs have also been applied to chiral separation. To separate D-Phe from a racemic mixture, Pickering and Chaudhuri [66,67] developed a chiral ELM using copper(II) N-decyl-L-hydroxyproline as a chiral carrier in a mixture of hexanol:decane (v/v = 1:1) as a membrane solvent. A maximum value of 2.4 was observed in the early stages of extraction when the pH was low in the source phase and high in the membrane phase. The authors concluded that this system could provide a low-cost means of producing large quantities of reasonably pure amino acid enantiomers from racemic mixtures.

Chiral resolution technology based on liquid membranes is an inexpensive approach with a relatively high selectivity and fast mass transfer rate. However, the extraction efficiency and selectivity are counteracted by the lack of stability and mechanical strength of liquid membranes. For example, a pressure difference is commonly employed as the driving force to enhance the mass transfer efficiency, but it cannot be effectively employed in liquid membrane systems except for SLMs. Large-scale production of reasonably optically pure compounds by liquid membrane systems awaits resolution of these shortcomings [59].

7.1.5 Theory

Numerous studies have been proposed to represent liquid membrane transport. A very detailed review on this topic is presented by Kislik [8]. Some results on the simulation of transport in ELMs are presented below. Ever since Li invented ELM [28], numerous mathematical models have been developed. Those models can be grouped into two main categories: diffusion type mass transfer models and carrier facilitated transport models [68].

A classical diffusion type mass transfer model was the advancing front model (AFM), which assumed that the solute reacts instantaneously and irreversibly with the internal receiving reagent at a reaction surface which advanced into the globules as the reagent was consumed (Figure 7.7) [69]. The resulting non-linear partial differential equations were solved by a perturbation method. The zero-order or pseudo-steady-state solution obtained provided reasonable prediction for the diffusion process.

The governing equations of the AFM are [10,70]:

Globules:

$$\frac{\partial c}{\partial t} = \frac{D_{\text{eff}}}{r^2}\frac{\partial}{\partial \text{r}}\left(r^2\frac{\partial c}{\partial r}\right) \quad (R_{\text{f}} < r < R) \tag{7.5}$$

$$t = 0 \quad c = 0 \quad (r < R) \tag{7.6}$$

$$r = R \quad c = Kc_{\text{r}} \quad (t > 0) \tag{7.7}$$

$$r = R_{\text{f}}(t) \quad c = 0 \quad (t > 0) \tag{7.8}$$

Continuous phase:

$$-V_{\text{e}}\frac{\text{d}c_{\text{e}}}{\text{d}t} = \frac{3}{R}(V_{\text{m}} + V_{\text{i}})D_{\text{eff}}\frac{\partial c}{\partial r}\bigg|_{\text{r}} - R \tag{7.9}$$

$$t = 0 \quad c_{\text{e}} = c_{\text{e0}} \tag{7.10}$$

Reaction front:

$$-\frac{V_{\text{i}}}{V_{\text{m}} + V_{\text{i}}}c_{\text{i0}}\frac{\text{d}R_{\text{f}}}{\text{d}t} = D_{\text{eff}}\frac{\partial c}{\partial r}\bigg|_{\text{r}} - R_{\text{f}} \tag{7.11}$$

$$t = 0 \quad R_{\text{f}} = R \tag{7.12}$$

where c is the solute concentration in the globule, c_{e} is its concentration in the continuous phase, R is the radius of the globule, R_{f} is the radius of the reaction front, V_{i} is the volume of the internal aqueous phase, V_{m} is the volume of the membrane phase, V_{e} is the volume of the continuous phase, K is the partition coefficient and D_{eff} is the effective diffusivity inside the globule.

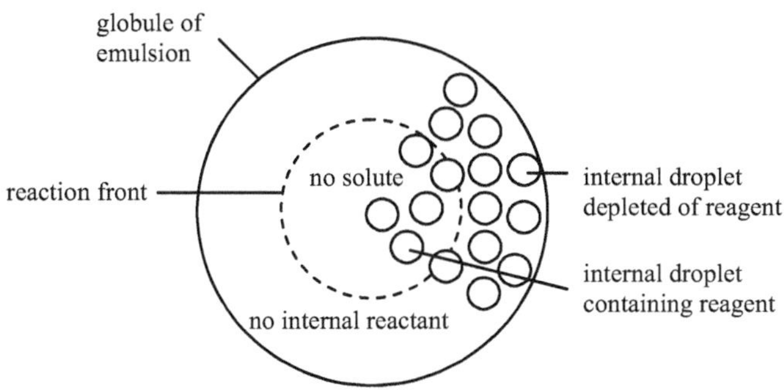

Figure 7.7 Advancing front model [from 69].

Extensions to the advancing front model include external mass transfer resistance [71,72], reaction reversibility in the internal phase [73–75] and leakage of the internal phase into the external phase [76,77]. Lorbach and Marr [78] developed a carrier facilitated transport model that takes into account the diffusion of the carrier and the carrier–solute complex in emulsion globules and reversible reactions at the external and internal interfaces. Although some simplifications have been made on constant summation of free and complex carrier, constant pH in the external phase and negligible resistance for the peripheral thin membrane, there were still four parameters dealing with extraction and stripping kinetics and equilibrium that were difficult to determine. Yan et al. [79] assumed that the reaction was not instantaneous but irreversible. This diffusion controlled mass transfer model considered the mass transfer inside and outside of the globule as well as the reaction between solute and internal phase reagent. Yan [80] proposed a reaction and diffusion controlled mass transfer model, considering the following resistances: mass transfer resistance inside the boundary layer of the globule and inside the globule, resistance due to chemical complexation reaction and chemical reaction of diffusion complex with internal phase reagent(s).

Other specific aspects have been taken into account. Lee et al. [45] developed a permeation model to examine the transport mechanism of penicillin G in an ELM system. The model took into account the mass transfer in the external aqueous film, the extraction reaction between penicillin G and ECA 4360J at the external interface, the diffusion of penicillin G in the emulsion phase, the stripping reaction at the internal interface and the pH change of internal aqueous solution containing Na_2CO_3 with penicillin G transported into the internal phase. The experimental data were well fitted with the model. Recently, Huang et al. [68] presented a closed-form solution for a mathematical model of ELM. The model included: (1) external phase mass transfer resistance from external phase to the external–membrane interface, (2) mass fluxes at the external–membrane interface, (3) chemical equilibrium of extraction reaction on the external–membrane interface, (4) simultaneous diffusion of the solute–carrier complex inside the emulsion globule phase, (5) stripping of the complex at the membrane–internal interface and (6) chemical equilibrium of the stripping reaction at the membrane–internal interface. The model was used to predict theoretically the effects of individual parameters on overall extraction rate, the concentration of solute in the external phase $c_e(t)$, the concentration profile inside membrane phase $c_m(r, t)$ and the interfacial concentration on external phase–membrane interface as well.

7.2 Membrane bioreactors

7.2.1 Introduction

Membrane bioreactors are alternative approaches to classical methods of immobilizing enzymes, microorganisms and antibodies. The biocatalysts are suspended in solution and compartmentalized by a membrane in a reaction vessel or immobilized within the membrane matrix itself. In the first method, the system consists of

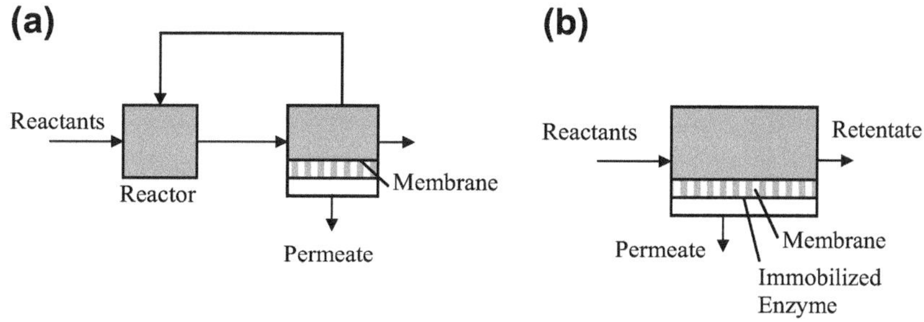

Figure 7.8 Membrane bioreactor configurations: (a) reactor combined with a membrane operation unit and (b) reactor with the membrane active as a catalytic and separation unit [81]. This article was published in Comprehensive Biotechnology, Second Edition, Vol. 2, C. Charcosset, Downstream Processing and Product Recovery/Membrane Systems and Technology, pp. 603–618, Copyright Elsevier (2011).

a traditional stirred tank reactor combined with a membrane separation unit, such as ultrafiltration (UF) and microfiltration (MF). In the second method, the membrane acts both as a support for the catalyst and as a separation unit (Figure 7.8). The biocatalyst can be flushed along a membrane module, segregated within a membrane module, or immobilized in or on the membrane by entrapment, gelification, physical adsorption, ionic binding, covalent binding or cross-linking.

The advantages of immobilizing enzymes are reported to be increased reactor stability and productivity, improved product purity and quality and reduction in waste. The efficiency of these systems depends on the biochemical (catalytic activity, reaction kinetics, concentration, viscosity of substrate and product, immobilization stability, etc), geometric parameters (membrane configuration, morphology and pore size distribution, etc) and hydrodynamics parameters (transmembrane pressure, flow velocity, etc). Membrane bioreactors have been reviewed previously by several authors [82–88].

7.2.2 Basic definitions

Catalytic activity of enzymes [89,90]

The theory of enzyme-catalysed reactions is based on the assumption that the enzyme (the catalyst, *E*) and the substrate (the reactant, *S*) form a complex (*ES*) by a reversible reaction. The complex is then converted into the product (*P*) with the reaction rate k_2:

$$S + E \xleftrightarrow{k_1, k_{-1}} ES \xrightarrow{k_2} P + E \tag{7.13}$$

The Michaelis–Menten equation shows the dependency of the reaction rate on the substrate concentration (first-order reaction):

$$v = \frac{V \times c_S}{K_M + c_S} \tag{7.14}$$

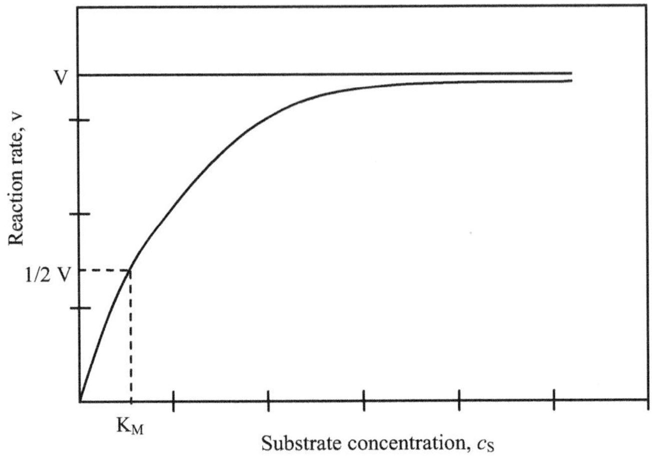

Figure 7.9 Reaction rate as a function of substrate concentration (enzyme at constant concentration).

where K_M is the Michaelis constant, v is the reaction rate, c_S is the substrate concentration and V is the maximum reaction rate, which is reached, when the entire enzyme is saturated with substrate.

From Equation 7.14, it can be demonstrated that the Michaelis constant K_M equals the substrate concentration at half the maximal reaction rate. The value K_M can be obtained by plotting the experimentally measured reaction rate versus the substrate concentration (Figure 7.9). Another approach is the plot using the reciprocal of the Michaelis–Menten equation:

$$\frac{1}{v} = \frac{K_M}{V} \times \frac{1}{c_S} + \frac{1}{V} \tag{7.15}$$

Michaelis constants for enzymes usually range from 10^{-2} to 10^{-5} mol L^{-1}. A low K_M indicates a high affinity between enzyme and substrate.

Membrane fouling and enzyme activity decay

The main limitations of enzymatic membrane reactor are associated with membrane fouling and enzyme activity decay. Membrane fouling leads to a decrease in permeate flux and has been frequently reported [85,86]. Membrane fouling mainly results from the deposition of retained particles on the membrane surface (external fouling) and the accumulation of small-size substances in membrane pores (internal fouling) [91]. In several cases, external fouling or fouling layer formation may be the major cause of membrane fouling. In addition, the internal fouling or pore-blocking can lead to the formation of irreversible fouling, which is harmful for the long-term operation of the membrane bioreactor. Fouling can eventually lead to higher operating cost arising from the requirement of higher transmembrane pressure, more intensive aeration and more frequent membrane cleaning and replacement [91].

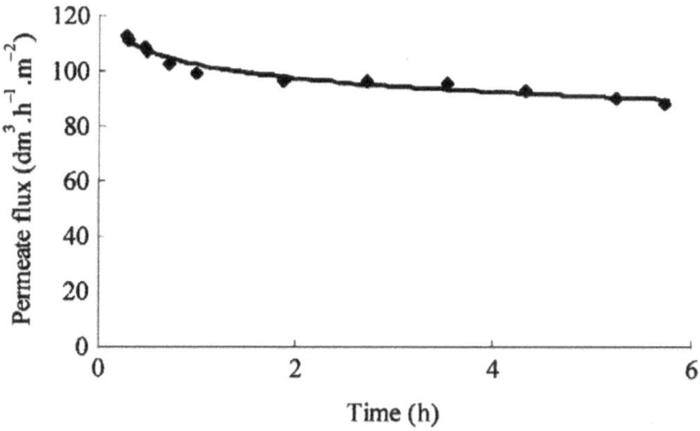

Figure 7.10 Permeate flux versus time during starch hydrolysis in a continuous recycle membrane reactor (CRMR) ($\theta = 80$ °C, $S_0 = 10\%$ w/w, $E = 5.5$ cm^3 kg^{-1}, $\tau = 1$ h, $V_R = 6$ dm^3) [93].
This article was published in Biochem. Eng. J., Vol. 5, D. Paolucci-Jeanjean, M.P. Belleville, G.M. Rios, N. Zakhia, The effect of enzyme concentration and space time on the performance of a continuous recycle membrane reactor for one-step starch hydrolysis, pp. 17–22, Copyright Elsevier (2000).

The presence of a pre-hydrolysis step before the continuous operation can minimize fouling. In the case of starch hydrolysis in a dead-end cell reactor, the lower viscosity of the initial solution made the implementation of the membrane reactor easier [92]. Paolucci-Jeanjean et al. [93] used a pre-hydrolysed solution for the hydrolysis of raw cassava starch into low-molecular weight compounds like glucose, maltose and maltotetraose, using Termamyl [93]. As enzymes cut starch into smaller-molecular weight compounds, no fouling was observed and a high flux (>80 L h^{-1} m^{-2}) was maintained during more than 10 h (Figure 7.10). On the contrary, in the absence of the pre-hydrolysis step, there was a very fast decrease in flux (<20 L h^{-1} m^{-2} in a few minutes) due to membrane fouling by high-molecular weight polysaccharides of native starch [87].

Enzyme activity decay is due to several phenomena such as catalyst leakage and enzyme denaturation under the effects of pH, temperature, shear effects or adsorption/deposit at the walls. Enzyme leakage is usually evaluated by continuous measurement of permeate activity [94–97]. Adsorption and shear effects are more difficult to characterize and less often reported [95,96]. Recently, Paolucci-Jeanjean et al. [97] have investigated the loss of enzyme activity during enzymatic starch hydrolysis in a continuous enzyme membrane reactor. These authors demonstrated that the effects of temperature and denaturation of catalyst by adsorption/deposit onto the membrane were the limiting phenomena. However, reducing temperature to overcome thermal denaturation was not efficient since the increase in viscosity led to a decrease in permeate flux and reactor performance.

7.2.3 Membranes and configurations

Membrane devices

Immobilized biocatalyst membrane reactors are frequently used in a hollow fibre configuration because of their high packing density (large surface area per unit volume of reactor space) (Figure 7.11). Other configurations included flat sheet and rotating bioreactor. Cells are grown in the extracapillary space with medium flow through the fibres, or grown within the fibres with medium flow outside or across the fibres. It was shown that a mass transfer limitation for oxygen and glucose was obtained and that the well-defined spacing between the fibres was a means of reducing this phenomenon [98].

Enzyme immobilization

For membrane bioreactors with immobilized enzymes, an important aspect is the chemistry of enzyme immobilization. It is often accomplished either directly on the

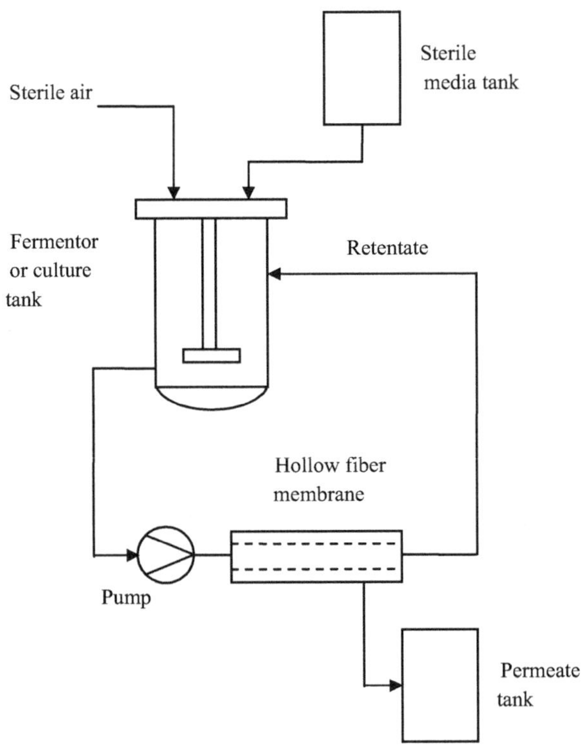

Figure 7.11 Typical experimental set-up of a bioreactor combined with a membrane operation unit [81].
This article was published in Comprehensive Biotechnology, Second Edition, Vol. 2, C. Charcosset, Downstream Processing and Product Recovery/Membrane Systems and Technology, pp. 603–618, Copyright Elsevier (2011).

membrane or via spacer arm, often through the ε-amino functionality of lysine residues on the protein [99]. Although immobilization of enzymes generally enhance their stability, one major disadvantage of random immobilization of enzymes onto polymeric MF type membranes is that the activity of the immobilized enzyme is often significantly decreased because the active site may be blocked from substrate accessibility, multiple point binding may occur or the enzyme may be denatured. Different approaches are developed in order to accommodate site-specific immobilization of enzymes with different structural characteristics, as gene fusion to incorporate a peptidic affinity tag at the N- or C-terminus of the enzyme; post-translational modification to incorporate a single biotin moiety on enzymes; and site-directed mutagenesis to introduce unique cysteines to enzymes [99].

7.2.4 Applications

Since research on membrane bioreactor technology began over 30 years ago, membrane bioreactor systems have been mostly applied in a large-scale to treat industrial wastewater, domestic wastewater and specific municipal wastewater [100]. In the food industry, the applications reported of biocatalytic membrane reactor concern the hydrolysis of pectins in fruit juices, the treatment of musts and wines by the conversion of polyphenolic compounds and anthocyanes, the reduction of the lactose content in milk and whey by its conversion into digestible sugar and the removal of peroxides from diary products [85]. In medical applications, important membrane bioreactors are the membrane bioartificial pancreas and the membrane bioartificial liver [82,86]. In the biotechnology and pharmaceutical fields, applications reported include chemical synthesis of various compounds and protein hydrolysis.

Chemical synthesis

There is a growing interest in applying enzymes to chemical synthesis processes. Lipases, in particular, have received much attention for a variety of applications, such as esterification and transesterification, ester hydrolysis and in the resolution of racemic mixtures to produce optically active compounds [101].

Membrane bioreactors have been used in the production of optically active compounds by lipase-catalysed hydrolysis [102–107]. After hydrolysis, one isomer is removed into an aqueous phase as the product and the other isomer remains in the organic phase. This enables the simultaneous recovery of both isomers with simpler downstream processing. For example, Sakaki and Itoh [108] applied a biphasic membrane reactor with lipase-immobilized hydrophilic/hydrophobic capillary membranes to the optical resolution of racemic 2-hydroxy octanoic acid methyl ester, and the optical purity and yield of recovered (R)- and (S)-isomers were measured. Table 7.2 shows the results of optical resolution of 2-hydroxy octanoic acid in a biphasic membrane reactor with a buffer–hexane system. (*S*)-2-hydroxy octanoic acid was recovered from the aqueous phase at 59–67% yield and 0.9–0.92 enantiomeric excess, and the ester of (R)-isomer was recovered from the organic phase at

Table 7.2 Results of optical resolution of racemic 2-hydroxy octanoic acid methyl ester in a lipase-immobilized hydrophilic UF membrane reactor using a buffer–hexane biphasic system [108]

	Product from aqueous phase				Residual substrate from organic phase			
Index	**Time (h)**	**Conversion (%)**	ee_P	**Yield of (*S*)-isomer[a] (%)**	**Time (h)**	**Conversion (%)**	ee_S	**Yield of (*R*)-isomer[b] (%)**
PAN-H	24	33	0.62	59	77.5	54	0.93	73
PAN-H1	11.5	36	0.92	67	34	55	0.99	75
PAN-H2	11	41	0.9	67	26	58	0.99	73

[a] (*S*) – Isomer was obtained as acid.
[b] (*R*) – Isomer was obtained as methyl ester.
This article was published in Biotechnol. Lett., Vol. 25, K. Sakaki, N. Itoh, Optical resolution of racemic 2-hydroxy octanoic acid by lipase-catalyzed hydrolysis in a biphasic membrane reactor, pp. 1591–1595, Copyright Springer (2003).

73–75% yield and 0.92–0.99 enantiomeric excess. The main losses were said to be due to the unrecovered 2-hydroxy octanoic acid produced after the first recovery of the product from the aqueous phase.

Membranes prepared by covalent attachment of *Candida antarctica* lipase B (CALB) onto a ceramic support previously coated by polymers (gelatine/polyethyleneimine PEI) were tested in a membrane reactor for the synthesis of butyl laurate, from butyl acetate and lauric acid, in hexane [109]. Butyl laurate synthesis was selected as a model reaction leading to high-added value esters with potential applications as cosmetic and pharmaceutical additives, emulsificants and antimicrobial agent. The ester production was found to be directly related to the permeation flux across the membrane. However, the permeate flux unexpectedly decreased when the Reynolds number values increased. This phenomenon was related to morphological changes, which occurred in the polymer layer when the tangential velocity varied.

Membrane bioreactors have also been used for the production of drugs. For example, lipase from *C. rugosa* immobilized on a nylon support has been used to synthesize lovastatin, a drug which lowers serum cholesterol levels [101]. The production of a derivative of kyotorphin (analgesic) was obtained in solvent media using α-chymotrypsin as catalyst and α-alumina mesoporous tubular support [110]. Low-molecular weight heparin was produced by an UF membrane bioreactor using maltose binding protein fused heparinase I [111]. Aglycon, which shows strong antioxidant activity, was obtained by hydrolysis of oleuropein in a membrane reactor containing immobilized β-glucosidase [112].

Protein hydrolysis

Enzymatic hydrolysis improves the functional, nutritional and immunological properties of proteins. Thus, protein hydrolysates are widely used as food ingredients in energy-providing drinks, hypoallergenic formulae and enteral diets for children and sick adults. Conventional batch hydrolysis of proteins has traditionally been used to obtain protein hydrolysates due to its simple operation and easy control [113].

However, there are several disadvantages associated to this method: high enzyme consumption, high labour costs and low yield if product inhibition takes place [114].

The enzymatic membrane bioreactor integrates enzymatic hydrolysis, product separation and catalyst recovery into a single operation [95,115–117]. In order to profit from the advantages of batch (easy operation and control) and membrane recycle reactor (enzyme reuse), a cyclic batch enzyme membrane reactor was recently proposed by Prieto et al. [114,118]. The cyclic reactor was successfully applied for the production of reduced antigenicity whey protein hydrolysates. Three steps, namely, hydrolysis, UF and enzyme recycling, were repeated a number of times. The reaction was carried out at 60 °C and alkaline conditions, with Subtilisin as the catalyst.

The enzymatic membrane bioreactor provides a number of advantages including the continuous removal of low molecular weight peptides as the hydrolysis process proceeds which reduces the problem of enzyme inhibition by the hydrolysates [119,120]. Also, the enzyme can be recycled which, unlike the batch process, leads to reduced cost on the process. In addition, a membrane technology eliminates side effects associated with other hydrolysis unit operations, because high temperatures are not required in order to inactivate the enzyme at the end of the batch hydrolysis.

The enzymatic membrane bioreactor has been widely applied to total conversion of food proteins of various origins in order to produce, in a continuous mode, hydrolysates with improved nutritional and/or functional properties. Cheison et al. [119] gave a detailed list of the various proteins hydrolysed in enzymatic membrane bioreactors. Several plant proteins have been hydrolysed including cottonseed protein, soy protein, corn gluten meal and alfalfa protein. Proteins from animal sources hydrolysed in enzymatic membrane bioreactors include marine animal protein sources such as fish skin gelatine, cod frame protein yellowfin sole (*Limanda aspera*) frame protein, Alaska Pollack skin and hoki frame protein. Mammalian proteins exploited through hydrolysis in enzymatic membrane bioreactors include beef rump (*Biceps femoris*), goat whey, casein, haemoglobin, bovine skin gelatine, bovine hoof and horn and bovine colostrum. Whey protein hydrolysates with specific properties like low antigenicity as well as bioactive peptides have also been produced in enzymatic membrane bioreactors.

In addition, some studies have reported the use of enzymatic membrane bioreactors as a tool for continuous production and extraction of specific peptidic sequences; that is bioactive peptides [121,122]. Prata-Vidal et al. [121] reported the use of an enzymatic membrane bioreactor for the hydrolysis of caseinomacropeptide (CMP) and showed the feasibility of the continuous process for the production of these antithrombotic peptides. CMP is the product of milk κ-casein hydrolysis by chymosin. This macropeptide with a molecular weight of about 7000 g mol^{-1} contains several sequences of amino acids, which show promising nutritional or biological properties.

7.2.5 Other applications of membranes for bioreactions

A number of other applications have been developed, which associate bioreactions and membranes. Two examples are presented below. In the first example, tubular membranes permit to supply oxygen to mammalian cell bioreactors with minimal shear introduction. In the second example, biomimetic membranes use a couple of reverse enzymes catalysing two opposite reactions to obtain a very specific separation under mild conditions (ambient temperature and pressure, and no solvent).

Membranes for bioreactor aeration

One of the most important factors in operating cell culture bioreactors is oxygen supply. Various small-scale methods such as surface aeration, membrane aeration and gas sparging are used [123]. Gas sparging is the preferred method for large-scale aeration because of great mass transfer rates and operational simplicity. However, gas sparging may damage cultured animal cells. Cell bubble interactions are generally considered as the likely cause of cell damage.

The use of tubular membranes for aeration of mammalian cell bioreactors permits to supply oxygen to the cells with minimal shear introduction. The process was introduced in the 80s and is termed 'bubble-free membrane aeration' [124–127]. The membrane functions as a gas distributor to control the gas–liquid interface area and the mass transfer coefficient. The tubular system consists of a stator, most frequently shaped as a cylinder, which is wrapped with a certain length of a tubular, gas-permeable membrane material, and a central mechanical agitator acting within the cylindrical space defined by the stator (Figure 7.12) [128]. However, the system is

Figure 7.12 Photo of the dynamic membrane aeration (DMA) concept (left) in contrast to a conventional rotor–stator system/membrane aeration setup (right). On the DMA, the membrane tubing is wrapped vertically around the rotor arms, which are arranged radially on the rotor shaft at the top and bottom of the bioreactor. The shaft is located in the centre of the reactor and performs an oscillating rotary movement driven by an external servo motor. The flow around the membrane tubing is initiated by the relative motion between the membrane and the liquid. The two concentric rings provide membrane gas supply and removal. On the conventional membrane aeration setup, the tubing is wrapped vertically on a stator with a concentric, steadily rotating anchor stirrer [128]. Original in colour.
This article was published in Cytotechnology, Vol. 59, B. Frahm, H. Brod, U. Langer, Improving bioreactor cultivation conditions for sensitive cell lines by dynamic membrane aeration, pp. 17–30, Copyright Springer (2009).

limited in overall mass transfer capacity. In addition, it is not properly scalable, because the tubing surface area (thus the oxygen supply capacity) increases with the linear dimension of a reactor system squared while the fermenter volume (i.e. the capacity demand) increases with the linear dimension cubed. Consequently, membrane aeration systems are mostly used at laboratory scale.

Other membrane aeration based techniques have then been proposed such as the generation of microbubbles using membrane pores. Nehring et al. [123] have produced very small microbubbles distributed equally across a surface area by using a ceramic membrane. The system was designed to keep a high oxygen transfer rate at low impeller agitation, in order to limit cell damage caused by cell bubble–surface interaction and impeller agitation. The system was applied to various scales of bioreactors, from 2 to 100 L, and oxygen transfer rates were measured under different conditions (Table 7.3).

The dynamic membrane aeration bioreactor uses another concept of bubble-free membrane aeration. It was originally developed by Bayer Technology Services for the oxygenation of sensitive animal cell cultures at low shear stress and low power input [128]. The system consists of a centrally arranged, oscillating rotor wrapped with silicone tubing in a star-like arrangement (Figure 7.13). It has been successfully developed up to the 200 L scale. Oxygenation occurs by diffusion of oxygen through the silicone tubing material from the gaseous phase inside the tubes directly into the aqueous phase of the reactor; hence, air bubbles are avoided and gentle agitation can be used for stirring. The system was applied to the oxidation of lactose to lactobionic acid by a bienzymatic system [129]. The membrane-aerated reactor was scaled up from a miniaturized membrane contactor to a 20 L dynamic membrane aeration reactor. The conversion was catalysed by an enzyme cascade consisting of cellobiose dehydrogenase as synthesizing enzyme and laccase as regenerating enzyme, coupled by a redox mediator. The dynamic membrane aeration reactor was successfully operated at very low power input and greatly reduced enzyme inactivation by eliminating the high gas/liquid interfacial area of conventionally aerated stirred reactors or bubble columns.

Table 7.3 Configuration and parameter set of long-term cell cultivation in different sizes of bioreactors [123]

	Reactor			
	Biostat MD	**Biostat B**	**30 DS**	**100 DS**
Volume (L)	2	5	30	100
Silicon tube (m/L)	2.5		2.5	2.5
No. of microspargers		1	2	2
Tip speed (min^{-1})	100	70	70	30
No. of impellers	1	1	2	2
Cells	MDCK	MDCK	MDCK	MDCK

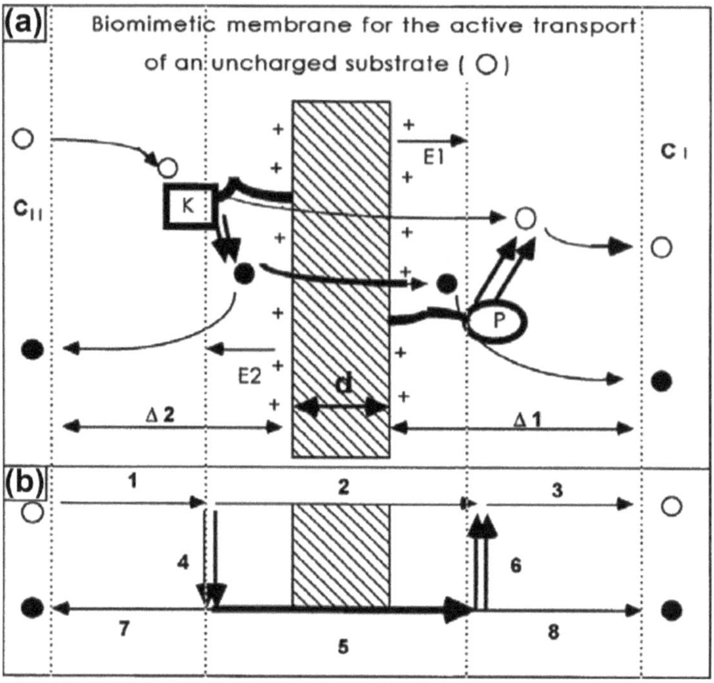

Figure 7.13 Biomimetic membrane for the active transport of an uncharged substrate (o). In (a), the membrane topography involves the action of a kinase (K) on the emitting side (C_{II}) and of a phosphatase (P) on the receiving side (C_I) acting on diffusion layers of thickness Δ on either side of a positively charged permeable membrane of thickness d; in (b) the different stages of diffusion and reaction resulting from the transport. The positive charges on the membrane favour the diffusion of the negatively charged product (●) compared with the diffusion of the uncharged substrate (○) [130].
This article was published in Enzyme Microb. Technol., Vol. 30, B. Maïsterrena, R. Couturier, B. Perrin, Artificial biomimetic membranes for the active and selective transport of small molecules, pp. 125–128, Copyright Elsevier (2002).

Biomimetic membranes

Biomimetic membranes are another example of membrane processes associated with bioreactions. They have been developed for the specific transport of small organic molecules by Maïsterrena et al. At present, they have been studied only on a laboratory scale. These composite membranes are composed of a permeable (+ or −) charged membrane inserted between two enzymatic membranes [130]. They are used to separate the donor compartment (C_I) from the receiver compartment (C_{II}) of a diffusion cell. An enzyme E_1 is immobilized on the membrane face adjacent to C_I and another enzyme E_2 on the face adjacent to C_{II}. E_1/E_2 constitutes a couple of 'reverse' enzymes catalysing two opposite reactions. The alkaline phosphatase (ALP)/kinase (K) enzyme couples or inverse couples (K/ALP), are good examples of reverse enzymes and were chosen to create these bioactive membranes. Depending on the reaction sequence of the immobilized enzymes (ALP/K or K/ALP) as well as the nature of the charge carried by the intermediary membrane (+ or −), the membrane permits either the active and selective transport of a phosphorylated molecule or the transport of a neutral molecule.

Figure 7.13a illustrates the principle of the mechanism of a biomimetic membrane capable of actively transporting an uncharged substrate. The permeable organic membrane, positively and uniformly charged, carries, on the transmitter side, the immobilized kinase (K) and, on the receptor side, the immobilized phosphatase (P). These two enzymes catalyse at alkaline pH (pH = 8.5) the following reactions:

$$S + (ATP/Mg)^{2-} \xrightarrow{(K)} S - P^{2-} + (ADP/Mg)^{-} + H^{+} \tag{7.16}$$

$$S - P^{2-} \xrightarrow{(P)} S + P^{2-} \tag{7.17}$$

Substrate S is taken as uncharged. The product S-P^{2-} generated by the kinase (which is also the phosphatase substrate) carries two negative charges. On either side of the membrane, the existence of immobilized layers of thickness Δ is assumed. Each layer is composed of an extremely thin diffuse layer (DL), adjacent to the membrane and within which there exists a constant electric field, generated by the fixed charges situated on the two membrane surfaces. The zone of the immobilized layer not subject to the influence of membrane charges corresponds to δ.

The different diffusion/reaction steps produced near the membrane during active transport of S from C_{II} to C_I are represented in Figure 7.13b. Step 1 corresponds to the passive diffusion of S into δ_2; step 2, the diffusion of S across the membrane and the two diffuse layers; step 3, the diffusion of S into δ_1; step 4, the phosphorylation of S by the kinase on the C_{II} side; step 5, diffusion of the product $(S\text{-}P)^{2-}$ across the membrane and both diffuse layers; step 6, the de-phosphorylation of the product on the C_I side by the phosphatase (and thus the regeneration of the substrate in the receiving compartment) and step 7, the retrodiffusion of the product into δ_2 on the C_{II} side; stage 8, the diffusion of the product into δ_1 on the C_I side. This specific transport was investigated using a model based on coupling between: (i) diffusion phenomena produced in unstirred layers (USLs) adjacent to the membrane surfaces; (ii) two opposite reactions produced on each side of the membrane; (iii) electrostatic interactions created between the charges carried by the intermediary membrane and at least one charged metabolite involved in the reaction sequence [131,132].

In addition, active and selective transport of glucose and glycerol was carried out by coupling biomimetic membranes and electrophoresis [133]. The positively charged composite membranes carry, on the face adjacent to the donor compartment of an electrophoresis module, a specific kinase (hexokinase or glycerokinase) and, on the opposite face, an alkaline phosphatase (ALP). Phosphorylation of the neutral substrate (glucose or glycerol) on the donor side by the kinase generates a negatively charged phosphorylated substrate, whose transmembrane migration is promoted by an electric field and by the membrane positive charge. De-phosphorylation of the phosphorylated substrate by ALP on the opposite face regenerates the neutral substrate, which accumulates in the receiver compartment of the electrophoresis module. Using an electrophoresis module specifically designed for this study, experiments were carried out enabling glucose and glycerol to be concentrated approximately 8- and 12-fold, respectively, in 8 h.

7.2.6 Theory

The behaviour of hollow fibre membrane bioreactors have been widely investigated from a theoretical point of view. For example, Schonberg and Belfort [134] showed that radial convection of the feed in a shell and tube hollow fibre reactor was useful to increase the catalyst efficiency and the production rates per unit volume. The solution of governing equations was obtained assuming Darcy flow, first-order kinetics, steady-state transport and irreversible reaction. Salzman et al. [135] studied hollow fibre enzyme reactors for a continuous and single-pass operation. In their model, the enzyme was physically confined within the shell section of the reactor and the substrate solution flowing in the fibre lumen; both substrate and reaction products diffused through the fibre wall. The solution of the equations yielded the conversion as a function of volumetric flow rate, kinetic constants, diffusion coefficients, geometric dimensions of the reactor, flow conditions and substrate concentrations. Calabrò et al. [136] developed a theoretical model for the analysis of steady-state behaviour of a hollow fibre membrane bioreactor with immobilized biocatalyst. The model permitted the evaluation of substrate consumption and product(s) formation as functions of dimensionless parameters.

Recently, Curcio et al. [137] improved this theoretical model by taking into account the behaviour of a membrane bioreactor operating in recycle. The model is described in details in the following. The model was based on the unsteady-state balance equations governing momentum and mass transfer within the regions that can be identified in a membrane reactor with immobilized enzyme, coupled to the unsteady-state mass balance on the components in the feed tank.

The system consisted of a feed tank and a hollow fibre bioreactor assembled in a tube and shell configuration. In a single fibre, three regions were distinguished: the membrane lumen (region 1), where substrate continuously flew mainly in the axial direction and partially in the radial direction; the membrane dense layer or 'skin' (region 2) and the membrane spongy layer (region 3) (Figure 7.14). The chemical reaction was supposed to take place solely in the membrane dense and spongy layers where the biocatalyst was confined and distributed. The substrate was forced to permeate through the membrane wall, where it came in contact with the immobilized enzyme for reaction. Products and unconverted substrate were continuously removed in the radial direction from the permeate stream, whereas the retentate stream still containing substrate and a certain amount of products was recycled back to the feed tank.

The reaction rate R_r was expressed by a modified Michaelis–Menten equation that accounted for the competitive product inhibition effects. R_r was written in terms of the actual substrate, S_E, and product, P_E, concentrations, evaluated at the interface with the enzyme:

$$R_r = \frac{V_{max} S_E}{K_M\left(1 + \frac{P_E}{K_i}\right) + S_E} \tag{7.18}$$

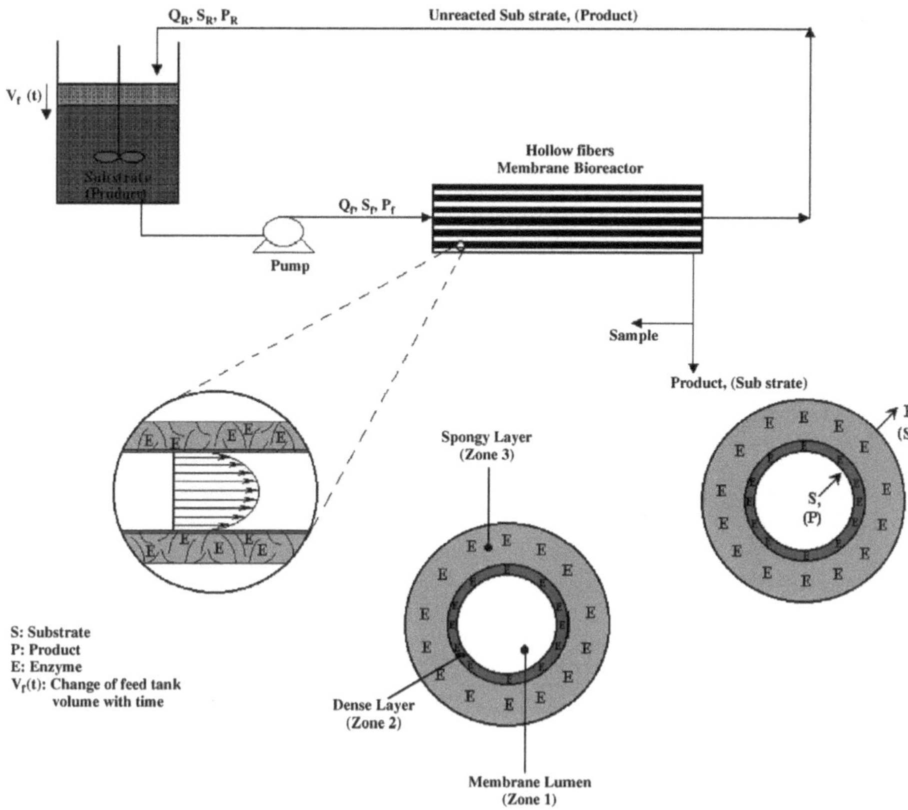

Figure 7.14 Schematic representation of a hollow fibre membrane bioreactor in recycle configuration [137].
This article was published in J. Membr. Sci., Vol. 273, S. Curcio, V. Calabrò, G. Iorio, A theoretical and experimental analysis of a membrane bioreactor performance in recycle configuration, pp. 129–142, Copyright Elsevier (2006).

where V_{max} and K_M were, respectively, the reaction-limiting velocity and the Michaelis constant and K_i was the product inhibition constant.

Momentum analysis

The general equation of motion and the continuity equation applied in the membrane lumen:

$$\rho\left[\frac{\partial u_1}{\partial t} + (u_1 \cdot \nabla)u_1\right] = -\nabla p_1 + \nabla \cdot \mu\left[\nabla u_1 + (\nabla u_1)^T\right] \tag{7.19}$$

$$\nabla \cdot u_1 = 0 \tag{7.20}$$

where ρ denotes the solution density, μ the viscosity, u_1 the velocity vector and p_1 the pressure.

The flow through both the dense and spongy layers was described by the Brinkman equation that applies for a porous media when the overall transport of momentum by shear stresses in the fluid is high. Therefore, the flow field in the membrane regions was described by:

$$\rho\frac{\partial u_\mathrm{i}}{\partial t}+\frac{\mu}{k_\mathrm{i}}u_\mathrm{i} = -\nabla p_\mathrm{i}+\nabla\bullet\mu\left[\nabla u_\mathrm{i}+(\nabla u_\mathrm{i})^\mathrm{T}\right] \tag{7.21}$$

$$\nabla\bullet u_\mathrm{i} = 0 \tag{7.22}$$

where the subscript $i = 2, 3$ referred to the dense and spongy layers, respectively. u_i was the corresponding superficial velocity vector that accounted for the porosity ε_i of the two regions, p_i was the pressure and k_i their permeability.

Mass transfer analysis

The mass transfer analysis accounts for the behaviour of both the feed tank and the hollow fibre membrane bioreactor. For the feed tank, an overall mass balance together with the substrate, S_f, and the product, P_f, mass balances expressed the time evolution of the feed tank volume and of the composition of the stream fed to hollow fibres.

Total mass balance:

$$\frac{\mathrm{d}V_\mathrm{f}}{\mathrm{d}t} = Q_\mathrm{R}-Q_\mathrm{f} \tag{7.23}$$

$$\text{Initial condition}\ :\quad V_\mathrm{f}(t=0) = V_0 \tag{7.24}$$

Mass balance of the substrate:

$$\frac{\mathrm{d}(V_\mathrm{f}S_\mathrm{f})}{\mathrm{d}t} = Q_\mathrm{R}S_\mathrm{R}-Q_\mathrm{f}S_\mathrm{f} \tag{7.25}$$

$$\text{Initial condition}\ :\quad S_\mathrm{f}(t=0) = S_0 \tag{7.26}$$

Mass balance of the product:

$$\frac{\mathrm{d}(V_\mathrm{f}P_\mathrm{f})}{\mathrm{d}t} = Q_\mathrm{R}S_\mathrm{R}-Q_\mathrm{f}P_\mathrm{f} \tag{7.27}$$

$$\text{Initial condition}\ :\quad P_\mathrm{f}(t=0) = 0 \tag{7.28}$$

where V_f was the volume of mixture contained in the feed tank, Q_R the volumetric flow rate coming out of the bioreactor in axial direction and recycled back to the tank, Q_f the volumetric flow rate fed to the bioreactor, and S_R and P_R were the average substrate and product concentrations flowing out from the bioreactor and recycled to the feed tank. Finally, V_0 and S_0 represented the initial volume and substrate concentration, respectively.

For the three different regions of the hollow fibre, the mass transfer was based on the classical time dependent convection–diffusion-reaction balance equation that was to be written for the substrate S, and the product P:

$$\frac{\partial \mathrm{S_i}}{\partial t} + \nabla\cdot\left(-\,D_{\mathrm{SB,eff,i}}\nabla \mathrm{S_i}\right) = -R_{\mathrm{ri}} + -u_{\mathrm{i}}\cdot\nabla \mathrm{S_i} \tag{7.29}$$

$$\frac{\partial \mathrm{P_i}}{\partial t} + \nabla\cdot\left(-D_{\mathrm{PB,eff,i}}\nabla \mathrm{P_i}\right) = -R_{\mathrm{ri}} + -u_{\mathrm{i}}\cdot\nabla \mathrm{S_i} \tag{7.30}$$

where subscript i ($i = 1$–3) indicated the fibre region. The reaction rate term R_{ri} ($i = 2$, 3) applied to the dense and spongy layers, and was based on the actual interfacial concentrations (S_{Ei} and P_{Ei}). $D_{\mathrm{SB,eff}}$ and $D_{\mathrm{PB,eff}}$ were the effective diffusion coefficients of both substrate and product in water, calculated accounting for the void fractions, ε_i, of each region. The interfacial concentrations, S_{Ei} and P_{Ei} ($i = 2, 3$), were evaluated from a balance between the amount of substrate reaching the biocatalyst surface, assuming that no accumulation occurred, and the amount of substrate converted by the chemical reaction, for both the dense and spongy layers:

$$K_{\mathrm{f,i}}A_{\mathrm{i}}(S_{\mathrm{i}} - S_{\mathrm{Ei}}) = \frac{k_2 E(E_{\mathrm{i}}/E)S_{\mathrm{Ei}}}{K_{\mathrm{m}}\left(1 + \dfrac{P_{\mathrm{Ei}}}{K_{\mathrm{i}}}\right) + S_{\mathrm{Ei}}}\,\varepsilon_{\mathrm{i}}V_{\mathrm{i}} \tag{7.31}$$

where $K_{\mathrm{f,i}}$ was the mass transport coefficient evaluated in the fluid phase surrounding the biocatalyst and estimated on the basis of velocity profiles in the membrane, V_{i} the geometrical volume of the ith membrane region and A_{i} the surface exposed to contact with the substrate solution.

The resulting system of non-linear partial differential equations was numerically solved. The evolution of three representative parameters, that is the effectiveness factor, the degree of conversion within the membrane and the bioreactor performance index, was calculated as functions of the operating conditions. The proposed transport model was said to be very general as it may describe the behaviour of different types of heterogeneous reacting systems by adapting either the reaction rate expression or the set of intrinsic kinetic constants.

A number of other models of membrane bioreactors can be found in the literature. For example, models have been developed to describe the enantioseparation of chiral compounds in a hollow fibre membrane configuration [138–140]. A mathematical model was also proposed to analyse the behaviour of a hollow fibre enzyme reactor as a function of operating temperature and of transmembrane temperature gradient [141].

References

[1] R.A. Bartsch, J.D. Way, Chemical separations with liquid membranes: an overview, in: R.A. Bartsch, J.D. Way (Eds.), Chemical Separations with Liquid Membranes, ACS Symposium Series. Am. Chem. Soc., Washington, 1996, pp. 1–10.

[2] R. Chiarizia, Application of supported liquid membranes for removal of nitrate, technetium(VII) and chromium(VI) from groundwater, J. Memb. Sci. 55 (1991) 39–64.
[3] J. Gega, W. Walkowiak, B. Gajda, Separation of Co(II) and Ni(II) ions by supported and hybrid liquid membranes, Sep. Purif. Technol. 22–23 (2001) 551–558.
[4] W. Ho, N. Li, Emulsion liquid membranes: definitions, in: W. Ho, K. Sirkar (Eds.), Membrane Handbook, van Mostrand Reinhold, New York, 1992, pp. 597–610.
[5] R.J. Marr, J. Draxler, Emulsion liquid membranes: applications, in: W. Ho, K. Sirkar (Eds.), Membrane Handbook, van Mostrand Reinhold, New York, 1992.
[6] M.E. Vilt, W.S.W. Ho, N.N. Li, Liquid membranes, in: E. Drioli, L. Giorno (Eds.), Comprehensive Membrane Science and Engineering, Elsevier, Kidlington, UK, 2010, pp. 79–106.
[7] J.M. Perera, G.W. Stevens, Use of emulsion liquid membrane systems in chemical and biotechnological separations, in: A.K. Pabby, S.S.H. Rizvi, A.M. Sastre (Eds.), Handbook of Membrane Separations, Chemical, Pharmaceutical, Food, and Biotechnological Applications, CRC Press, Taylor & Francis Group, 2009, pp. 709–740.
[8] V.S. Kislik, Liquid Membranes, Principles and Applications in Chemical Separations and Wastewater Treatment, Elsevier, Amsterdam, The Netherlands, 2010.
[9] J. Pellegrino, R.D. Noble, Enhanced transport and liquid membranes in bioseparations, Trends Biotechnol. 8 (1990) 216–224.
[10] P.I. Patnaik, Liquid emulsion membranes: principles, problems and application in fermentation processes, Biotechnol. Adv. 13 (1995) 175–208.
[11] A.M. Sastre, A. Kumar, J.P. Shukla, R.K. Singh, Improved techniques in liquid membrane separations: an overview, Sep. Purif. Methods 27 (1998) 213–298.
[12] N.M. Kocherginsky, Q. Yang, L. Seelam, Recent advances in supported liquid membrane technology, Sep. Purif. Technol. 53 (2007) 171–177.
[13] M.F. San Román, E. Bringas, R. Ibañez, I. Ortiz, Liquid membrane technology: fundamentals and review of its applications, J. Chem. Technol. Biotechnol. 85 (2010) 2–10.
[14] V.S. Kislik, Hybrid liquid membrane processes with organic water-immiscible carriers (OHLM): application in chemical and biochemical separations, in: A.K. Pabby, S.S.H. Rizvi, A.M. Sastre (Eds.), Handbook of Membrane Separations, Chemical, Pharmaceutical, Food, and Biotechnological Applications, CRC Press, Taylor & Francis, 2009, pp. 371–408.
[15] J. Berrios, D.L. Pyle, G. Aroca, Gibberellic acid extraction from aqueous solutions and fermentation broths by using emulsion liquid membranes, J. Memb. Sci. 348 (2010) 91–98.
[16] Y.S. Mok, K.H. Lee, W.K. Lee, Control of swelling in liquid emulsion membranes employed for lactic acid separation, J. Chem. Technol. Biotechnol. 65 (1996) 309–316.
[17] Y. Wan, X. Zhang, Swelling determination of W/O/W emulsion liquid membranes, J. Memb. Sci. 196 (2002) 185–201.
[18] J. Yan, R. Pal, Isotonic swelling behavior of W/O/W emulsion liquid membranes under agitation conditions, J. Memb. Sci. 213 (2003) 1–12.
[19] S.C. Lee, Comparison of extraction efficiencies of penicillin G at different w/o ratios in the emulsion liquid membrane systems with dilute polymer solutions, J. Memb. Sci. 237 (2004) 225–232.
[20] R. Molinari, P. Argurio, F. Pirillo, Comparison between stagnant sandwich and supported liquid membranes in copper(II) removal from aqueous solutions: flux, stability and model elaboration, J. Memb. Sci. 256 (2005) 158–168.

[21] W.S.W. Ho, T.K. Poddar, New membrane technology for removal and recovery of chromium from waste waters, Environ. Prog. 20 (2001) 44–52.

[22] W.S.W. Ho, B. Wang, Strontium removal by new alkyl phenylphosphonic acids in supported liquid membranes with strip dispersion, Ind. Eng. Chem. Res. 41 (2002) 381–388.

[23] I. Ortiz, M.F. San Román, S.M. Corvalán, A.M. Eliceche, Modeling and optimization of an emulsion pertraction process for removal and concentration of Cr(VI), Ind. Eng. Chem. Res. 42 (2003) 5891–5899.

[24] E. Bringas, M.F. San Román, I. Ortiz, Separation and recovery of anionic by the emulsion pertraction technology. Remediation of polluted groundwaters with Cr(VI), Ind. Eng. Chem. Res. 45 (2006) 4295–4303.

[25] J.V. Sonawane, A.K. Pabby, A.M. Sastre, Au(I) extraction by LIX- 79/n-heptane using the pseudo-emulsion-based hollow-fiber strip dispersion (PEHFSD) technique, J. Memb. Sci. 300 (2007) 147–155.

[26] Z.W. Ren, Zhang, H. Li, W. Lin, Mass transfer characteristics of citric acid extraction by hollow fiber renewal liquid membrane, Chem. Eng. J. 146 (2009) 220–226.

[27] Z. Guo, L. Wang, C. Song, L. Zhang, Extraction of nicotine from local tobacco using double-supported liquid membranes technique, J. Memb. Sci. 356 (2010) 105–109.

[28] N.N. Li, Liquid surfactant membranes, US Patent No. 3,410,794 (1968).

[29] M. Vasudevan, J.M. Wiencek, Mechanism of the extraction of proteins into Tween 85 nonionic microemulsions, Ind. Eng. Chem. Res. 35 (1996) 1085–1089.

[30] H. Ikeda, A. Matsuhisa, A. Ueno, Efficient transport of saccharides through a liquid membrane mediated by a cyclodextrin dimer, Chem. Eur. J. 9 (2003) 4907–4910.

[31] J.A. Calzado, C. Palet, M. Valiente, Facilitated transport and separation of aromatic amino acids through activated composite membrane, Anal. Chim. Acta. 431 (2001) 59–67.

[32] P. Wieczorek, J.Å Jönsson, L. Mathiasson, Concentration of amino acids using supported liquid membranes with di-2-ethylhexyl phosphoric acid as a carrier, Anal. Chim. Acta 346 (1997) 191–197.

[33] M.P. Thien, T.A. Hatton, D.I.C. Wang, Separation and concentration of amino-acids using liquid emulsion membranes, Biotechnol. Bioeng. 32 (1988) 604–615.

[34] H. Itoh, M.P. Thien, T.A. Hatton, D.I.C. Wang, A liquid emulsion membrane process for the separation of amino acids, Biotechnol. Bioeng. 35 (1990) 853–860.

[35] M. Teramoto, T. Yamashiro, A. Inoue, A. Yamamoto, H. Matsuyama, Y. Miyake, Extraction of amino acids by emulsion liquid membranes containing di(2-ethylhexyl) phosphoric acid as a carrier biotechnology; coupled, facilitated transport; diffusion, J. Memb. Sci. 58 (1991) 11–32.

[36] S.A. Hong, H.J. Choi, S.W. Nam, Concentration of amino acids by a liquid emulsion membrane with a cationic extractant, J. Memb. Sci. 70 (1992) 225–235.

[37] T. Kaghazchia, A. Kargaria, R. Yegania, A. Zare, Emulsion liquid membrane pertraction of L-lysine from dilute aqueous solutions by D2EHPA mobile carrier, Desalination 190 (2006) 161–171.

[38] T. Shinbo, T. Yamaguchi, H. Yangishita, K. Sakaki, D. Kitamoto, M. Sugiura, Supported liquid membranes for enantioselective transport of amino-acid mediated by chiral crown-ether – effect of membrane solvent on transport rate and membrane stability, J. Memb. Sci. 84 (1993) 241–248.

[39] P. Wieczorek, Factors influencing the transport of tryptophan hydrochloride through supported liquid membranes containing macrocyclic carriers, J. Memb. Sci. 127 (1997) 87–92.

[40] R. Molinari, L. De Bartolo, E. Drioli, Coupled transport of amino-acids through a supported liquid membrane. 1. Experimental optimization, J. Memb. Sci. 73 (1992) 203–215.
[41] M. Ma, B. Chen, X. Luo, H. Tan, D. He, Q. Xie, S. Ya, Study on the transport selectivity and kinetics of amino acids through di(2-ethylhexyl) phosphoric acid-kerosene bulk liquid membrane, J. Memb. Sci. 234 (2004) 101–109.
[42] S.K. Chang, H.S. Hwang, H. Son, J. Yonk, Y.S. Kang, Selective transport of amino acid esters through an chloroform liquid membrane by a calix[6]arene-based ester carrier, J. Chem. Soc. Chem. Commun. 4 (1991) 217–218.
[43] T. Oshima, K. Inoue, S. Furusaki, M. Goto, Liquid membrane transport of amino acids by a calix[6]arene carboxylic acid derivative, J. Memb. Sci. 217 (2003) 87–97.
[44] T. Scheper, Z. Likidis, K. Makryaleas, C. Nowottny, K. Schugerl, Different examples of enzymatic bioconversion in liquid membrane reactors, Enzyme Microb. Technol. 9 (1987) 625–631.
[45] S.C. Lee, J.H. Chang, B.S. Ahn, W.K. Lee, Mathematical modeling of penicillin G extraction in an emulsion liquid membrane system containing only a surfactant in the membrane phase, J. Memb. Sci. 149 (1998) 39–49.
[46] S.C. Lee, Continuous extraction of penicillin G by emulsion liquid membranes with optimal surfactant compositions, Chem. Eng. J. 79 (2000) 61–67.
[47] S.C. Lee, Development of a more efficient emulsion liquid membrane system with a dilute polymer solution for extraction of penicillin G, J. Ind. Eng. Chem. 14 (2008) 207–212.
[48] W.S.W. Ho, Removal and recovery of metals and other materials by supported liquid membranes with strip dispersion, Ann. N. Y. Acad. Sci. 984 (2003) 97–122.
[49] Z. Lazarova, B. Syska, K. Schügerl, Application of large-scale hollow fiber contactors for simultaneous extractive removal and stripping of penicillin G, J. Memb. Sci. 202 (2002) 151–164.
[50] J. Kawasaki, R. Egashira, T. Kawai, H. Hara, L. Boyadzhiev, Recovery of erythromycin by a liquid membrane, J. Memb. Sci. 112 (1996) 209–215.
[51] H. Habaki, R. Egashira, G.W. Stevens, J. Kawasaki, A novel method improving low separation performance for W/O/W ELM permeation of erythromycin, J. Memb. Sci. 208 (2002) 89–103.
[52] S.C. Boey, M.C. Garcia Del Cerro, Extraction of citric acid by liquid membrane extraction, Chem. Eng. Res. Des. 65 (1987) 218–223.
[53] B. Yordanov, L. Boyadzhiev, Pertraction of citric acid by means of emulsion liquid membranes, J. Memb. Sci. 238 (2004) 191–197.
[54] R.S. Juang, L.J. Chen, Analysis of the transport rates of citric acid through a supported liquid membrane containing tri-n-octylamine, Ind. Eng. Chem. Res. 35 (1996) 1673–1679.
[55] C.Z. Guo, J.M. Zhang, Mass transfer condition of citric acid extraction by polypropylene supported liquid membrane, Technol. Water Treat. 33 (2007) 33–36.
[56] S.C. Lee, K.S. Hyun, Development of an emulsion liquid membrane system for separation of acetic acid from succinic acid, J. Memb. Sci. 350 (2010) 333–339.
[57] Y.K. Hong, W.H. Hong, D.H. Han, Application of reactive extraction to recovery of carboxylic acids, Biotechnol. Bioprocess Eng. 6 (2001) 386–394.
[58] Y.K. Hong, W.H. Hong, Removal of acetic acid from aqueous solutions containing succinic acid and acetic acid by tri-n-octylamine, Sep. Purif. Technol. 42 (2005) 151–157.
[59] R. Xie, L.Y. Chu, J.G. Deng, Membranes and membrane processes for chiral resolution, Chem. Soc. Rev. 37 (2008) 1243–1263.

[60] P. Hadik, L.P. Szabo, E. Nagy, D, L-lactic acid and D, L-alanine enantio separation by membrane process, Desalination 148 (2002) 193–198.
[61] P. Hadik, L.P. Szabo, E. Nagy, Z. Farkas, Enantio separation of D, L-lactic acid by membrane techniques, J. Memb. Sci. 251 (2005) 223–232.
[62] B.B. Lakshmi, C.R. Martin, Enantioseparation using apoenzymes immobilized in a porous polymeric membrane, Nature 388 (1997) 758–760.
[63] A. Skolaut, J. Retey, Use of enzymes deactivated by site-directed mutagenesis for the preparation of enantioselective membranes, Angew. Chem. Int. Ed. 41 (2002) 2960–2962.
[64] H.M. Krieg, J. Lotter, K. Keizer, J.C. Breytenbach, Enrichment of chlorthalidone enantiomers by an aqueous bulk liquid membrane containing beta-cyclodextrin, J. Memb. Sci. 167 (2000) 33–45.
[65] D. Stella, J.A. Calzado, A.M. Girelli, S. Canepari, R. Bucci, C. Palet, M. Valiente, Liquid membranes for chiral separations. Application of cinchonidine as a chiral carrier, J. Sep. Sci. 25 (2002) 229–238.
[66] P.J. Pickering, J.B. Chaudhuri, Emulsion liquid membranes for chiral separations: selective extraction of rac-phenylalanine enantiomers, Chirality 9 (1997a) 261–267.
[67] P.J. Pickering, J.B. Chaudhuri, Enantioselective extraction of (D)-phenylalanine from racemic (D/L)-phenylalanine using chiral emulsion liquid membranes, J. Memb. Sci. 127 (1997b) 115–130.
[68] C.R. Huang, H. Fan, D. Zhou, A closed-form solution for a mathematical model of emulsion liquid membrane, J. Memb. Sci. 339 (2009) 233–238.
[69] W.S. Ho, T.A. Hatton, E.N. Lightfoot, N.N. Li, Batch extraction with liquid surfactant membrane: a diffusion control model, AIChE J. 28 (1982) 662–670.
[70] R.N. Chilamkurti, C.T. Rhodes, Transport across liquid membranes, effect of formulation variables, J. Appl. Biochem. 2 (1980) 7–16.
[71] J.L. Fales, P. Strove, A perturbation solution for batch extraction with double emulsion: role of continuous-phase mass transfer resistance, J. Memb. Sci. 21 (1984) 35–53.
[72] P. Stroeve, P.P. Varanasi, Transport process in liquid membrane: double emulsion separation systems, Sep. Purif. Methods 11 (1982) 29–69.
[73] M. Teramoto, H. Takihana, M. Shibutani, T. Yuasa, N. Hara, Modeling of the permeation of copper through liquid surfactant membrane by continuous operation, Sep. Sci. Technol. 18 (1983a) 985–997.
[74] M. Teramoto, T. Sakai, K. Yanagann, M. Ohsuga, Y. Miyake, Modeling of the permeation of copper through liquid surfactant membrane, Sep. Sci. Technol. 18 (1983b) 735–764.
[75] A.L. Burge, R.D. Noble, A diffusion model for reversible consumption in emulsion liquid membrane, J. Memb. Sci. 21 (1984) 55–71.
[76] C.C. Chan, C.J. Lee, A mass transfer model for the extraction of weak acids/bases in emulsion-liquid membrane systems, Chem. Eng. Sci. 42 (1987) 83–95.
[77] R.P. Borwankar, C.C. Chan, D.T. Wasan, R.M. Kurzeja, Z.M. Gu, N.N. Li, Analysis of the effect of internal phase leakage on liquid membrane separation, AIChE J. 34 (1988) 753–762.
[78] D. Lorbach, R.J. Mar, Emulsion liquid membrane. Part II. Modeling mass transfer of zinc with bis(2-ethylhexyl)dithiophosphoric acid, Chem. Eng. Process. 21 (1987) 83–93.
[79] N. Yan, Y. Shi, Y.F. Su, A mass transfer model for type I facilitated transport in liquid membranes, Chem. Eng. Sci. 47 (1992) 4365–4371.
[80] N. Yan, A mass transfer model for type-II-facilitated transport in liquid membranes, Chem. Eng. Sci. 48 (1993) 3835–3843.

[81] C. Charcosset, Downstream processing and product recovery/membrane systems and technology, in: M.Y. Murray (Ed.), Comprehensive Biotechnology, second ed., vol. 2, Elsevier, 2011, pp. 603–618.

[82] G. Belfort, Membranes and bioreactors: a technical challenge in biotechnology, Biotechnol. Bioeng. 33 (1989) 1047–1066.

[83] G. Belfort, C.A. Heath, New developments in membrane bioreactors, in: J.G. Crespo, K.W. Böddeker (Eds.), Membrane Processes in Separation and Purification, NATO ASI Series E: Applied Science, vol. 272, Kluwer Academic Publishers, Dordrecht, 1993, pp. 127–149.

[84] M. Cheryan, M.A. Mehaia, Membrane bioreactors, in: W.C. McGregor (Ed.), Membrane Separation in Biotechnology, Marcel Dekker, New York, 1986, pp. 255–301.

[85] L. Giorno, E. Drioli, Biocatalytic membrane reactors: applications and perspectives, Trends Biotechnol. 18 (2000) 339–349.

[86] L. Giorno, L. De Bartolo, E. Drioli, Membrane bioreactors for biotechnology and medical applications, in: D. Bhattacharyya, D.A. Butterfield (Eds.), New Insights into Membrane Science and Technology: Polymeric and Biofunctional Membranes, Elsevier, 2003, pp. 187–218 (Chapter 9).

[87] G.M. Rios, M.P. Belleville, D. Paolucci, J. Sanchez, Progress in enzymatic membrane reactors—a review, J. Memb. Sci. 242 (2004) 189–196.

[88] R. Mazzei, E. Drioli, L. Giorno, Biocatalytic membranes and bioreactors, in: E. Drioli, L. Giorno (Eds.), Chemical and Biochemical Transformations in Membrane Systems, Elsevier, 2010, pp. 195–210.

[89] W. Aehle, Enzymes in Industry, Productions and Applications, third ed., Wiley, Weinheim, Germany, 2007.

[90] H. Bisswanger, Enzyme Kinetics, Principles and Methods, second ed., Wiley-VCH, Weinheim, Germany, 2008.

[91] F. Meng, B. Liao, S. Liang, F. Yang, H. Zhang, L. Song, Morphological visualization, componential characterization and microbiological identification of membrane fouling in membrane bioreactors (MBRs), J. Memb. Sci. 361 (2010) 1–14.

[92] D. Darnoko, M. Cheryan, W.E. Artz, Saccharification of cassava starch in an ultrafiltration reactor, Enzyme Microb. Technol. 11 (1989) 154–159.

[93] D. Paolucci-Jeanjean, M.P. Belleville, G.M. Rios, N. Zakhia, The effect of enzyme concentration and space time on the performance of a continuous recycle membrane reactor for one-step starch hydrolysis, Biochem. Eng. J. 5 (2000) 17–22.

[94] P. Bressolier, J.M. Petit, R. Julien, Enzyme hydrolysis of plasma proteins in a CSTR UF reactor: performances and modelling, Biotechnol. Bioeng. 31 (1988) 650–658.

[95] A. Mannheim, M. Cheryan, Continuous hydrolysis of milk protein in a membrane reactor, J. Food Sci. 55 (1990) 381–385.

[96] F. Sannier, J.M. Piot, P. Dhustler, D. Guillochon, Stability of a mineral membrane ultrafiltration reactor for peptic hydrolysis of hemoglobin, J. Chem. Technol. Biotechnol. 61 (1994) 43–47.

[97] D. Paolucci-Jeanjean, M.P. Belleville, G.M. Rios, A comprehensive study of the loss of enzyme activity in a continuous membrane reactor. Application to starch hydrolysis, J. Chem. Technol. Biotechnol. 76 (2001) 273–278.

[98] C.A. Heath, G. Belfort, Immobilization of suspended mammalian cells: analysis of hollow fiber and microcapsule bioreactors, Adv. Biochem. Eng. Biotechnol. 34 (1988) 1–31.

[99] D.A. Butterfield, D. Bhattacharyya, S. Daunert, L. Bachas, Catalytic biofunctional membranes containing site-specifically immobilized enzyme arrays: a review, J. Memb. Sci. 181 (2001) 29–37.
[100] W. Yang, N. Cicek, J. Ilg, State-of-the-art of membrane bioreactors: worldwide research and commercial applications in North America, J. Memb. Sci. 270 (2006) 201–211.
[101] F. Yang, T.W. Weber, J.L. Gainer, G. Carta, Synthesis of lovastatin with immobilized Candida rugosa lipase in organic solvents: effects of reaction conditions on initial rates, Biotechnol. Bioeng. 56 (1997) 671–680.
[102] D.R. Wu, S.M. Cramer, G. Belfort, Kinetic resolution of racemic glycidyl butyrate using a multiphase membrane enzyme reactor: experiments and model verification, Biotechnol. Bioeng. 41 (1993) 979–990.
[103] L. Giorno, R. Molinari, E. Drioli, D. Bianchi, P. Cesti, Performance of a biphasic organic/aqueous hollow fibre reactor using immobilized lipase, J. Chem. Technol. Biotechnol. 64 (1995) 345–352.
[104] J. Ceynowa, I. Koter, Lipase-catalyzed kinetic resolution of (R, S)-1-phenylethyl propionate in an enzyme membrane reactor, Acta. Biotechnol. 17 (1997) 253–263.
[105] J.L. Lopez, S.L. Matson, A multiphase/extractive enzyme membrane reactor for production of diltiazem chiral intermediate, J. Memb. Sci. 125 (1997) 189–211.
[106] K. Sakaki, L. Giorno, E. Drioli, Lipase-catalyzed optical resolution of racemic naproxen in biphasic enzyme membrane reactors, J. Memb. Sci. 184 (2001) 27–38.
[107] K. Sakaki, S. Hara, N. Itoh, Optical resolution of racemic 2-hydroxy octanoic acid using biphasic enzyme membrane reactor, Desalination 149 (2002) 247–252.
[108] K. Sakaki, N. Itoh, Optical resolution of racemic 2-hydroxy octanoic acid by lipase-catalyzed hydrolysis in a biphasic membrane reactor, Biotechnol. Lett. 25 (2003) 1591–1595.
[109] T. Gumí, J. Fernandez-Delgado Albacete, D. Paolucci-Jeanjean, M.P. Belleville, G.M. Rios, Study of the influence of the hydrodynamic parameters on the performance of an enzymatic membrane reactor, J. Memb. Sci. 311 (2008) 147–152.
[110] M.P. Belleville, P. Lozano, J.L. Iborra, G.M. Rios, Preparation of hybrid membranes for enzymatic reaction, Sep. Purif. Technol. 25 (2001) 229–233.
[111] F. Ye, Y. Kuang, S. Chen, C. Zhang, Y. Chen, X.H. Xing, Characteristics of low molecular weight heparin production by an ultrafiltration membrane bioreactor using maltose binding protein fused heparinase I, Biochem. Eng. J. 46 (2009) 193–198.
[112] R. Mazzei, L. Giorno, E. Piacentini, S. Mazzuca, E. Drioli, Kinetic study of a biocatalytic membrane reactor containing immobilized beta-glucosidase for the hydrolysis of oleuropein, J. Memb. Sci. 339 (2009) 215–223.
[113] P.M. Nielsen, H.S. Olsen, Enzymatic modification of food protein, in: R. Whitehurst, B.A. Law (Eds.), Enzymes in Food Technology, Sheffield Academic Press, Sheffield, 2002, pp. 109–143.
[114] C.A. Prieto, E.M. Guadix, A. Guadix, Influence of temperature on protein hydrolysis in a cyclic batch enzyme membrane reactor, Biochem. Eng. J. 42 (2008) 217–223.
[115] W.D. Deeslie, M. Cheryan, A CSTR-hollow fiber system for continuous hydrolysis of proteins: performance and kinetics, Biotechnol. Bioeng. 23 (1981) 2257–2271.
[116] W.D. Deeslie, M. Cheryan, A CSTR-hollow-fiber system for continuous hydrolysis of proteins: factors affecting long-term stability of the reactor, Biotechnol. Bioeng. 24 (1982) 69–82.
[117] A. Perea, U. Ugalde, Continuous hydrolysis of whey proteins in a membrane recycle reactor, Enzyme Microb. Technol. 18 (1996) 29–34.

[118] C.A. Prieto, A. Guadix, P. González-Tello, E.M. Guadix, A cyclic batch membrane reactor for the hydrolysis of whey protein, J. Food Eng. 78 (2007) 257–265.
[119] S.C. Cheison, Z. Wang, S. Xu, Hydrolysis of whey protein isolate in a tangential flow filter membrane reactor. I. Characterisation of permeate flux and product recovery by multivariate data analysis, J. Memb. Sci. 283 (2006) 45–56.
[120] A. Guadix, F. Camacho, E.M. Guadix, Production of whey protein hydrolysates with reduced allergenicity in a stable membrane reactor, J. Food Eng. 72 (2006) 398–405.
[121] M. Prata-Vidal, S. Bouhallab, G. Henry, P. Aimar, An experimental study of caseinomacropeptide hydrolysis by trypsin in a continuous membrane reactor, Biochem. Eng. J. 8 (2001) 195–202.
[122] J.T. Wei, B.H. Chiang, Bioactive peptide production by hydrolysis of porcine blood proteins in a continuous enzymatic membrane reactor, J. Sci. Food Agric. 89 (2009) 372–378.
[123] D. Nehring, P. Czermak, J. Vorlop, H. Lübben, Experimental study of a ceramic microsparging aeration system in a pilot-scale animal cell culture, Biotechnol. Prog. 20 (2004) 1710–1717.
[124] P. Côté, J.L. Bersillon, A. Huyard, G. Faup, Bubble-free aeration using membranes: process analysis, J. Water Pollut. Control Fed. 60 (1988) 1986–1992.
[125] P. Côté, J.L. Bersillon, A. Huyard, A bubble-free aeration using membranes: mass transfer analysis, J. Memb. Sci. 47 (1989) 91–108.
[126] F. Hesse, M. Ebel, N. Konisch, R. Sterlinski, W. Kessler, R. Wagner, Comparison of a production process in a membrane-aerated stirred tank and up to 1000-L airlift bioreactors using BHK-21 cells and chemically defined protein-free medium, Biotechnol. Prog. 19 (2003) 833–843.
[127] S. Lütz, N. Rao, Membrane applications in red and white biotechnology, in: K.V. Peinemann, S. Pereira-Nunes (Eds.), Membranes for Life Science, Wiley-VCH, Germany, 2008, pp. 155–174.
[128] B. Frahm, H. Brod, U. Langer, Improving bioreactor cultivation conditions for sensitive cell lines by dynamic membrane aeration, Cytotechnology 59 (2009) 17–30.
[129] W. van Hecke, D. Haltrich, B. Frahm, H. Brod, J. Dewulf, H. Van Langenhove, R. Ludwig, A biocatalytic cascade reaction sensitive to the gas–liquid interface: modeling and upscaling in a dynamic membrane aeration reactor, J. Mol. Catal. B 68 (2011) 154–161.
[130] B. Maïsterrena, R. Couturier, B. Perrin, Artificial biomimetic membranes for the active and selective transport of small molecules, Enzyme Microb. Technol. 30 (2002) 125–128.
[131] K. Fiaty, C. Charcosset, B. Perrin, R. Couturier, B. Maïsterrena, ATP-dependent active transport simulations based on a phosphatase-channel-kinase membrane structure, J. Comput. Chem. 25 (2004) 1264–1276.
[132] C. Charcosset, K. Fiaty, B. Perrin, R. Couturier, B. Maïsterrena, Key role of enzyme positions and membrane surface potentials in the properties of biomimetic membranes, Arch. Biochem. Biophys. 424 (2004) 235–245.
[133] B. Perrin, R. Couturier, K. Fiaty, C. Charcosset, B. Maïsterrena, Enzymatic membranes for the selective transport of neutral molecules by electrophoresis, Electrophoresis 29 (2008) 2288–2292.
[134] J.A. Schonberg, G. Belfort, Enhanced nutrient transport in hollow fiber perfusion bioreactors: a theoretical analysis, Biotechnol. Prog. 3 (2) (1987) 80–89.
[135] G. Salzman, R. Tadmor, S. Guzy, S. Sideman, N. Lotan, Hollow fiber enzymatic reactors for a two substrate process: analytical modeling and numerical simulations, Chem. Eng. Process. 38 (1999) 289–299.

[136] V. Calabrò, S. Curcio, G. Iorio, A theoretical analysis of transport phenomena in a hollow fiber membrane bioreactor with immobilized biocatalyst, J. Memb. Sci. 206 (2002) 217–241.
[137] S. Curcio, V. Calabrò, G. Iorio, A theoretical and experimental analysis of a membrane bioreactor performance in recycle configuration, J. Memb. Sci. 273 (2006) 129–142.
[138] S. Bhatia, W.S. Long, A.H. Kamaruddin, Enzymatic membrane reactor for the kinetic resolution of racemic ibuprofen ester: Modeling and experimental studies, Chem. Eng. Sci. 59 (2004) 5061–5068.
[139] L.W. Wang, Y.C. Cheng, S.W. Tsai, Process modeling of the lipase-catalyzed dynamic kinetic resolution of (R, S)-suprofen 2,2,2-trifluoroethyl thioester in a hollow-fiber membrane, Bioprocess Biosystem. Eng. 27 (2004) 39–49.
[140] S.Y. Lau, M.H. Uzir, A.H. Kamaruddin, S. Bhatia, Lipase-catalyzed dynamic kinetic resolution of racemic ibuprofen ester via hollow fiber membrane reactor: modeling and simulation, J. Memb. Sci. 357 (2010) 109–121.
[141] N. Diano, V. Grano, S. Rossi, U. Bencivenga, M. Portaccio, U. Amato, F. Canfora, M. Lepore, F.S. Gaeta, D.G. Mita, Hollow-fiber enzyme reactor operating under nonisothermal conditions, Biotechnol. Prog. 20 (2004) 457–466.

8 Some perspectives

Chapter Outline

This last chapter presents two recent membrane processes which are increasingly reported for pharmaceutical and biotechnological applications. These processes show a wide range of potential opportunities; however, they still experience unresolved points that slow down their applications. The first process is solvent-resistant nanofiltration (SRNF). New developments in membrane materials and techniques allow these membranes to be used in a wide range of industrial solvents. The second technique uses a membrane for crystallization. Several configurations were developed such as osmosis and reverse osmosis (RO), evaporative membrane crystallization and membrane contactor. SRNF and membrane crystallization are presented in this chapter.

8.1 Solvent-resistant nanofiltration

8.1.1 Introduction

Nanofiltration (NF) is a pressure-driven membrane process, involving pressures between 5 and 20 bar, used to separate molecules in the molecular weight range of

Membrane Processes in Biotechnology and Pharmaceutics. DOI: 10.1016/B978-0-444-56334-7.00008-3

200–2000 g mol^{-1}. NF is performed at near ambient temperatures, which is a considerable advantage compared to conventional thermal separation techniques such as steam distillation and vacuum distillation, which generally require an increase in temperature and a high energy consumption, which can be inappropriate for heat-sensitive products. These methods may also result in a loss of compounds of low molecular weight, which can be removed together with the solvent during evaporation. Since many active pharmaceutical ingredients and high-value natural molecules are susceptible to thermal degradation, the operating temperature of NF can minimize loss of activity and/or nutritive properties.

NF dates back to the 1970s when RO membranes with a relatively high water flux operating at relatively low pressures were developed. Such low pressure RO membranes were termed NF membranes. Large-scale applications of NF currently exist in water treatment. NF recently emerged from this application area with the development of SRNF, also termed organic solvent nanofiltration (OSN) [1,2]. The application of NF in organic solvents is still limited, despite a wide range of potential opportunities in the pharmaceutical and natural products industries. Actual limitations of SRNF are mainly related to membrane stability and lifetime, and lack of fundamental understanding and modelling.

Several reviews, articles or books present main features of NF. Schäfer et al. [2] proposed a comprehensive and up-to-date book which covers the principles of NF, its potential and actual applications. Various chapters are provided including history of NF membranes from 1960 to 1990, materials, module design and operation, preparation, membrane characterization, modelling, fouling, pre-treatment and hybrid processes. Applications include water treatment, food industry, chemical processing industry, pulp and paper industry, metal and acid recovery and so on. A specific chapter is devoted to non-aqueous applications of NF [3]. Hilal et al. [4] presented a review of NF in water treatments, including applications in treating water as well as in the pre-treatment process for desalination. Mechanisms, minimization of NF membrane fouling problems and theories for modelling and transport of salt, charged and non-charged organic compounds in NF membranes were also presented. In addition, the application of AFM in studying the morphology of the NF membrane surfaces was summarized. Van der Bruggen et al. [5] focused on unresolved problems that slow down large-scale NF applications. Six key points were identified: (1) avoiding membrane fouling, and possibilities to remediate, (2) improving the separation between solutes, (3) further treatment of concentrates, (4) chemical resistance and limited lifetime of membranes, (5) insufficient rejection of pollutants in water treatment and (6) need for modelling and simulation tools. Silva et al. [6] provided a recent and complete review on NF in organic solvents. Organic solvent NF membranes were presented including polymeric membranes (integral asymmetric polymeric membranes and thin-film composite membranes) and inorganic membranes. The major drawback of the inorganic membranes was underlined to be their relatively high MWCO $>$1000 Da, which made them unsuitable for NF. SRNF transport mechanisms were presented as well as applications including fine chemical and pharmaceutical synthesis, food and beverage and refining.

In NF, basic parameters are mostly the same as in UF and MF: permeate flux, solute-sieving coefficients and rejection coefficients. The configurations are also those used in UF and MF, dead-end and cross-flow and the diafiltration mode. The following sections present data on membranes, governing phenomena, the specific diafiltration mode, applications and some theoretical considerations.

8.1.2 Membranes

Several SRNF membranes are currently commercially available. Some of their properties are summarized in Table 8.1. The STARMEM® membrane series was commercialized until recently by Membrane Extraction Technology Ltd. (UK). These asymmetric membranes have an active layer of polyimide (PI). They are available as flat sheet with MWCO of 220 and 400. SelRO (MPS series) are pH stable flat sheet hydrophobic membranes commercialized by Koch Membrane Systems (USA) for industrial NF applications. The allowable pH is between 1 and 13. These composite membranes are made of an active layer of cross-linked polydimethylsiloxane (PDMS) mounted on a poly(acrylonitrile) support. The membranes are available with 200 and 1000 MWCO.

Both DuraMem® and PuraMem® (Evonik Industries AG, Germany) membranes are made of P84 PI. The DuraMem membranes are reported to be stable in a wide

Table 8.1 Properties of some commercial solvent-resistant NF membranes

Membranes	Manufacturer	Material, configuration, MWCO	Stable in solvents
STARMEM®	Membrane Extraction Technology Ltd (UK)	Polyimide, flat sheet, 220 and 400 Da	Alcohols, alkanes, aromatics, ethers, ketones
DuraMem®	Evonik Industries AG (Germany)	Modified polyimide, flat sheet and spiral-wound modules from 150 to 900 Da	Acetone, ethanol, methanol, tetrahydrofuran, dimethylformamide, dimethylacetamide, isopropanol, acetonitrile, methylethyl ketone, ethyl acetate and more
PuraMem®	Evonik Industries AG (Germany)	Modified polyimide, flat sheet and spiral-wound modules, 280 Da	Toluene, heptane, hexane, methylethyl ketone, ethyl acetate methylisobutyl ketone and more
MPS series	Koch Membrane Systems (USA)	200 and 1000 Da	Acid and base

range of industrial solvents including solvents of the polar aprotic solvent family. They are available with MWCO from 150 to 900 Da. Due to their excellent stability and availability with a range of different MWCO, these membranes open up new opportunities of molecular separations. They are commercialized in flat sheet or spiral-wound modules. The PuraMem membrane series is stable in apolar hydrocarbon-type solvents. These membranes are available with 280 Da MWCO in flat sheet or spiral-wound modules.

Other recent SRNF membranes have been proposed to bypass actual limitations [7–9]. Skinned asymmetric membranes made from PI were prepared by immersion precipitation from a casting solution containing dimethylformamide (DMF) as the solvent and 1,4-dioxane (dioxane) as the co-solvent [10]. The variation of the ratio between DMF and dioxane in the dope solution used for membrane formation provided control over the MWCO curves of the membrane. Coupling this methodology with chemical cross-linking provided good solvent stability, enabling application of the SRNF membrane to separations in DMF.

PI membranes cross-linked with diamines and modified polypyrrole (PPy) have been shown to be stable in DMF [11–13]. New SRNF membranes with PPy modified top layer were prepared on different type of support by in situ pyrrole polymerization [13]. Due to the poor solubility of PPy, an in situ polymerization method was adopted to modify commercial membranes. Three different membrane classes were selected to be modified and were tested in several solvents for different solutes. The filtration experiment in strong aprotic DMF of PPy modified membranes showed a stable permeability and retention over 30 h. In addition, the PPy modified membranes showed a much higher flux in tetrahydrofuran (THF) systems than for earlier reported cross-linked PI membranes.

Polyaniline (PANI) skinned asymmetric NF membranes were prepared using a process of acid doping [14]. Further treatment by thermal cross-linking produced a membrane stable in most organic solvents. Filtration experiments showed that these membranes had a low MWCO with reasonably good solvent fluxes. Further studies have shown that these membranes can be chemically cross-linked using α,α'-dichloro-*p*-xylene and glutaraldehyde [15]. These membranes were also stable in a wide variety of organic solvents including harsh polar aprotic solvents. Filtration tests have been performed with methanol, acetone, THF and DMF, even at temperatures of up to 65 °C.

Recently, the same authors have prepared PANI membranes in a spiral-wound configuration [16]. Continuous casting allowed to prepare large membrane area as required for the spiral-wound membrane module. Filtrations carried out in acetone, THF and DMF gave stable permeate fluxes and good separation performances. MWCO of the PANI membranes in different solvents was found between 150 and 300 g mol^{-1} at 30 °C. Figure 8.1 presents the performance of the spiral-wound module in DMF at 30, 40 and 65 °C. For each temperature, the module was tested over 2 days. A shift in the MWCO from 300 to 500 g mol^{-1} was observed when the temperature was increased from 30 to 65 °C. The increase in MWCO was possibly attributed to the increase in PANI polymer chain mobility at higher temperatures. The authors concluded that these devices could be potentially useful for large-scale applications.

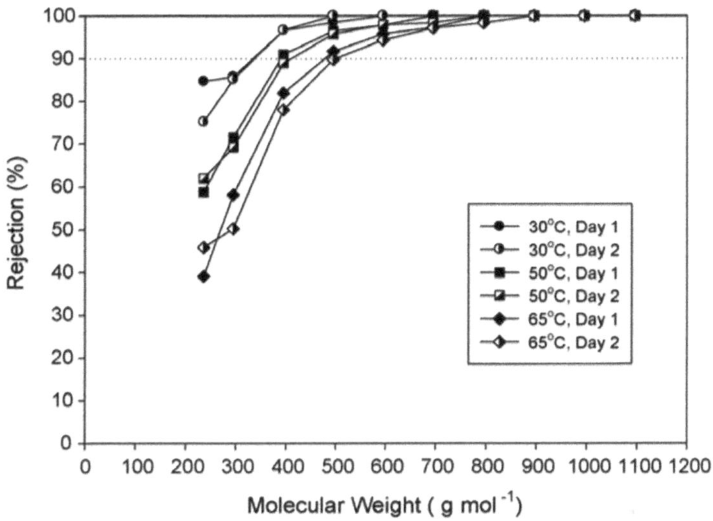

Figure 8.1 Nanofiltration performance of the spiral-wound module in DMF at different system temperatures at 30 bar [16].
This article was published in J. Membr. Sci., Vol. 349, M. Sairam, X.X. Loh, Y. Bhole, I. Sereewatthanawut, K. Li, A. Bismarck, J.H.G. Steinke, A.G. Livingston, Spiral-wound polyaniline membrane modules for organic solvent nanofiltration (OSN), pp. 123–129, Copyright Elsevier (2010).

The preparation of other SRNF membranes was described recently. Volkov et al. [17] described the preparation of composite membranes comprising a poly[1-(trimethylsilyl)-1-propyne] top layer of about 1 μm on a porous poly(acrylonitrile) support. The composite membrane had ethanol permeability of 3.8 kg m^{-2} h^{-1} bar^{-1} and 90% retention of the negatively charged dye Remazol Brilliant Blue R (MW 626.5) at 5 bar. The permeability of methanol, ethanol or acetone through the composite membranes was reported to be higher than many commercial available NF membranes. Li et al. [18] prepared SRNF membranes from multilayered polyelectrolyte (PE) complexes obtained by an alternating adsorption of polycations and polyanions. NaCl ions were added to the PE dipping solutions to control the charge density and the configuration of the PE adsorbed, hence, of the membrane morphology and performance. The membranes showed a IPA permeate flux from 0.06 to 0.98 L m^{-2} h^{-1} bar^{-1} with a very high rejection of Rose Bengal (1017 Da). The resulting membranes were claimed to have comparable selectivity properties with commercial membranes, better solvent stability and higher fluxes.

8.1.3 Governing phenomena

The performance of a SRNF process is usually characterized by the solvent permeability and the solute rejection. Solvent permeation through SRNF membranes has been realized for a wide range of solvents, solvent mixtures and solutes [19–25].

In SRNF, the membrane–solvent interactions play a determining role in the value of the solvent flux. Flux measurements indicate that exposure of SRNF membranes to organic solvents causes important changes in the polymeric structure and therefore a deterioration of the membrane performance [8]. Hydrophilic membranes showed a decrease of the pure water flux after solvent exposure. The decrease of the degree of surface hydrophilicity was explained as the cause of the lower water fluxes. Geens et al. [25] used contact angle measurements and surface tension calculations on new and solvent-treated membranes. Their results confirmed that organic solvents change the hydrophilicity of polymeric membrane surfaces, leading to different water permeability before and after solvent treatment. Swelling measurements were realized to confirm that a reorganization of the membrane material takes place, leading to differences in porosity. Swelling of a polymer can be quantified by determining the weight difference between a wet and a dry sample. Swelling is a measure for the volume of liquid absorbed by the membrane and is defined as

$$Q = \frac{1}{\rho_s} \frac{W_{wet} - W_{dry}}{W_{dry}} \tag{8.1}$$

with Q the degree of swelling, ρ_s the liquid density, W_{wet} the weight of a wet membrane sample and W_{dry} the weight of a dry membrane sample.

Other studies examined the rejection behaviour of SRNF membranes [26–33]. Whu et al. [26] and Yang et al. [21] found lower rejections with a commercial PDMS made membrane (MPF-60) for Vitamin B12, brilliant blue R and safranin O in methanol than expected from the specified MWCO. White [22] suggested that the pore structure of the membranes and polymer–solvent interactions have a strong influence on the separation performance. In organic solvent systems, the MWCO is solvent dependent and can no longer be used for the prediction of solute rejection.

Key phenomena governing separation in SRNF were underlined by Darvishmanesh et al. [34] who measured in a systematic way the rejection of dyes (Sudan II, Sudan Black, Sudan 408) from common organic solvents (methanol, ethanol, acetone,

Table 8.2 Solvent permeability (l h^{-1} m^{-2} bar^{-1}) and rejection R (%) of Sudan II, Sudan Black and Sudan 408 [34]

Solvent	Methanol		Ethanol		Acetone		Methyl ethyl ketone		Toulene		*n*-Hexane	
Solutes	L_p	Rejection (%)	L_p	Rejection (%)	L_p	Rejection (%)	L_p	Rejection (%)	L_p	Rejection (%)	L_p	Rejection (%)
Sudan II	5.48	31.3	2.41	39.1	4.65	33.2	8.51	44.7	1.37	62.4	0.08	−70.6
Sudan Black		92.8		94.1		41.4		53.2		79.2		−33.5
Sudan 408		91.4		92.6		13.8		50.3		78.7		−35.9

Phys. Chem. Chem. Phys., Vol. 12, S. Darvishmanesh, J. Degrève, B. Van der Bruggen, Mechanisms of solute rejection in solvent resistant nanofiltration: the effect of solvent on solute rejection, pp. 13333–13342 - Reproduced by permission of the PCCP Owner Societies.

methylethyl ketone, toluene and *n*-hexane) through a PI-based SRNF membrane, STARMEM™122 (Table 8.2). The separation in SRNF was highly dependent on interaction of solvents with the membrane. Depending on the solvent and the membrane, a high interaction may exist within each other. In this case, two phenomena occur. First, the solvent passes through the membrane with a high flux (alcohol) without rearrangement of the membrane structure, and the rejection is relatively high. Secondly, the solvent changes the membrane structure by swelling, and a low separation performance is obtained at a high flux. Solvents affect the separation performances in two ways: solvent–membrane interaction and solvent–solute interaction. Solvent–membrane interaction has a dominant effect on the separation (high rejection in alcohol while negative rejection occurs in *n*-hexane). Solvent–solute interactions are more pronounced when solvents from the same chemical family are compared (methanol, ethanol, isopropanol).

8.1.4 Diafiltration

Solvent exchange is a common unit operation in the pharmaceutical industry since many sequential organic synthesis reactions must be carried out in different organic solvents. Conventional solvent exchange via distillation is able to swap solvents with differing boiling points, when the solvent to be removed has a lower boiling point than the replacing solvent. SRNF was proposed recently for solvent exchange [35,36].

Sheth et al. [36] employed diafiltration using methanol as the solvent needed in the next synthesis step to drastically reduce the concentration of ethyl acetate used as the solvent in the previous synthesis step. Ethyl acetate was reduced to the level of a low concentration impurity in methanol by both batch and continuous diafiltration using SRNF membranes MPF-50 and MPF-60 (Koch Membrane Systems). The MPF-60 membrane has a high rejection of around 96% for the solute of interest, erythromycin, which is a widely used broad-spectrum macrolide antibiotic.

Sereewatthanawut et al. [37] reported the use of SRNF for diafiltration separations of two solutes in organic solvent solutions with two case studies. In Case Study A, two dyes, *Solvent Yellow 7* (198 g mol^{-1} MW) and *Brilliant Blue R* (826 g mol^{-1} MW), were used as model product and impurity compounds, respectively. Case Study B was an actual separation challenge at Janssen Pharmacuetica NV (Belgium) involving an intermediate of a new active pharmaceutical ingredient (API-INT 675 g mol^{-1} MW) and its oligomeric impurities (1000 g mol^{-1} MW). SRNF spiral-wound membrane modules were used for diafiltration at kilo scale with typical organic solvents, DMF and THF. For Case Study A, a final product with 99.7% purity of SY7 was generated from an initial starting solution of 91% purity, while recovering 90% of SY7. For Case Study B, 99% of the higher MW oligomeric impurities (i.e. tetramer and higher) have been removed from API-INT, whilst reducing the content of oligomeric impurities in the synthesis solution from 6.8 wt to 2.4 wt%, which was below the allowed limit of 3 wt% oligomeric impurities, and recovering more than 99% of API-INT.

However, the SRNF diafiltration processes required a large volume of solvent to achieve the target purity and yield of the desired compound. Sereewatthanawut et al. [37] proposed recently to couple the purification process with a downstream OSN-solvent recovery system (named 'dual membrane diafiltration' DMD). This process did not generate large volumes of waste and/or did not provide a dilute product solution that would require further processing. The dual membrane diafiltration process combined two membrane stages, that is a purification stage combined with a solvent recovery stage (Figure 8.2). Instead of adding fresh solvent to the process, recovered solvent from the solvent recovery stage was returned to the purification stage to provide further purification. The feasibility of the DMD process was demonstrated through Case Study B.

Lin and Livingston [38] proposed to use a counter-current membrane cascade using SRNF membranes. The effect of process parameters, such as number of stages and flow rate ratio of replacing solvent to initial solvent, on solvent exchange performance were tested through simulations and experiments. Experimental results were 47.8%, 59.2% and 75.3% solvent exchange for single-stage, two-stage and three-stage cascades, values which were close to the 50.0%, 66.6% and 75.0% values predicted by the simulations. It was acknowledged that this configuration would not be suitable for complete exchange of two solvents as this would require an infinite numbers of stages. Integration into other solvent exchange processes or a direct modification of this configuration was claimed to be needed.

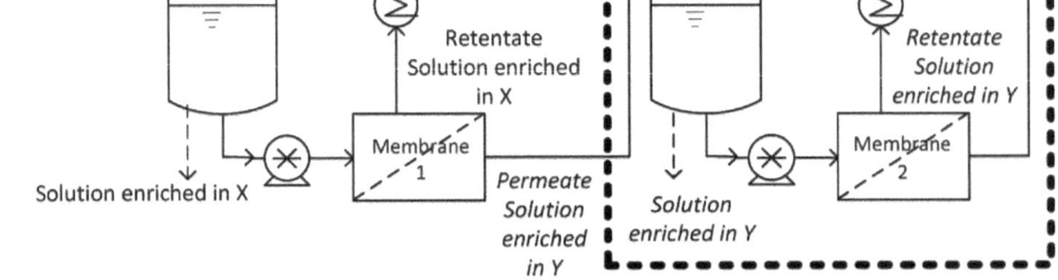

Figure 8.2 Schematic diagram of dual membrane diafiltration (DMD) process, compound Y (lower MW) is separated from compound X (higher MW) in the primary stage using Membrane 1. In the secondary stage, compound Y is retained using Membrane 2, allowing solvent to be recycled back into the primary stage [37].
Reprinted with permission from Org. Process Res. Dev., Vol. 14, I. Sereewatthanawut, F.W. Lim, Y.S. Bhole, D. Ormerod, A. Horvath, A.T. Boam, A.G. Livingston, Demonstration of molecular purification in polar aprotic solvents by organic solvent nanofiltration, pp. 600–611, Copyright (2010) American Chemical Society.

8.1.5 Applications

Some applications have been reported above such as the separation of an intermediate of a new drug candidate and its oligomeric impurities using SRNF spiral-wound membrane modules in a diafiltration mode [37]. One of the first applications was reported by Sheth et al. [36] who employed SRNF membranes in a diafiltration configuration for exchanging the solvent methanol, needed in the next synthesis step, for ethyl acetate, used as the solvent in the previous synthesis step. Ethyl acetate was reduced to the level of a low concentration impurity in methanol by both batch and continuous diafiltration using SRNF membranes MPF-50 and MPF-60 (Koch Membrane Systems). The latter membrane had a high rejection of around 96% for the solute, erythromycin, a widely used broad-spectrum macrolide antibiotic.

Recently, SRNF was tested to recover the solvent from an aqueous ethanol solution of soybean isoflavones [39]. Soybean isoflavones are reported to play an important role in protecting against several common diseases, including cancer, heart disease and osteoporosis. In industry, the purification of soybean isoflavones is usually carried out by macroporous resin chromatography, desorption of soybean isoflavones and evaporation of the desorption solution. Among the five SRNF membranes tested, the Starmem™ 122 membrane was shown to give high permeate flux and retention of total soybean isoflavones.

In addition, Tylkowski et al. [40] reported results on the extraction of propolis compounds with an ethanol–water mixture and sequential concentration of the extract by NF using two membranes: Starmem™ 122 and Duramem™ 200. Propolis is a resinous natural product, produced by bees (*Apis mellifera*) from vegetable secretions. Propolis contains more than 300 bioactive compounds. Among them, flavonoids and phenolic acids as well as their derivatives have been used since ancient times as a medicine because of their biological properties such as antibacterial, antifungal, antiprotozoan, antiviral, immunomodulatory, anti-inflammatory, antioxidant and antitumour activity. The content of flavones, flavonols, flavanones, dihydroflavonols as well as total phenolic substances was determined spectrophotometrically in the feed and permeate obtained during dead-end NF experiments. Rejections higher than 95% were obtained, and extract concentration of over three times was achieved with the Duramem™ 200 membrane.

So et al. [41] combined SRNF with solution-phase peptide synthesis. The process was termed ‘membrane enhanced peptide synthesis (MEPS)’. The benefits of the process from the solution-phase synthesis were claimed to avoid the purification steps that make solution-phase synthesis practically difficult. A first amino acid was linked to a soluble polyethylene glycol anchor. Through subsequent repeated coupling and deprotection steps, the peptide was extended to the desired length. The residual by-products and excess reagents after each reaction were removed by diafiltration through a solvent-stable membrane which retained the peptide. Two pentapeptides were produced using this new technology. The purity of the peptides produced by MEPS was higher than that of peptides produced by solid-phase synthesis, under the same conditions. The process was presented as an alternative route for peptide production at industrial scale, especially as a production method for PEGylated peptides.

8.1.6 Theory

The development of transport models through SRNF is particularly challenging because SRNF shows unusual performance compared to traditional NF, due to the interaction between solvents and membranes. Effects such as swelling of the polymeric network, the polymeric chain reorganization and surface energy varying in different solvent have been frequently reported [19,21,25,42,43]. A number of transport models through SRNF have been developed using solvent and membrane parameters which influence transport. Some of these models are summarized below [24].

Bhanushali et al. [20] proposed the following equation based on a traditional solution-diffusion model [44]:

$$J_{\text{solvent}}\ \alpha\ A \tag{8.2}$$

$$A\ \alpha \left(\frac{V_{\text{m}}}{\mu}\right)\left(\frac{1}{\phi^{\text{n}}\gamma_{\text{SV}}}\right) \tag{8.3}$$

A was the pure solvent permeability coefficient and consisted of a diffusivity and solubility term in the original form of the solution-diffusion model. Four physical properties were further included in the model: the viscosity μ, the molar volume V_{m}, the surface energy of the solid membrane material γ_{SV}, and a sorption value ϕ [20,22,45]. Although the model showed good correlation with experimental data, several limitations were pointed out [24]. In the case of high affinity between the solvent and the membrane, a higher flux would be expected, but according to the model the flux decreased. In addition, the model did not cover the whole range of solvents from polar to non-polar ones.

Some models used the Hagen–Poiseuille equation to predict and fit experimental data of transport through SRNF membranes [23,26]. Other approaches include, for example, a resistance in series model based on viscous flow [46]:

$$A\ \alpha \frac{1}{\lambda[\gamma_{\text{C}} - \gamma_{\text{L}} + f_1\mu] + f_2\mu} \tag{8.4}$$

The membrane was modelled as a combination of a NF top layer with a UF sublayer. Solvent independent parameters were introduced to characterize the NF (f_1) and UF (f_2) sublayers. λ was a solvent–membrane dependence parameter that shows the porosity of the membrane surface. Affinity of the solvent and membrane has been considered by measuring the difference between γ_{C}, the critical surface tension of the membrane material, and γ_{L}, the surface tension of the solvent.

Another semi-empirical model was introduced by Geens et al. [25]. Their model used three parameters: the solvent viscosity, the solvent molar volume and the difference in surface tension between the solid membrane material and the liquid solvent. Combination of these parameters led to the introduction of a new pure solvent permeability coefficient:

$$A\ \alpha \left(\frac{V_{\text{m}}}{\mu\Delta\gamma}\right) \tag{8.5}$$

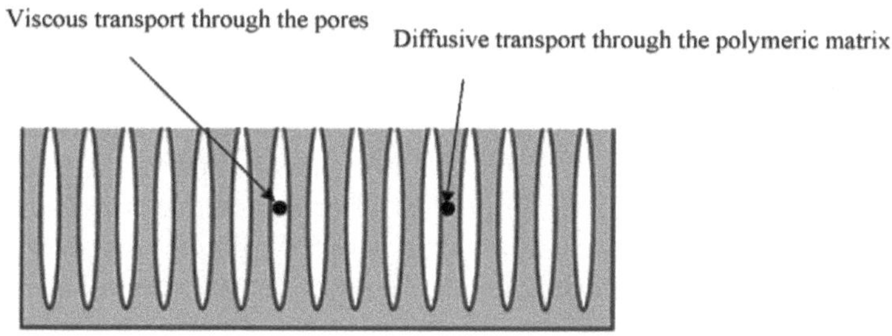

Figure 8.3 Schematic representation of solvent transport through the NF membrane [24]. This article was published in J. Membr. Sci., Vol. 334, S. Darvishmanesh, A. Buekenhoudt, J. Degrève, B. Van der Bruggen, General model for prediction of solvent permeation through organic and inorganic solvent resistant nanofiltration membranes, pp. 43–49, Copyright Elsevier (2009).

The model was in good correlation with experimental data. However, a number of limitations were underlined [24]. In a mathematical point of view, the difference in surface tension might be zero so that the flux reached infinity. In addition, the model predicted that by increasing the molecular volume of solvent, V_m (representative of the solvent size), the flux increased, although the contrary would be expected.

Recently, Darvishmanesh et al. [24] developed a new semi-empirical model based on the solution-diffusion model. The SRNF membrane was considered as a one layer membrane, with two different permeabilities (Figure 8.3). The solvent molecules permeated through the membrane polymeric matrix and through the pores. For the molecules that diffused through the matrix, the diffusional permeability was considered, and for those that passed through the pores, viscous transport was used. In the model, the solvent permeability was dependent on viscosity, and on two non-dimensional parameters defined from the surface tension and dielectric constant.

$$J_V = \frac{a_0 \alpha}{\mu \exp(1-\beta)}(\Delta p - \Delta \pi) + \frac{b_0}{\mu \exp(1-\beta)} \Delta p \qquad (8.6)$$

where J_V was the volume flux, α a non-dimensional polarity coefficient, β the ratio of the surface tension of the membranes, a_0 and b_0 specific diffusivity and permeability values. The authors demonstrated that the model fitted well with experimental permeation of various solvents from different chemical families through a hydrophobic polymeric membrane (MPF-50) and a hydrophilic ceramic membrane (HITK 275).

8.2 Membrane crystallization

8.2.1 Introduction

Crystallization is widely used in the chemical and pharmaceutical industry for the separation, purification and production of chemicals [47,48]. In particular, most pharmaceutical manufacturing processes include a series of crystallization steps to achieve high purity and to produce the desired final crystal form [49,50]. The characteristic features of the crystallized product are purity, polymorphic form, crystal shape and crystal size distribution (CSD), which should be as narrow as possible. These properties determine the efficiency of downstream operations, such as filtration, drying and formulating, and the product effectiveness, such as bioavailability and shelf-life. For pharmaceuticals that exhibit various polymorphs, crystallization also affects the polymorph produced and the extent of chiral separation [51].

Crystal nucleation and growth are induced by supersaturation which is generated by increasing solute concentration above its solubility limits or by adding an antisolvent which reduces the solubility of the solute in the mixed solvent/antisolvent solution. The kinetics of the crystallization process involves the nucleation and growth rates of the crystals, which are directly dependent on the supersaturation ratio. In crystallization, the rate and extent of supersaturation have a direct effect on the quality of the crystals obtained. Uncontrolled supersaturation can lead to undesirable CSDs, which, in turn, negatively affect the downstream operations. Regarding the level of impurities, crystallization at low supersaturation levels usually results in better crystal quality, that is larger and better-shaped crystals, with less inclusions and therefore greater purity. Furthermore, for compounds exhibiting polymorphism, the operating conditions required to obtain the desired form, either spontaneously or by seeding may require a limited supersaturation range. These features are governed by a variety of factors such as the crystallization process, crystallizer technology and operating conditions.

Membrane processes provide recent alternatives to other crystallization processes. In a membrane crystallization system, the membrane does not act as a sieving barrier to select the transport of specific components like in UF and MF, but as a physical support able to generate and control the supersaturated environment in which crystals can nucleate and grow. The main features of membrane crystallization systems are to control and limit the maximum level of supersaturation due to defined mass transfer across the membrane; to act as heterogeneous nucleation-inducing substrates; to control solid features such as size, polymorphic form, shape and purity, and to reduce energy consumption compared to cooling or evaporative crystallization. To achieve these goals, several configurations have been examined: osmosis and RO, evaporative membrane crystallization and membrane contactor.

Di Profio et al. [52] recalled that the main features of 'membrane crystallization' are (1) the use of membranes to control the composition of the crystallizing solution, by opposing a well defined and tunable resistance to mass flow occurring in the vapour phase; (2) the action of the porous surface of the membrane as a suitable support to activate heterogeneous nucleation mechanisms; (3) the possibility to induce nucleation and crystal growth in separate sites, thus reducing the risk of membrane fouling even when the same membrane supports heterogeneous nucleation. Other processes involving membranes for crystallization were termed 'membrane-assisted crystallization processes'. In this paper, Di Profio et al. [52] provided a review on the developments and perspectives of membrane crystallization. A book chapter was proposed by the same authors focussing on operational configurations (solvent evaporation and antisolvent membrane crystallizers), control of supersaturation (effect on crystal morphology and purity, influence of polymorphism), heterogeneous nucleation above the membrane surface and finally advantages of membrane crystallization processes [53]. Another review paper on membrane crystallization and membrane-assisted crystallization processes described the various techniques: osmosis and RO, evaporative membrane crystallization, membrane contactor and membrane templates [54].

8.2.2 Basic definitions

The mechanisms involved in crystallization are usually nucleation, growth, agglomeration and breakage [47,48,55,56]. Nucleation means that particles (nuclei) of size L_0 are created from a supersaturated solution. The rate of nucleation is J. Growth means an increase of crystal size. The speed of growth is $G = \mathrm{d}L/\mathrm{d}t$, and may be a function of L. Agglomeration implies that two particles stick together and create a bigger particle. The rate of agglomeration is measured by r_A, the net rate of creation of crystals per unit volume of crystallizer. Breakage means that a big crystal divides into smaller ones. r_B is the net rate of disappearing of crystals per unit volume of crystallizer. Details are given below on supersaturation, and the main mechanisms of nucleation and growth.

Supersaturation

The driving force for nucleation and growth is supersaturation. After dissolving the chemical species in a solvent, the solution must be supersaturated in order to observe nucleation or growth. Supersaturation is the difference between the chemical potential of the solute molecules in the supersaturated (μ) and saturated (μ_s) states respectively. For one molecule the expression of this difference is:

$$\mu = \mu - \mu_s = kT \log \beta \tag{8.7}$$

where k is the Boltzmann constant, T is the temperature and β is the supersaturation ratio. Activities are usually considered equal to the concentrations; therefore, the supersaturation is expressed as:

$$\beta = \frac{C_i}{C_s} \tag{8.8}$$

where C_i is the concentration of the solute in solution and C_s its saturated or equilibrium concentration. If $\beta > 1$, the crystal grows; if $\beta < 1$, the crystal dissolves; and if $\beta = 1$, the crystals and the solution are at equilibrium.

Supersaturation is usually created by cooling, evaporation and/or addition of antisolvent, including changing the pH by addition of acid or base. A schematic batch crystallizer is shown in Figure 8.4. In evaporative crystallization, the solution is heated which causes the solvent to evaporate. This loss of solvent makes the solution more concentrated which simultaneously causes the generation of supersaturation. Cooling crystallization relies on the fact that most solutes experience a decrease in solubility as temperature drops. In antisolvent crystallization, supersaturation is

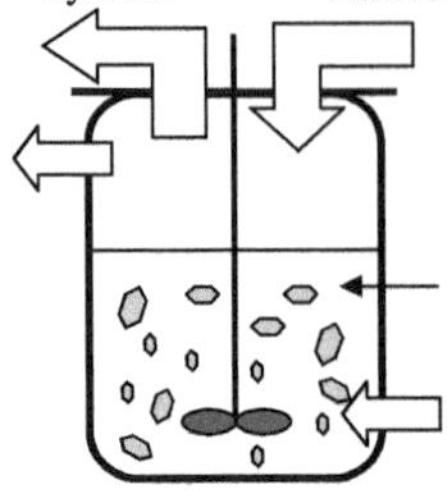

Figure 8.4 A schematic batch crystallizer. The crystals are not drawn to scale.

generated by the addition of an antisolvent that reduces the solubility of the solute in the original solvent. This antisolvent may either be a liquid, gas or a supercritical fluid. The limited temperature stability of most pharmaceuticals precludes the use of evaporation and limits the temperature range that can be used during cooling crystallization. Moreover, it is important to control the supersaturation since the size, shape and solid-state phase of the crystals are dependent on the supersaturation profile.

Nucleation

When a solution is supersaturated, the solid phase forms more or less rapidly depending on the growth conditions such as temperature, supersaturation, chemical conditions and hydrodynamics. Primary nucleation occurs in a solution that is clear, without crystals. Nucleation is said to be homogeneous if the nuclei form in the bulk of the solution. Nucleation is said to be heterogeneous if the nuclei preferentially form on substrates such as the wall of the crystallizer, the stirrer or dust particles. Finally, secondary nucleation may be induced by the presence of existing crystals of the same phase.

The nucleation rate, J, is the number of crystals that form in a supersaturated solution per unit of time and unit of volume. J is expressed as:

$$J = nN_0\nu \exp\left(-\frac{f\Omega^2\gamma^3}{(kT)^3\ln^2\beta}\right) \tag{8.9}$$

where f the nuclei form factor. J is proportional to n times the solubility expressed in number of molecules per unit of volume, N_0. ν is the frequency with which nuclei of critical size r^* become supercritical by addition of a molecule and develop into crystals. Equation (8.9) shows that the frequency of nucleation depends on the supersaturation β and on the concentration of molecules nN_0. For systems with low solubility, the solute molecules are separated by larger distances and by a greater number of solvent molecules. The probability that the molecules come into contact and form a nucleus is thus lower. On the contrary, systems with high solubility show a higher probability of contact, and thus easier nucleation.

Growth

Once the nuclei are formed and exceed the critical size, they become crystals. A crystal is limited by its faces. The set of equivalent faces resulting from the crystal symmetry is a form. All the forms present on a crystal represent the morphology of the crystal. The concept of habit includes the notion of face extension. In addition, the growth form of the crystal only includes the faces with the slowest growth rates. The growth form thus depends on the crystallization conditions.

Growth kinetics and mechanisms depend on external factors (chemical conditions, temperature, supersaturation and hydrodynamics) and on internal factors (structure, bonds and defects). The growth medium influences the growth kinetics of the faces in different ways. First, the solvent is more or less adsorbed by the faces and selectively slows down their growth rates. Solubility also plays a key role, the higher the solubility, the higher the growth rate. The growth medium also influences solvation, desolvation and complex formation. If not predetermined by the process, variations in temperature also produce extremely different growth rates. Lastly, hydrodynamics, or more precisely the relative velocity of the solution compared to the crystal, is an important parameter.

8.2.3 Configurations and applications

Osmosis and reverse osmosis

RO is a pressure-driven process involving two solutions with different concentrations on either side of a semi-permeable membrane [57]. The rate at which fresh water crosses the membrane is proportional to the pressure differential that exceeds the natural osmotic pressure differential. In osmotic dewatering, the RO membrane is used to separate two solutions having different solute concentrations [58,59]. The lower concentration solution contains the substance to be crystallized. Due to the osmotic pressure, resulting from the difference in osmolarity across the membrane, water from the low-concentration solution permeates through the membrane into the high-concentration solution. This dehydration, at a controllable rate, gradually concentrates the low-concentration solution and causes crystallization to occur in the presence of a suitable precipitant (Figure 8.5).

Todd et al. [60] used RO membranes to obtain high-quality crystals of biological macromolecules. Their method called 'osmotic dewatering crystallization' was successfully applied to growing lysozyme crystals. The use of dialysis membranes was described by Zeelen and Wierenga [61] for the crystallization of biological molecules on the microscale. Using the osmotic dewatering technique, Lee et al. [58,59] successfully crystallized an oligonucleotide. X-ray diffraction analysis showed a maximum resolution of 2.46 Å for one crystal, comparable to that of crystals grown by conventional methods. Cuellar et al. [51] evaluated the use of RO and NF for water removal to achieve selective crystallization of L-phenylalanine anhydrate from aqueous solution. Water removal and crystallization were studied separately, and models were used to define the operating limits for the coupled system so that the target crystal form and production rate could be achieved. A narrow

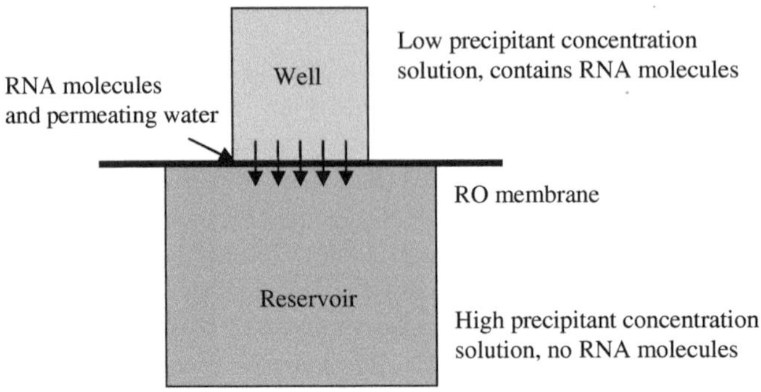

Figure 8.5 Schematic diagram of the principle of osmotic dewatering [from 58].

window of operation was obtained. The authors underlined that it could make the technique difficult to operate in practice.

Recently, Kuhn et al. [62] characterized membrane performance in solvent removal from an aqueous adipic acid solution under RO conditions. The presence of adipic acid was shown to strongly reduce the flux, as compared to the flux from a pure water feed. This effect was attributed to adsorption on the membrane surface and to concentration polarization. In addition, the optimal operational conditions resulted in a six-fold lower energy conversion as compared to evaporative crystallization, with a final mean size that corresponded closely to the maximum mean size that could be achieved in a crystallizer under ideal supersaturation control.

Concentration polarization effects were also observed in a crystallization vessel and a separate membrane module operated in RO mode for an ammonium sulphate water system [63,64]. The process was termed 'membrane-assisted crystallization using reverse osmosis'. A stable mass flux around 2.5 kg m^{-2} h^{-1} was measured in five successive experiments with an accumulated process time of 30 h. The very low mass flux was attributed to concentration polarization on the permeate side of the membrane. The authors concluded that future research should focus on increasing the mass flux across the membrane by minimizing the polarization effects, rather than by simply increasing the membrane surface area. In a later study [65], practical and theoretical feasibility of membrane-assisted crystallization using RO was investigated with an emphasis on the influence of solubility characteristics. The authors concluded that membrane-assisted crystallization using RO had potential for many applications. This method should be practically most suitable for components with a moderate solubility, high molecular weight and a strong dependency of solubility on temperature.

Evaporative membrane crystallization

Evaporative membrane crystallization is based on membrane distillation (MD), a thermally driven membrane process in which a hydrophobic microporous membrane separates hot and cold solvent streams [66,67]. The hydrophobic nature of the membrane prevents the passage of liquid solvent through the pores while

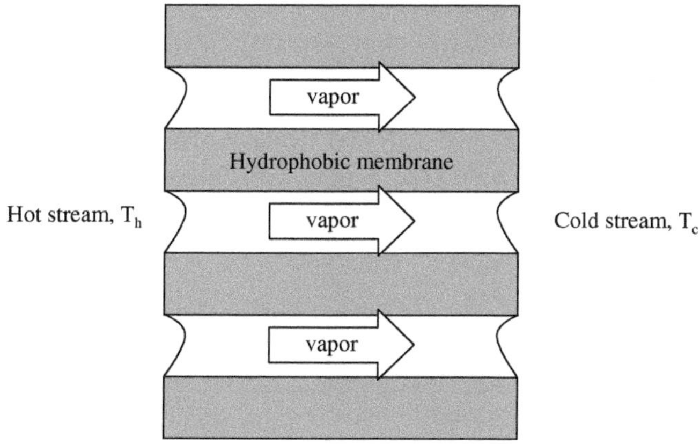

Figure 8.6 Vapour–liquid interface in membrane distillation.

allowing the passage of solvent vapour (Figure 8.6). The temperature difference produces a vapour pressure gradient which causes solvent vapour to pass through the membrane and condense on the colder surface. The result is a distillate of very high purity which, unlike in conventional distillation, does not suffer from non-volatile contaminants.

In evaporative membrane crystallization, a hydrophobic porous hollow fibre membrane device is used to maintain tight control of supersaturation while crystallization takes place in a circulating crystallizer [68,69]. Two different operative configurations can be used [70]. (1) Static conditions, where both solutions in contact are quiescent. In this configuration, several experiments can be performed simultaneously, even using small amounts of crystallizing solution (a few tenths of a microlitre). (2) Dynamic conditions, where solutions are recirculated. In this configuration, the amount of solution is of the order of a few millilitres or tens of millilitres (with the possibility of further scale-up).

Evaporative membrane crystallization has been applied mainly to protein crystallization, which today plays a crucial role. The 3D structure elucidation of biomacromolecules is essential for an understanding of their complex biological functions. In the pharmaceutical industry, knowledge of the complete atomic 3D structure is expected to speed-up the process of designing new molecules as potential ligands or inhibitors of proteins involved in pathological processes [71]. X-ray crystallography is the method of choice for determining a protein structure at atomic resolutions but requires highly ordered crystals of adequate size (≥30–50 μm). Macromolecular crystallization also represents an important separation/purification step in a number of chemical and biotechnological processes, as well as the first step in the production of cross-linked enzyme crystals, with high surface-to-volume ratios.

Evaporative membrane crystallization has been used for the preparation of fumaric acid crystals in aqueous L-malic acid solutions [72], single lysozyme crystals suitable for X-ray diffraction analysis with sodium chloride as precipitant [68,73] and trypsin crystals with ammonium sulphate as precipitant [74,75]. For example, lysozyme crystallization was carried out in a microporous hydrophobic membrane device [73]. The high quality of the tetragonal lysozyme crystals obtained was confirmed by diffraction analysis using synchrotron radiation. Crystallization kinetics was investigated using a turbidity method. The shorter induction periods (ranging between 1.2 and 10 h), compared to those measured using conventional crystallization techniques, as well as the constant high growth rates at low supersaturation ratios, demonstrated the effect of the membrane surface as a heterogeneous support for crystal nucleation and growth.

Evaporative membrane crystallization has also been used to control polymorphism. The various polymorphs of the same substance, having different crystalline lattices, are characterized by diverse physicochemical properties such as solubility, bioavailability and stability. This is a key point for organic molecules for pharmaceutics. Commercial paracetamol, dissolved in distilled water, was crystallized in a static membrane crystallizer [76,77]. The driving force of the membrane-assisted evaporation mechanism was varied by changing the rate of solvent evaporation, J, and, hence, the rate of variation of the supersaturation. The morphology of the paracetamol crystals obtained varied depending on J: a needle-like shape, typical of Form II, and either an elongated prismatic shape, usual for Form I when grown at low supersaturation, or a crystal habit in which all the faces were well developed and distinguishable from Form I when grown at higher supersaturations. The selective crystallization of either α or γ polymorph of glycine was also obtained by both static and dynamic membrane crystallization [78]. Depending on the operative parameters (stripping solution concentration and recirculation velocity in the dynamic configuration), either one of two crystal morphologies shown in Figure 8.7 was selectively observed. The two forms, named Form I and Form II, appeared in a time lag ranging from 2 to 10 days.

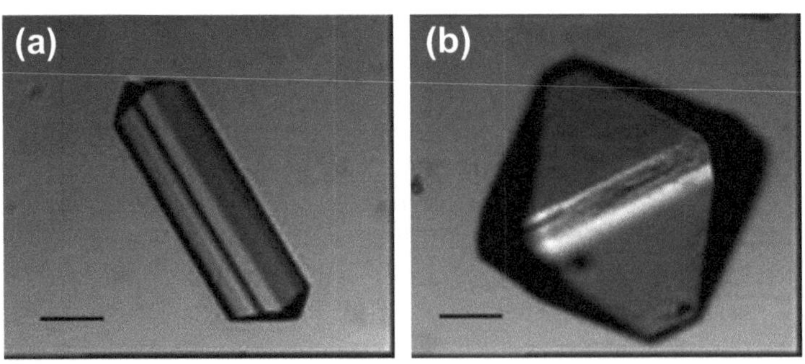

Figure 8.7 Optical images of the two crystal morphologies obtained during the membrane crystallization tests of glycine: (a) Form I; (b) Form II (the unit bar inside the figure corresponds to 50 μm) [78].
Reprinted with permission from Cryst. Growth Des., Vol. 7, G. Di Profio, S. Tucci, E. Curcio, E. Drioli, Selective glycine polymorph crystallization by using microporous membranes, pp. 526–530, Copyright (2007) American Chemical Society.

Antisolvent membrane crystallization

Antisolvent membrane crystallization makes use of a membrane contactor technique. The membrane contactor involves one phase introduced through the membrane pores coming into contact with a second phase which flows tangentially to the membrane surface (Figure 8.8). If the two phases are not miscible, then droplets are formed at the pore outlets and are detached by the flowing continuous phase. The process is called 'membrane emulsification' and w/o, o/w and multiphase emulsions are prepared [79] (Chapter 5). When the two phases are miscible, mixing and reaction occur between the two phases inside the membrane device [80]. The membrane contactor may be used in a tangential configuration (one phase circulates tangentially to the membrane surface) or in a dead-end (normal flow) configuration.

Membrane contactors have been used for precipitation operation. Particle formation by precipitation is an important unit operation for the production of fine solids, such as catalysts, pigments and pharmaceuticals [47,48]. Precipitation is increasingly being used as an alternative to cooling and evaporative crystallization processes for the isolation and separation of organic fine chemicals, especially for active pharmaceutical ingredients whose biological activity might be degraded when used at high-temperature conditions. Precipitation typically consists of mixing two liquid streams to create supersaturation, which then induces nucleation, particle growth and (very often)

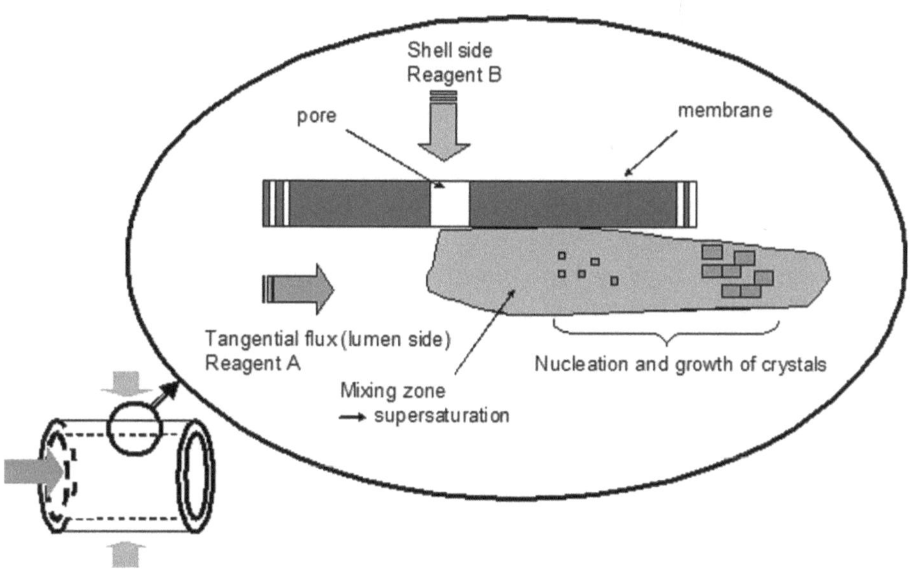

Figure 8.8 Schematic diagram showing the principle of a hollow fibre membrane device for precipitation [84]. Original in colour.
This article was published in Chem. Eng. Sci., Vol. 64, R. Kieffer, D. Mangin, F. Puel, C. Charcosset, Precipitation of barium sulphate in a hollow fibers membrane contactor: Part I Investigation of fouling, pp. 1759–1767, Copyright Elsevier (2009).

agglomeration. The hydrodynamic factors governing the initial contact and mixing of the reacting fluids play a major role in controlling particle quality due to their influence on the supersaturation distribution in the reactor volume.

A recent study presented results in relation to product properties using several crystallization techniques and membrane processes: antisolvent membrane crystallization and evaporation membrane crystallization [81,82]. Membrane-based crystallization techniques using microporous hydrophobic membranes were applied to a broad range of substances: an inorganic bulk crystallization product (NaCl), an organic bulk product (benzoic acid) and a protein (lysozyme). Parameters studied were the influence of the supersaturation conditions and consequently the applied crystal growth rate on the product shape and size parameters. In all model systems, larger, well-shaped crystals with fewer inclusions, and therefore of higher purity, were obtained by applying gentle supersaturation rates.

Antisolvent crystallization of L-asparagine was investigated in porous hollow fibre devices [83]. Two configurations were investigated: the first involved the addition of antisolvent to a crystallizing solution flowing on the lumen side of a hollow fibre membrane, the second involved the addition of the crystallizing solution on the lumen side to the antisolvent flowing over the shell side by keeping the lumen side pressure higher. The experimental results showed that the CSDs obtained had shifted to smaller sizes compared to data recorded in the literature for batch-stirred crystallizers. However, a decrease in antisolvent flux was measured with the first configuration, which was blamed primarily on fibre blockage by newly formed crystals. It was pointed out that optimization of the operating conditions and of the device was necessary to increase the performance of this membrane crystallization process.

Recently, Kieffer et al. [84,85] investigated the precipitation of $BaSO_4$ particles in a hollow fibre membrane device. A solution of barium chloride was flowed tangentially to the membrane surface and reacted with a solution of potassium sulphate introduced through the membrane pores. The supersaturation thus created induced nucleation and particle growth on the lumen side of the hollow fibres. Fouling was observed using X-ray images with $BaSO_4$ as a radiocontrast agent. Fouling was blamed primarily on fibre blockage by newly formed crystals rather than on pore blockage. Moreover, a larger inner diameter of hollow fibre, 250 10^{-6} m, was shown to slow down the fouling of the device considerably, as observed for $D = 110\ 10^{-6}$ m. A proper methodology was developed to measure crystal size directly at the outlet of the hollow fibre device by using EDTA as a neutralizing agent of crystal nucleation and growth. Concentrations of barium chloride and potassium sulphate were shown to be the main influence on the CSD, the effect of inlet flow rates on the lumen and shell sides being less pronounced. The chemical conversions measured were between 3.7 and 20.5%. Again, it was emphasized that an improvement in the design of the hollow fibre device would be necessary to give higher chemical conversions.

Contrary to the above-mentioned configurations, Di Profio et al. [86] proposed forcing the antisolvent through the membrane in the vapour phase, instead of the liquid phase. Two configurations were proposed: (a) solvent/antisolvent demixing crystallization, where the solvent was removed at a higher flow rate than the antisolvent to promote supersaturation and (b) antisolvent addition, where antisolvent was added to the crystallizing solution. In both configurations, accurate control of supersaturation was achieved with effects on crystal morphology of two organic molecules of industrial interest, paracetamol and glycine. The antisolvent membrane crystallization process has also been used to influence the polymorphic composition in the crystallization of L-histidine [87]. When dosing the amount of ethanol in the crystallizing solution, by generating the proper transmembrane flux, the composition of the precipitate shifted from the prevalence of the thermodynamic stable polymorph A toward the predominance of the metastable phase B.

8.2.4 Theory

Few models have been proposed to describe membrane crystallization processes. They are often limited to flux through the membrane and supersaturation prediction, and do not provide results on crystal characteristics such as size and shape. Kieffer et al. [84] proposed a simulation of the supersaturation profile inside a hollow fibre for an antisolvent membrane crystallization configuration. Di Profio et al. [52] provided a theoretical modelling of flux through the membrane in an evaporative membrane configuration. Both simulations are summarized in the following.

Antisolvent membrane crystallization

In the antisolvent membrane crystallization operation, solution A flows tangentially to the membrane surface and reacts with solution B introduced through the membrane pores [84]. The resulting supersaturation induces nucleation and particles grow on the lumen side of the hollow fibres. C_A and C_B are concentrations of components A and B, respectively. The crystal growth rate, G, and the primary nucleation rate, J, are calculated as functions of the supersaturation ratio, S, with k_g, the crystal growth rate constant, A and B, kinetic constants of nucleation, $\gamma_\pm$, the mean ionic activity coefficient, and K_{sp}, the solubility product:

$$G = k_g(\text{S-1})^2 \tag{8.10}$$

$$J = A\exp\left[-\frac{B}{\ln^2 S}\right] \tag{8.11}$$

$$S = \gamma_\pm\left[\frac{C_A + C_B}{K_{sp}}\right]^{0.5} \tag{8.12}$$

The model was developed to calculate the supersaturation ratio and the number of crystals in the lumen side of the hollow fibre. In particular, the model was used to predict the location of fouling in the fibre (corresponding to a high supersaturation ratio). In case of crystallization of $BaSO_4$ from a solution of $BaCl_2$ (A) and a solution of K_2SO_4 (B), fouling was attributed primarily to fibre blockage by newly formed crystals at the inlet of the hollow fibres.

Evaporative membrane crystallization

In evaporative membrane crystallization, macroporous membranes (pore size >50 nm) are used to provide sufficient high flux through the membrane. The approximate mean free path of water molecule passing through the membrane in vapour phase is in the same order of the pore size, considering its diameter at room temperature and pressure (2.75 Å). Therefore, the reduced Knudsen-molecular diffusion transition form of the dusty-gas model (neglecting surface diffusion) can be used to describe vapour flux J_i across the membrane [52]:

$$J_i = -\frac{1}{RT_{avg}}\left(\frac{D_w^k D_{w-a}^0}{D_{w-a}^0 + p_a D_w^k}\right)\frac{\Delta p}{\delta}M \tag{8.13}$$

$$D_w^k = \frac{2\varepsilon r_p}{3\tau}\sqrt{\frac{8RT_{avg}}{\pi M}} \tag{8.14}$$

$$D_{w-a}^0 = 4.46 \times 10^{-6}\,\frac{\varepsilon}{\tau}\,T_{avg}^{2.334} \tag{8.15}$$

where Δp is the vapour pressure gradient across the membrane, r_p is the pore radius, δ is the membrane thickness, τ is the tortuosity factor, M the molecular weight, T_{avg} the average temperature, and D_w^k is the Knudsen diffusion coefficient. In the frequent case of non-ideal mixtures, the vapour–liquid equilibrium is described in terms of partial pressure p_i, vapour pressure of pure i, p_i^0 , and activity coefficient ξ_i according to the relationship:

$$p_i = Py_i = p_i^0\,a_i = p_i^0\,a_i = p_i^0\,\zeta_i\,x_i \tag{8.16}$$

where x_i and y_i are the liquid and vapour mole fraction, respectively. The vapour pressure p^0 of a pure substance varies with temperature according to the Clausius–Clapeyron equation:

$$\frac{dp^0}{dT} = \frac{\lambda}{RT^2} \tag{8.17}$$

where λ is the latent heat of vapourization. The expression for the activity coefficient in diluted aqueous ionic solutions can be derived from the Debye–Hückel theory:

$$\log \xi_\pm = -|z_+ z_l|\,\psi\,\sqrt{I} \tag{8.18}$$

Here, $\xi_\pm$ is the activity coefficient of the electrolyte, ψ is a constant which depends on the temperature and solution permittivity, $z_\pm$ is the ion valence and I the ionic strength of the solution, given by

$$I = \frac{1}{2}\sum z_i^2 c_i \tag{8.19}$$

According to equations (8.13–8.19), the flux through the membrane of a component from one phase to the other depends on both the driving force, Δp, and the membrane characteristics: ε, r_p, τ and δ. Migrating molecules experience a resistance which depends on these properties, which are fixed once a specific membrane is chosen, while the strength of the driving force is fixed by the combination of operating conditions affecting Δp. As Δp is directly proportional to the activity gradient Δa of the volatile component(s) and is dependent on temperature, the driving force is created by a temperature or an activity difference between the two contacted solutions.

References

[1] D. Bhanushali, D. Bhattacharyya, Advances in solvent-resistant nanofiltration membranes, Adv. Membr. Technol. 984 (2003) 159–177.

[2] A.I. Schäfer, A.G. Fane, T.D. Waite, Nanofiltration: Principles and Applications, Elsevier, Oxford, UK, 2005.

[3] F.P. Cuperus, K. Ebert, Non-aqueous application of NF, in: A.I. Schaefer, A.G. Fane, T.D. Waite (Eds.), Nanofiltration: Principles and Applications, Elsevier Science, New York, 2004, pp. 521–536. Chapter 21.

[4] N. Hilal, H. Al-Zoubi, N.A. Darwish, A.W. Mohammad, M. Abu Arabi, A comprehensive review of nanofiltration membranes: treatment, pretreatment, modelling, and atomic force microscopy, Desalination 170 (2004) 281–308.

[5] B. Van der Bruggen, M. Mänttäri, M. Nyström, Drawbacks of applying nanofiltration and how to avoid them: a review, Sep. Purif. Technol. 63 (2008) 251–263.

[6] P. Silva, L.G. Peeva, A.G. Livingston, Nanofiltration in organic solvents, in: N.N. Li, A.G. Fane, W.S.W. Ho, T. Matsuura (Eds.), Advanced Membrane Technology and Applications, Wiley, Hoboken, NJ, 2008, pp. 451–468.

[7] P. Vandezande, L.E.M. Gevers, I.F.J. Vankelecom, Solvent resistant nanofiltration: separating on a molecular level, Chem. Soc. Rev. 37 (2008) 365–405.

[8] B. Van der Bruggen, J. Greens, C. Vandecasteele, Influence of organic solvents on the performance of polymeric nanofiltration membranes, Sep. Sci. Technol. 37 (2002) 783–797.

[9] J.T. Scarpello, D. Nair, L.M.F. Dos Santos, A.G. Livingston, The separation of homogeneous organometallic catalysts using solvent resistant nanofiltration, J. Membr. Sci. 203 (2002) 71–85.

[10] Y.H. See Toh, M. Silva, A. Livingston, Controlling molecular weight cut-off curves for highly solvent stable organic solvent nanofiltration (OSN) membranes, J. Membr. Sci. 324 (2008) 220–232.

[11] Y.H. See Toh, F.W. Lim, A.G. Livingston, Polymeric membranes for nanofiltration in polar aprotic solvents, J. Membr. Sci. 301 (2007) 3–10.

[12] K. Vanherck, P. Vandezande, S.O. Aldea, I.F.J. Vankelecom, Cross-linked polyimide membranes for solvent resistant nanofiltration in aprotic solvents, J. Membr. Sci. 320 (2008) 468–476.

[13] X. Li, P. Vandezande, I.F.J. Vankelecom, Polypyrrole modified solvent resistant nanofiltration membranes, J. Membr. Sci. 320 (2008) 143–150.

[14] M. Sairam, X.X. Loh, K. Li, A. Bismarck, J.H.G. Steinke, A.G. Livingston, Nanoporous asymmetric polyaniline films for filtration of organic solvents, J. Membr. Sci. 330 (2009) 166–174.

[15] X.X. Loh, M. Sairam, A. Bismarck, J.H.G. Steinke, A.G. Livingston, K. Li, Crosslinked integrally skinned asymmetric polyaniline membranes for use in organic solvents, J. Membr. Sci. 326 (2009) 635–642.

[16] M. Sairam, X.X. Loh, Y. Bhole, I. Sereewatthanawut, K. Li, A. Bismarck, J.H.G. Steinke, A.G. Livingston, Spiral-wound polyaniline membrane modules for organic solvent nanofiltration (OSN), J. Membr. Sci. 349 (2010) 123–129.
[17] A.V. Volkov, V.V. Parashchuk, D.F. Stamatialis, V.S. Khotimsky, V.V. Volkov, M. Wessling, High permeable PTMSP/PAN composite membranes for solvent nanofiltration, J. Membr. Sci. 333 (2009) 88–93.
[18] X. Li, W. Goyens, P. Ahmadiannamini, W. Vanderlinden, S. De Feyter, I. Vankelecom, Morphology and performance of solvent-resistant nanofiltration membranes based on multilayered polyelectrolytes: Study of preparation conditions, J. Membr. Sci. 358 (2010) 150–157.
[19] D.R. Machado, D. Hasson, R. Semiat, Effect of solvent properties on permeate flow through nanofiltration membranes. Part I. Investigation of parameters affecting solvent flux, J. Membr. Sci. 163 (1999) 93–102.
[20] D. Bhanushali, S. Kloos, C. Kurth, D. Bhattacharyya, Performance of solvent-resistant membranes for nonaqueous systems: solvent permeation results and modelling, J. Membr. Sci. 189 (2001) 1–21.
[21] X.J. Yang, A.G. Livingston, L. Freitas dos Santos, Experimental observations of nanofiltration with organic solvents, J. Membr. Sci. 190 (2001) 45–55.
[22] L.S. White, Transport properties of a polyimide solvent resistant nanofiltration membrane, J. Membr. Sci. 205 (2002) 191–202.
[23] J.P. Robinson, E.S. Tarleton, C.R. Millington, A. Nijmeijer, Solvent flux through dense polymeric nanofiltration membranes, J. Membr. Sci. 230 (2004) 29–37.
[24] S. Darvishmanesh, A. Buekenhoudt, J. Degrève, B. Van der Bruggen, General model for prediction of solvent permeation through organic and inorganic solvent resistant nanofiltration membranes, J. Membr. Sci. 334 (2009) 43–49.
[25] J. Geens, B. Van der Bruggen, C. Vandecasteele, Characterisation of the solvent stability of polymeric nanofiltration membranes by measurement of contact angles and swelling, Chem. Eng. Sci. 59 (2004) 1161–1164.
[26] J.A. Whu, B.C. Baltzis, K.K. Sirkar, Nanofiltration studies of larger organic microsolutes in methanol solutions, J. Membr. Sci. 170 (2000) 159–172.
[27] D. Bhanushali, S. Kloos, D. Bhattacharyya, Solute transport in solvent-resistant nanofiltration membranes for non-aqueous systems: experimental results and the role of solute-solvent coupling, J. Membr. Sci. 208 (2002) 343–359.
[28] J. Geens, A. Hillen, B. Bettens, B. Van der Bruggen, C. Vandecasteele, Solute transport in non-aqueous nanofiltration: effect of membrane material, J. Chem. Technol. Biotechnol. 80 (2005) 1371–1377.
[29] Y. Zhao, Q. Yuan, A comparison of nanofiltration with aqueous and organic solvents, J. Membr. Sci. 279 (2006) 453–458.
[30] K. Ebert, J. Koll, M.F.J. Dijkstra, M. Eggers, Fundamental studies on the performance of a hydrophobic solvent stable membrane in non aqueous solutions, J. Membr. Sci. 285 (2006) 75–80.
[31] F. Zheng, C. Li, Q. Yuan, F. Vriesekoop, Influence of molecular shape on the retention of small molecules by solvent resistant nanofiltration (SRNF) membranes: a suitable molecular size parameter, J. Membr. Sci. 318 (2008) 114–122.
[32] E.S. Tarleton, J.P. Robinson, J.S. Low, Nanofiltration: a technology for selective solute removal from fuels and solvents, Chem. Eng. Res. Des. 87 (2009) 271–279.
[33] R. Valadez-Blanco, A.G. Livingston, Solute molecular transport through polyimide asymmetric organic solvent nanofiltration (OSN) membranes and the effect of membrane-formation parameters on mass transfer, J. Membr. Sci. 326 (2009) 332–342.

[34] S. Darvishmanesh, J. Degrève, B. Van der Bruggen, Mechanisms of solute rejection in solvent resistant nanofiltration: the effect of solvent on solute rejection, Phys. Chem. Chem. Phys. 12 (2010) 13333–13342.

[35] A. Livingston, L. Peeva, S.J. Han, D. Nair, S.S. Luthra, L.S. White, L.M.F. Dos Santos, Membrane separation in green chemical processing – solvent nanofiltration in liquid phase organic synthesis reactions, Ann. N. Y. Acad. Sci. 984 (2003) 123–141.

[36] J.P. Sheth, Y. Qin, K.K. Sirkar, B.C. Baltzis, Nanofiltration-based diafiltration process for solvent exchange in pharmaceutical manufacturing, J. Membr. Sci. 211 (2003) 251–261.

[37] I. Sereewatthanawut, F.W. Lim, Y.S. Bhole, D. Ormerod, A. Horvath, A.T. Boam, A.G. Livingston, Demonstration of molecular purification in polar aprotic solvents by organic solvent nanofiltration, Org. Process Res. Dev. 14 (2010) 600–611.

[38] J.C.T. Lin, A.G. Livingston, Nanofiltration membrane cascade for continuous solvent exchange, Chem. Eng. Sci. 62 (2007) 2728–2736.

[39] Q. Zhang, Q. Yuan, Modeling of nanofiltration process for solvent recovery from aqueous ethanol solution of soybean isoflavones, Sep. Sci. Technol. 44 (2009) 3239–3257.

[40] B. Tylkowski, B. Trusheva, V. Bankova, M. Giamberini, G. Peev, A. Nikolova, Extraction of biologically active compounds from propolis and concentration of extract by nanofiltration, J. Membr. Sci. 348 (2010) 124–130.

[41] S. So, L.G. Peeva, E.W. Tate, R.J. Leatherbarrow, A.G. Livingston, Organic solvent nanofiltration: a new paradigm in peptide synthesis, Org. Process Res. Dev. 14 (2010) 1313–1325.

[42] E.M. Tsui, M. Cheryan, Characteristics of nanofiltration membranes in aqueous ethanol, J. Membr. Sci. 237 (2004) 61–69.

[43] D.F. Stamatialis, N. Stafie, K. Buadu, M. Hempenius, M. Wessling, Observations on the permeation performance of solvent resistant nanofiltration membranes, J. Membr. Sci. 279 (2006) 424–433.

[44] J.G. Wijmans, R.W. Baker, The solution–diffusion model: a review, J. Membr. Sci. 107 (1995) 1–21.

[45] L.G. Peeva, E. Gibbins, S.S. Luthra, L.S. White, R.P. Stateva, A.G. Livingston, Effect of concentration polarisation and osmotic pressure on flux in organic solvent nanofiltration, J. Membr. Sci. 236 (2004) 121–136.

[46] D.R. Machado, D. Hasson, R. Semiat, Effect of solvent properties on permeate flow through nanofiltration membranes. Part II. Transport model, J. Membr. Sci. 166 (1) (2000) 63–69.

[47] A. Mersmann, Crystallization Technology Handbook, second ed., Marcel Dekker, New York, 2001.

[48] A.S. Myerson, Handbook of Industrial Crystallization, second ed., Butterworth-Heinemann, Boston, MA, 2002.

[49] B.Y. Shekunov, P. York, Crystallization processes in pharmaceutical technology and drug delivery design, J. Cryst. Growth 211 (2000) 122–136.

[50] M. Fujiwara, Z.K. Nagy, J.W. Chew, R.D. Braatz, First-principles and direct design approaches for the control of pharmaceutical crystallization, J. Process Control 15 (2005) 493–504.

[51] M.C. Cuellar, S.N. Herreilers, A.J.J. Straathof, J.J. Heijnen, L.A.M. van der Wielen, Limits of operation fort the integration of water removal by membranes and crystallization of L-phenylalanine, Ind. Eng. Chem. Res. 48 (2009) 1566–1573.

[52] G. Di Profio, E. Curcio, E. Drioli, Supersaturation control and heterogeneous nucleation in membrane crystallizers: facts and perspectives, Ind. Eng. Chem. Res. 49 (2010) 11878–11889.

[53] G. Di Profio, E. Curcio, E. Drioli, Membrane crystallization technology, in: E. Drioli, L. Giorno (Eds.), Comprehensive Membrane Science and Engineering, Elsevier, Kidlington, UK, 2010.

[54] C. Charcosset, R. Kieffer, D. Mangin, F. Puel, Coupling between membrane processes and crystallization operations, Ind. Eng. Chem. Res. 49 (2010) 5489–5495.

[55] R. Boistelle, J.P. Astier, Crystallization mechanisms in solution, J. Cryst. Growth 90 (1988) 14–30.

[56] D. Mangin, F. Puel, S. Veesler, Polymorphism in processes of crystallization in solution: a practical review, Org. Process Res. Dev. 13 (2009) 1241–1253.

[57] C. Fritzmann, J. Löwenberg, T. Wintgens, T. Melin, State-of-the-art of reverse osmosis desalination, Desalination 216 (2007) 1–76.

[58] C.Y. Lee, M.G. Sportiello, S.P. Cape, S. Ferree, P. Todd, C.E. Kundrot, S. Lietzke, C. Barnes, Characterization and applications of osmotic dewatering to the crystallization of oligonucleotides, Biotechnol. Prog. 13 (1997) 77–81.

[59] C.Y. Lee, S.R. McEntyre, P. Todd, K. Schaefer, C.E. Kundrot, Control of nucleation in oligonucleotide crystallization by the osmotic dewatering method with kinetic water removal rate control, J. Cryst. Growth 187 (1998) 490–498.

[60] P. Todd, S.K. Sikdar, C. Walker, Application of osmotic dewatering to the controlled crystallization of biological macromolecules and organic compounds, J. Cryst. Growth 110 (1991) 283–292.

[61] J.P. Zeelen, R.K. Wierenga, The growth of yeast thiolase crystals using a polyacrylamide-gel as dialysis membrane, J. Cryst. Growth 122 (1992) 194–198.

[62] J. Kuhn, R. Lakerveld, H.J.M. Kramer, J. Grievink, P.J. Jansens, Characterization and dynamic optimization of membrane-assisted crystallization of adipic acid, Ind. Eng. Chem. Res. 48 (2009) 5360–5369.

[63] R. Lakerveld, J. Kuhn, M.A. Bosch, H.J.M. Kramer, P.J. Jansens, J. Grievink, Membrane assisted evaporative crystallization: Optimization of task supersaturation generation, in: 17th International Symposium on Industrial Crystallization, 14–17 September 2008, Maastricht (The Netherlands).

[64] R. Lakerveld, H.J.M. Kramer, P.J. Jansens, J. Grievink, The application of a task-based concept for the design of innovative industrial crystallizers, Comp. Chem. Eng. 33 (2009) 1692–1700.

[65] R. Lakerveld, J. Kuhn, H.J.M. Kramer, P.J. Jansens, J. Grievink, Membrane assisted crystallization using reverse osmosis: influence of solubility characteristics on experimental application and energy saving potential, Chem. Eng. Sci. 65 (2010) 2689–2699.

[66] K.W. Lawson, D.R. Lloyd, Membrane distillation, J. Membrane Sci. 124 (1997) 1–25.

[67] C.M. Tun, A.G. Fane, J.T. Matheickal, R. Sheikholeslami, Membrane distillation crystallization of concentrated salts – flux and crystal formation, J. Membr. Sci. 257 (2005) 144–155.

[68] E. Curcio, G. Di Profio, E. Drioli, A new membrane-based crystallization technique: tests on lysozyme, J. Cryst. Growth 247 (2003) 166–176.

[69] M. Gryta, Concentration of NaCl solution by membrane distillation integrated with crystallization, Sep. Sci. Tech. 37 (2002) 3535–3558.

[70] G. Di Profio, E. Curcio, E. Drioli, Progresses on membrane crystallizers, in: 17th International Symposium on Industrial Crystallization, 14–17 September 2008, Maastricht (The Netherlands)

[71] Y. Wu, E. Drioli, The behaviour of membrane distillation of concentrated aqueous solution, Water Treatment 4 (1989) 399–415.
[72] G. Di Profio, E. Curcio, E. Drioli, Protein crystallization by membrane-based techniques, Membr. News 66 (2004) 14–16.
[73] G. Di Profio, E. Curcio, A. Cassetta, D. Lamba, E. Drioli, Membrane crystallization of lysozyme: kinetic aspects, J. Cryst. Growth 257 (2003) 359–369.
[74] G. Di Profio, E. Curcio, E. Drioli, Trypsin crystallization by membrane-based techniques, J. Struct. Biol. 150 (2005) 41–49.
[75] G. Di Profio, G. Perrone, E. Curcio, A. Cassetta, D. Lamba, E. Drioli, Preparation of enzyme crystals with tunable morphology in membrane crystallizers, Ind. Eng. Chem. Res. 44 (2005) 10005–10012.
[76] G. Di Profio, S. Tucci, E. Curcio, E. Drioli, Controlling polymorphism with membrane-based crystallizers: application to form I and II of paracetamol, Chem. Mater 19 (2007) 2386–2388.
[77] G. Di Profio, E. Curcio, S. Ferraro, C. Stabile, E. Drioli, Effect of supersaturation control and heteregeneous nucleation on porous membrane surfaces in the crystallization of L-glutamic acid polymorphs, Cryst. Growth Des. 9 (2009) 2179–2186.
[78] G. Di Profio, S. Tucci, E. Curcio, E. Drioli, Selective glycine polymorph crystallization by using microporous membranes, Cryst. Growth Des. 7 (2007) 526–530.
[79] G.T. Vladisavljević, R.A. Williams, Recent developments in manufacturing emulsions and particulate products using membranes, Adv. Colloid Interface Sci. 113 (2005) 1–20.
[80] R. Kieffer, C. Charcosset, F. Puel, D. Mangin, Numerical simulation of mass transfer in a liquid-liquid membrane contactor for laminar flow conditions, Comput. Chem. Eng. 32 (2008) 1333–1341.
[81] D. Weckesser, A. König, Membrane based crystallization for product design, in: 17th International Symposium on Industrial Crystallization, 14–17 September 2008, Maastricht (The Netherlands).
[82] D. Weckesser, A. König, Particle shape and purity in membrane based crystallization, Chem. Eng. Technol. 31 (2008) 157–162.
[83] D.M. Zarkadas, K.K. Sirkar, Antisolvent crystallization in porous hollow fiber devices, Chem. Eng. Sci. 61 (2006) 5030–5048.
[84] R. Kieffer, D. Mangin, F. Puel, C. Charcosset, Precipitation of barium sulphate in a hollow fibers membrane contactor: part I Investigation of fouling, Chem. Eng. Sci. 64 (2009) 1759–1767.
[85] R. Kieffer, D. Mangin, F. Puel, C. Charcosset, Precipitation of barium sulphate in a hollow fibers membrane contactor: part II Influence of process parameters, Chem. Eng. Sci. 64 (2009) 1885–1891.
[86] G. Di Profio, C. Stabile, A. Caridi, E. Curcio, E. Drioli, Antisolvent membrane crystallization of pharmaceutical compounds, J. Pharm. Sci. 98 (2009) 4902–4913.
[87] G. Di Profio, A. Caridi, R. Caliandro, A. Guagliardi, E. Curcio, E. Drioli, Fine dosage of antisolvent in the crystallization of L-Histidine: Effect on polymorphism, Cryst. Growth Des. 10 (2010) 449–455.

Index

Note: Page numbers followed by f indicate figures and t indicate tables.

FSC
www.fsc.org
MIX
Papier | Fördert
gute Waldnutzung
FSC® C083411

Kolibri 360 GmbH
Ferdinand-Jühlke-Straße 7
99095 Erfurt, Deutschland
produktsicherheit@kolibri360.de